Rock and Pop Venues

Niels Werner Adelman-Larsen

Rock and Pop Venues

Acoustic and Architectural Design

 Springer

Niels Werner Adelman-Larsen
Flex Acoustics
Kongens Lyngby
Denmark

ISBN 978-3-642-45235-2   ISBN 978-3-642-45236-9 (eBook)
DOI 10.1007/978-3-642-45236-9
Springer Heidelberg New York Dordrecht London

Library of Congress Control Number: 2014931753

© Springer-Verlag Berlin Heidelberg 2014
This work is subject to copyright. All rights are reserved by the Publisher, whether the whole or part of the material is concerned, specifically the rights of translation, reprinting, reuse of illustrations, recitation, broadcasting, reproduction on microfilms or in any other physical way, and transmission or information storage and retrieval, electronic adaptation, computer software, or by similar or dissimilar methodology now known or hereafter developed. Exempted from this legal reservation are brief excerpts in connection with reviews or scholarly analysis or material supplied specifically for the purpose of being entered and executed on a computer system, for exclusive use by the purchaser of the work. Duplication of this publication or parts thereof is permitted only under the provisions of the Copyright Law of the Publisher's location, in its current version, and permission for use must always be obtained from Springer. Permissions for use may be obtained through RightsLink at the Copyright Clearance Center. Violations are liable to prosecution under the respective Copyright Law.
The use of general descriptive names, registered names, trademarks, service marks, etc. in this publication does not imply, even in the absence of a specific statement, that such names are exempt from the relevant protective laws and regulations and therefore free for general use.
While the advice and information in this book are believed to be true and accurate at the date of publication, neither the authors nor the editors nor the publisher can accept any legal responsibility for any errors or omissions that may be made. The publisher makes no warranty, express or implied, with respect to the material contained herein.

Printed on acid-free paper

Springer is part of Springer Science+Business Media (www.springer.com)

Preface

My interest in acoustics for performance spaces originates from some 1,200 concerts as a former rock and jazz drummer and a B.Sc. in mechanical engineering right after high school. While still touring about 80 concerts a year I took a few courses in acoustics at the Technical University of Denmark. I was immediately drawn into the field and my master thesis is in many ways basis of this very book. Ironically, after the defence I got kicked out of the band! Then I started to rather design the music halls than play in there. The book, which I was asked to write already in 2008, is dedicated to my engineer and freedom-fighter father, Knud W. Larsen, for passing on to me a sense of courageousness and curiosity as well as to my musical mother Therese Adelman Larsen who gave me the loving ballast needed to employ these attributes.

The 9½ week 2010 tour around Europe, during which the 55 music halls presented in this book were measured, was financed by: d&b audiotechnik, Flex Acoustics, Oticon Fonden, Cowi Fonden, Brødrene Hartmanns Fond.

Apart from these companies and foundations the author would like to thank (in random order): all the great people in the halls visited who opened their doors and made me feel welcome; my friend "Dr. Schnips" Teis Schnipper, Leo L. Beranek, Michael Barron for inspiring research, illustrations etc., Bård Støfringsdal, Jens Jørgen Dammerud, Magne Skålevig, Rasmus Rosenberg, Christoph Baumann, Finn Jacobsen, Hugo Fastl, Eric R. Thompson, Jan Voetmann, Jens Holger Rindel, Per Brüel, Cheol-Ho Jeong and Jens Cubick for their direct and profound contributions. You have made this book better. Special thanks to the great crew at d&b audiotechnik. Many more, friends and family (you know who you are), have supported me during this time too—thank you!

Contents

1 Principles of Acoustics and Hearing 1
  Sound Propagation 3
  Sound in Rooms 5
  Human Hearing 9
  Acoustic Defects: Echo 11
  Scattering 12
  Acoustic Vocal Sound 13
  Absorber Types 15
  Audience Absorption 17
  Air Absorption 18
  Critical Distance and Level of Reverberation 19
  Reverberation Time Design 20
  Background Noise 21
  Sound Levels and Amplified Events 21

2 Auditorium Acoustics: Terms, Language, and Concepts 25
  Objective Parameters 29
  EDT, Reverberation Time, Liveliness, and Reverberance 29
  C_{80}, D_{50}, Early Reflections, Clarity, and Intimacy ... 30
  LF, Envelopment, and Lateral Reflections 31
  G, Strength, and Room Gain 31
  Bass Ratio, Warmth, and Bass Response 32
  ST, Support, and Ensemble 33

3 Reinforcement of Sound Sources 35
  The Sound of a Rock Band 38
  Providing Amplified Direct Sound from Loudspeakers 39
  Single Point Sources 40
  Virtual Point Sources 40
  Directional Subwoofer Arrays 41
  Split Mono Systems 43
  Line Arrays 45

4	**Assessments of 20 Halls**	49
	Amager Bio	50
	Forbrændingen	53
	Godset	56
	Lille VEGA	59
	Loppen	62
	Magasinet	65
	Palletten	68
	Pumpehuset	71
	Rytmeposten	74
	Musikhuzet	77
	Skråen	80
	Slagelse Musikhus	83
	Stars	86
	Store VEGA	90
	Sønderborghus	93
	Tobakken	96
	Torvehallerne	99
	Train	102
	Viften	105
	Voxhall	108
5	**Recommended Acoustics for Pop and Rock Music**	111
	The Basis of the Recommendations	112
	Results of the Interviews	114
	The First Page of the Questionnaire	114
	PA System Versus Omnidirectional Source Measurements	116
	General Ratings of the Halls	117
	Musicians' Preferences	119
	Sound Engineers' Preference	120
	Debate	121
	Spectral Analysis of Surveyed Data	122
	Recommended Reverberation Time for a Given Hall Volume	125
	Acceptable Tolerances of T_{30} in Pop Rock Venues	125
	Suitable Reverberation Times in Larger Halls and Arenas	128
6	**Design Principles**	129
	Hall Size	130
	Hall Shape	132
	Stage and Its Surroundings	133
	Surface Materials	135
	Balconies and Overhangs	137
	Floor	137
	Stage	138

Seating.. 138
Platforms.. 139
Sound Insulation... 139
Interior Noise Sources................................... 139
Multipurpose Halls....................................... 140
Music Schools.. 142

7 Gallery of Halls that Present Pop and Rock Music Concerts........ 143
Ancienne Belgique (AB)................................... 143
L'Aeronef.. 151
Alcatraz... 157
Apolo La [2]... 163
Apolo.. 168
Astra.. 174
Bikini... 179
Cavern... 185
La Coopérative de Mai.................................... 192
Le Chabada... 197
Cirkus... 203
Le Confort Moderne....................................... 209
Debaser Medis.. 214
Elysée Montmartre.. 220
Festhalle.. 226
Forest National.. 233
Globen... 238
Grosse Freiheit.. 243
Hallenstadion.. 249
HMV Hammersmith Apollo................................... 253
Heineken Music Hall...................................... 261
Hanns-Martin-Schleyer-Halle.............................. 266
Jyske Bank BOXEN... 271
Kaiser Keller.. 276
Live Music Club.. 280
LKA/Langhorn... 286
Mediolanum Forum... 292
Melkweg—The Max.. 297
MEN Arena.. 303
Nosturi.. 309
O_2 Berlin... 315
O_2 World Hamburg...................................... 321
O_2 London... 325
O_{13} Tilburg... 331
Olympia.. 337
Oslo Spektrum Arena...................................... 343

Palau Sant Jordi... 348
Paradiso... 353
Porsche Arena 360
Rote Fabrik, Aktionshalle 365
Rote Fabrik Clubraum... 371
Rockefeller 376
Rockhal 381
Razzmatazz 1... 386
Razzmatazz 2... 391
Sala Barcelona'92/Sant Jordi Club 396
Scala 400
Tunnel 406
Vega 412
Wembley Arena.. 419
Werk 425
Zeche.. 430
Zeche Carl "Kaue" 435
Zénith Paris—La Villette..................................... 440
Zenith Strasbourg 446

**Appendix A: Measurements of the 55 Venues Presented
    in the Gallery in Chapter 7** 453

Appendix B 463

Appendix C 467

Appendix D: Two Sound Engineers' Statements 469

Introduction

Please imagine an outdoor rock concert. Visualize a flat field, a stage with a band, a crowd listening to the music mainly coming from PA speakers placed at the stage front. That's a rock concert without any room acoustics. Imagine then a huge box being placed around the whole event—a hall. Surely the whole atmosphere changes, but what effect should we allow it to have on the sound? Should the inside surfaces of the building reflect or absorb the sound? How much and at what frequencies?

What is "good acoustics"? The present book answers this question with regards to pop and rock music performances on the basis of the author's research that has been conducted in the field. The knowledge gathered origins from studies based on interviews with musicians and sound engineers.

Of course such a terminology as "good acoustics" only makes sense when referring to *the use* of a particular room. The acoustical demands of a room to be filled by children in a kindergarten must be very different from those favoring a musical choir. The noise aspect of room acoustics is employed in factories, offices and institutions where the sound is regarded largely as unwanted or simply too loud and therefore to be attenuated by acoustical means. Other sounds, in other rooms, with other purposes, are indeed wanted and must be affected by the acoustics of the room in a positive and enhancing way. This desired sound contains a message that must come across to the listener. The room must therefore carry, or transfer, this message to the receiving persons, such as an audience, in a favorable way. But preferably, also the person sending the message, such as a speaker or a musician, should be content with the way the room affects the sound as he or she perceives it. A musician and his/her performance can certainly be negatively affected by undesirable acoustics.

For pop and rock acoustics, in halls larger than club-size, acoustic absorptive materials are not used to correct the sound level of the music. The level is predominantly controlled by the sound engineer sliding faders on the mixer boards. The frequency content of the music that meets the ears of the audience origins of course from the instruments, and is then controlled and enhanced by the sound engineer by electronic equalization means etc. However, part of the sound perceived by the audience, does not come directly from the loudspeakers, but has bounced off of surfaces in the room first. Acoustic absorption should predominantly be employed to control the speed by which these sound reflections decay—the so called reverberation—of the room—and to shape the frequency content of these reflections.

Pop and rock concerts are unique by the fact that they are heavily depending on amplification of the sound that the band produces. A significant ingredient in the music is often a highly amplified and well-articulated bass line supported by a more or less syncopated, staccato of nature, bass drum rhythm. The precise timing of both, down to few hundreds of a second, is what professional orchestras are depending on in order to make hard grooving music, not only satisfactory for the orchestra itself but also, most notably, to the audience. Many other instruments are active in the bass frequency domain. The room's acoustics should enable for this message to be transmitted to the audience with a high degree of intelligibility. It is evident that if a hall allows the bass tones to last for long time, then this rhythmic backbone of that type of music is easily lost. The start and stop of bass sounds are rather converted into a long, dull, legato continuum that can hardly be identified as "rhythmic" by performers or audience. What was expected to be a musical experience is transformed into sound with little sense sometimes close to noise.

Unlike speech the bass level at rock concerts cannot simply be turned down in the overall mix in the PA speakers in order to achieve intelligibility. Then it would not be perceived as a pop or rock concert any longer. Loud bass sound is just as well a prerequisite of amplified music. It is part of the excitement of the performance and simply sounds right. Since the low frequency sound cannot really be aimed or directed at the audience like higher frequency sound, and the audience do not absorb nearly as much bass as higher pitched tones, the largest share of the audience will suffer from the legato-like bass sound if the low frequency sound is not very well controlled and not allowed to last too long. This undefined reverberant sound will partly shadow for, or *partially mask,* the defined, direct bass and higher frequency sound, significantly reducing the understanding of the content of the music. While the level of the direct loudspeaker-based sound decreases with distance, the reflected and later arriving sounds remain more or less constant throughout the venue. This is one reason why considerations regarding the shape of a room are critical in early phases of the design process of a pop and rock venue in order to achieve perfection.

On stage the musicians use open monitor speakers or in-ear monitoring systems to hear themselves and their colleagues. The audience listens to the music mainly through PA speakers operated by a sound engineer. These two sound systems depend on each other to some extend and on the acoustics in the two environments of stage and audience area.

So what is good acoustics for such concerts?—and according to who? Who should determine this? The audience? The musicians? The sound engineer? Do the three groups agree on what kind of room acoustics they can accept and maybe even find recommendable? The sound engineers of course want for the audiences to get value for money and experience the incredible ambience and moods of a concert. They feel obligated to present the audience with a clean and transparent sound. This can lead them to ask for an acoustically very dead stage-area that does not allow the unprocessed sound from instruments and monitor speakers to be

Introduction

reflected to the audience or irrelevant microphones on stage. However, this may contradict the needs of the musicians and the audience.

As we shall see, the investigations among people attending thousands of pop and rock performances lead us to understand that appropriate acoustics for pop and rock, rather than facilitate only a high fidelity sound experience, also shall support the concert as a social event where people meet to obtain a sensation of being together and share an experience in sound. Not only audiences in between but also musicians who need to be *involved* and therefor to be in the *same acoustic climate* as their audience. The musicians too want to experience *togetherness* with their audience as long as they are able to clearly hear what music they are creating for themselves as individuals and as a group. And that cannot be obtained through use of monitors alone but also requires some reflections from stage surroundings as well as from the space of the audience.

Which acoustics will fulfill these different demands? And how is *rock intelligibility* and *togetherness* achieved at the same time? What effects must certainly be avoided? Strangely, before the research paper *"suitable reverberation times for halls for rock and pop music"* from 2010 by the author, no proper research had been carried out in the field. Yes, acoustics for pop and rock is regarded a science. It must, more specifically, be treated as the science of quantifying the fine sensations of people as encountered in the ecstatic moments at pop and rock concerts around the Globe. And these sensations must, through an understanding of sound and general room acoustics, be translated into practical recommendations for the building of venues. That is precisely what this book does. The book is easily understood and is equally informative for sound engineers, rock aficionados as well as acoustic consultants and architects who are often, at the end of the day, the ones responsible for the acoustics in new and renovated venues. There are chapters on the basics of sound and room acoustics, the actual recommendations resulting from research as well as specific comments on 20 rock venues and why they received the ratings they did from musicians and sound engineers. As further examples 55 European venues, some of them being world famous, from tiny basements to enormous arenas, are presented with acoustic measurement results, architectural drawings and photos. It is the hope of the author that it will lead to even better musical experiences.

Endorsements

Acoustics are important within pop and rock venues to ensure a great experience for audiences and performers. This book fills an important gap of knowledge on the acoustics of venues. It will be of value to sound engineers as well as building owners and operators and building design professionals.

Rob Harris, Arup Acoustics

Niels' efforts to gather and analyse data and make this available for others, is highly appreciated and will lead to better sounding concerts in future. The book presents among other things data on the acoustical qualities of many existing venues including information on the crucial low frequencies. This is important knowledge for future research, improvement of existing halls and design of new venues. Designing the acoustics for halls for amplified music takes some knowledge that was not written in books prior to this one.

Martijn Vercammen, Peutz

Everybody thinks they know about acoustics, very few people actually do. Books like this can and should make a difference, and will make the world a better sounding place. For anybody who cares about what they hear in a live setting, it's essential reading.

Simon Honywill, Live Sound Engineer for Jeff Wayne's Musical Version of the War of The Worlds, Katherine Jenkins, Chris Rea, Jose Carreras etc.

For me, the acoustics of a room are of primary concern for a concert, secondary only to the quality of the musicians and their instruments. Amplified music is much harder to manage in spaces with longer reverberation times, especially at faster tempos and with denser instrumentation. The loss of definition in the bass frequencies really blurs the groove and feeling of music that depends on amplified bass and drums.

Ben Surman, FOH Engineer for Jack DeJohnette, John Scofield and many others, USA.

Chapter 1
Principles of Acoustics and Hearing

What is a sound wave? A visual interpretation can be obtained by imagining the membrane of a speaker that moves back and forth. Air molecules immediately next to the speaker will co-oscillate with the speaker. These air molecules will push and pull their neighbors, who in turn will push and pull their neighbors, and so on, forming a longitudinal wave of oscillating molecules. This is in essence a sound wave: tiny pockets of air, oscillating around an equilibrium position, causing small air pressure variations imposed on the static air pressure. A sound wave travels with a speed of 343 m/s in air at 20 °C whereas in solids and fluids the speed is faster and the total distance it can travel is longer, mainly due to the higher density of molecules.

The magnitude of the wave determines the amplitude of the sound given by a pressure maximum and a pressure minimum as graphically interpreted in Fig. 1.1. Because the human ear is capable of detecting sounds levels ranging from 1 unit to sounds one million times louder, the *decibel* (dB) scale has been introduced. Named after Alexander Graham Bell (1877–1922) the decibel is a logarithmic unit and overcomes thereby the handling of very large numbers. The decibel is used to measure sound levels and of course therefore also sound level differences.

The quietest sound that the human ear can detect is about 0 dB and the least noticeable difference is about 1 dB. Doubling the number of bass players in a symphonic orchestra will increase the sound level by 3 dB, however, a double brick wall will cause an isolation of the sound pressure of approximately 55 dB. The sound level in the open drops by 6 dB per doubling of distance. A symphony orchestra produces only approximately 2.5 W of acoustic power when playing *fortissimo* (very loud). The dynamic range of a symphony orchestra is as much as 70 dB, that of a pop and rock band considerably less, whereas jazz performances can span over a large dynamic interval from *ppp* to *fff*, in musical terms, as well.

The decibel is hence based on a ratio and is therefore not a unit like a meter or a watt. In order to express the absolute value of a sound pressure the ratio is taken between a given level and a reference level being 2.10^{-7} mbar or 20 μPa corresponding roughly to the lowest audible sound at 1,000 Hz.

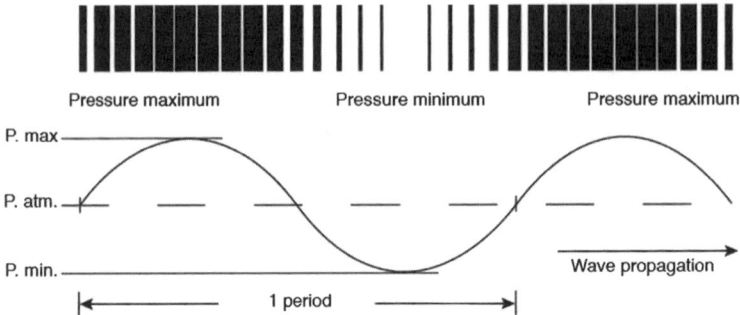

Fig. 1.1 Graphic representation of a pure tone. One period equals the wavelength of the tone

Frequency, f, and wavelength, λ, are two parameters expressing a wave. In, for instance, musical tones, the wavelength equals one period of the sound wave implying that the shape of the wave is periodic. The distance between pressure maxima (or minima) in the sound wave is the wavelength. The wave travels with the speed of sound c, and the three parameters interrelate as expressed by the equation

$$c = \lambda f \qquad (1.1)$$

which is in good agreement with one's intuitive understanding: the shorter the wave, the more often will one pass a certain point in as much as all waves pass with a constant speed, the speed of sound. The unit of frequency is s^{-1} (cycles/s) which is written as Hertz, Hz, named after Heinrich Rudolf Hertz (1857–1894). The wavelength unit is meters.

A pure tone has a single frequency associated with it. Musical instruments as well as the human voice, however, produce a number of overtones creating a unique sound where the lowest tone normally determines the pitch. A doubling of frequency corresponds to an octave in musical terms. Acoustic measurements are traditionally often made over octave intervals with center frequencies being: 63, 125, 250, 500, 1,000 Hz, and so on.

Many instruments depend on *resonance* in order to create their sound. The length and diameter of a flute, the volume of a drum and the tension of its head, and the body volume of an acoustic guitar are examples of resonators and ways to achieve a particular resonance frequency. But in amplified music, electric instruments are often used and the vibration of a string on an electric bass or guitar, for example, is picked up and amplified by the use of a transducer and amplifiers. The lowest note is called the *fundamental frequency* (Fig. 1.2) and the overtones are referred to as *harmonics* of the instrument. The A above middle C on a piano is often tuned to 440 Hz and the lowest E string on an acoustic or electric bass is about 41 Hz. The shapes of such complex waves are still close to periodic. The overtones that create the spectrum of the instrument and thereby its characteristic sound together with the sound created where the hand, stick, bow, or another tool touches the instrument, in many cases extend all the way to the highest sound possible for the human ear to hear above 15–20 kHz.

Sound Propagation

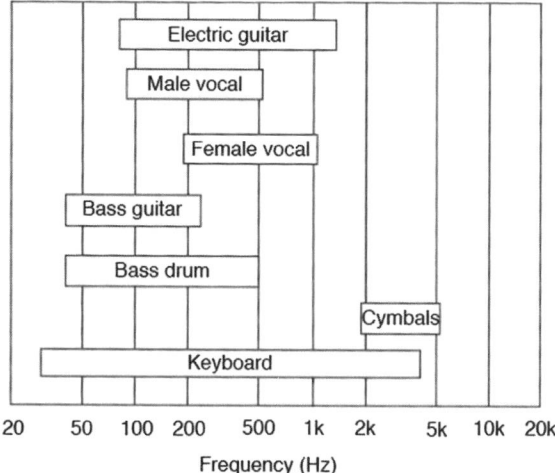

Fig. 1.2 Fundamental frequencies of different instruments used in pop and rock music

Sound Propagation

Sound rarely propagates in a uniform pattern away from the sound source. The pattern by which sound is distributed is dependent on, among other things, the frequency produced. Such patterns can be drawn for each frequency band and are called directivity patterns.

Sound propagates in its simplest form outdoor in the open, where the waves are not affected by any obstacles. Without reflections from walls or ceiling the unamplified human voice is capable of reaching approximately at least 50 m, which is also the longest distance from a performer to a listener in, for instance, the ancient Greek theatres. In these theatres, though, the proscenium walls behind the performers reflect the part of the performers' sound that propagates away from the audience to be redirected towards the audience permitting a higher sound level for the performer and audience.

Indoors, the level of the direct sound of course decreases farther away from the sound source. From pop and rock concerts, most of us have experienced being placed behind taller audience members standing in front of us blocking us visually. Actually we are also blocked from the direct sound if the placement of the sound system does not compensate for this by being placed high enough or if the floor is not sloped.

Only frequencies with wavelengths comparable to the size of the obstacle in front or smaller are blocked because large wavelengths bend around obstacles of inferior size. This phenomenon is called diffraction (Fig. 1.3). The obstacle is simply being surrounded by lower frequency sound. Likewise, it takes a very large and heavy object such as a concrete wall to reflect a low-frequency sound precisely. When the reflecting surface is large compared to the wavelength of an incoming sound wave, the sound will be reflected just like light in a mirror. This is called *specular reflection*.

Fig. 1.3 Sound waves with larger wavelengths bend around smaller objects and higher frequency sounds are blocked

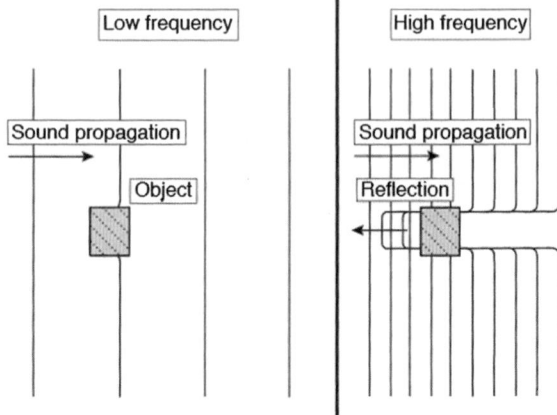

Fig. 1.4 The reflected sound will be affected differently depending on the shape of the boundary the incoming sound wave meets

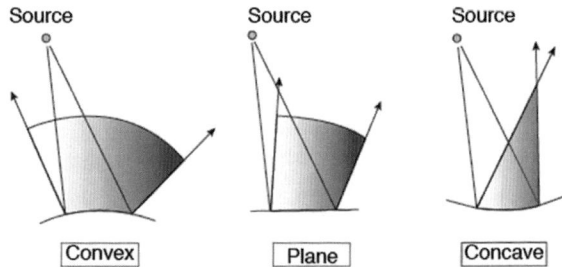

As light spreads scattered by a matte white piece of paper, sound can also be scattered more or less universally out in all directions from the scattering surface. The term *diffusing surface* is also used to account for the same effect and should not be confused with the diffuse sound field discussed later. Scattering is a very important property that is used extensively in auditorium design.

Conversely, black matte paper will not let any light get away, and in the same way, for instance, a thick layer of mineral wool will absorb incoming sound. Any building material, and the way it is mounted, will determine how much sound it absorbs and reflects and at which frequencies it does so.

Sound is reflected in different ways depending on its frequency and the shape of the boundary it meets. At mid and high frequencies sound waves can be considered as rays or beams, and the incoming angle of the wave is equal to the angle at which it bounces away from the boundary, very similar to a billiard ball. This situation with a plane boundary is sketched in the middle drawing of Fig. 1.4. If the surface is convex, however, the beam will be scattered, whereas a concave surface will lead to a situation where the sound is reflected from several places on the surface back to the listener's ear with small time differences. This causes a so-called focusing effect, highly disruptive for the intelligibility of the sound.

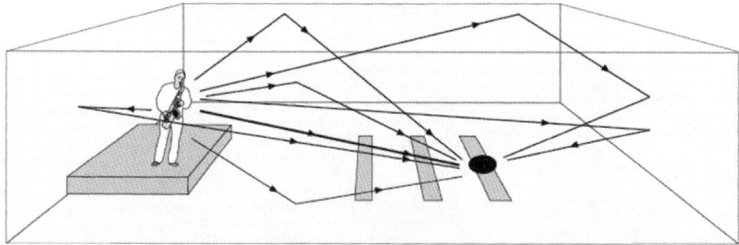

Fig. 1.5 Direct sound and first reflections from various surfaces in a room

Fig. 1.6 Reflectogram indicating how sound reflections soon multiply into reverberation

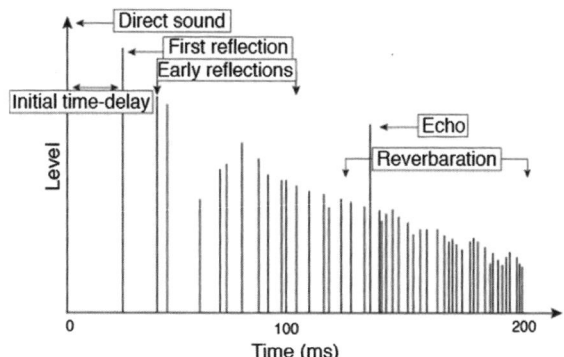

Sound in Rooms

Soon after the direct sound arrives at the listener, a series of early reflections follows from the side walls, ceiling, and so on (Fig. 1.5). This sound evidently will arrive later at the listener because it travels longer. The initial sound wave can spread in many directions; it hits numerous surfaces in the room and soon multiplies into hundreds of reflections. The later sound arriving after about 0.1 s (depending on hall size, etc.) is called the *reverberant sound* or simply *reverberation*. The characteristic sound at a given location in any hall is thus a result of thousands of reflections and where, at which frequencies, and to what degree they are reflected, scattered, or absorbed. Luckily the human hearing system is capable of distinguishing subtle differences and acknowledging the unique quality of different halls.

A diagram showing the amplitude of reflections versus time is called the *impulse response* or a *reflectogram* (Fig. 1.6). The term "impulse response" can also mean an actual recording of the room's response to a sound.

Any enclosed space has normal modes given by its dimensions. These modes are also referred to as the *standing waves* of the room when it is exposed to sound. The room modes are responsible for there being places in the room where a given frequency is louder than in other places of the room.

For a rectangular room with dimensions l_x, l_y, l_z it can be shown that the modes, also called *natural frequencies*, are

$$f_{l,m,n} = \frac{c}{2}\left[\left(\frac{\ell}{l_x}\right)^2 + \left(\frac{m}{l_y}\right)^2 + \left(\frac{n}{l_z}\right)^2\right]^{\frac{1}{2}} \quad (1.2)$$

where l, m, and n are integers that indicate the number of nodal planes perpendicular to the *x*-, *y*-, and *z*-axes, the three dimensions of the room. They are, for example, [1, 0, 1] or [7, 4, 3] or any other integers. The important point here is that the natural frequencies are given by the room dimensions. This is also true in odd-sized rooms whose natural frequencies can be calculated by use of computer programs.

It can be proven that the density of natural frequencies is proportional to the frequency squared, f^2, indicating that at higher frequencies the density is high, in fact so high that the modes can be regarded as a continuum. Likewise the size of the room plays a role and in small rooms the modes are very separated indeed at low frequencies, leading to severe colorations of the sound at certain frequencies. Such colorations will be even bigger if the room dimensions are a low integer multiple of one another because the modal frequencies will then coincide. So at higher frequencies, the sound energy at a given frequency is most likely more uniformly distributed in the room. It is said that the sound field is more diffuse at these higher frequencies.

In any room with a continuous (stationary) sound source, the sound field will be formed by standing waves. The decay of sound "from" the standing waves at the time right after a sound source has stopped, is one feature of reverberation.

But reverberation can occur without standing waves; think of a hand clap in a room. There will be reverberation because of hundreds of echoes repeated soon after one another, but standing waves will not have had time to build up. The same thing happens in a forest: a sudden percussive sound creates a reverb because of reflections from the many trees, but the sound waves would propagate or "run," probably not "stand".

It is easy to understand intuitively that the more the sound reflections hit some sound-absorptive surface, the faster the sound dies out in a given room. And likewise: the bigger the room is, the longer a reverberation. The time it takes for the sound in a room to attenuate 60 dB is called the *reverberation time* of the room. Despite many advances in acoustics over the latest decades this remains the single most important parameter in room acoustics. It is essential to note that the reverberation time is usually not the same in different frequency bands for a given room. It can be shown that the time T, in seconds, it takes for the sound to attenuate by 60 dB in a perfectly diffuse room proposed by Sabine (Wallace Clement Sabine 1868–1919), in a given frequency band, is approximately

$$T_{60} \approx 0.16 \frac{V}{A} \quad (1.3)$$

This equation is called Sabine's formula and is rightfully the most well-known relation of parameters in room acoustics. In reality, especially smaller rooms are

Fig. 1.7 Graphic representation of reverberation time (*RT*) and T_{30}

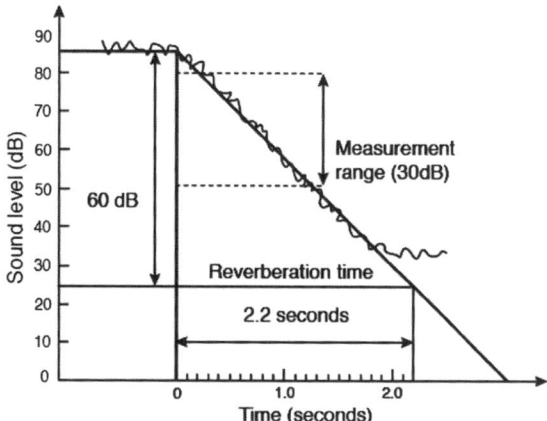

not perfectly diffuse and certainly not in all frequency bands. *V* is the volume of the hall in cubic meters and *A* is the absorption area, meaning the number of equivalent square meters of 100 % sound-absorbing material in a given frequency band. The reverberation time is sometimes, also in this book, denoted simply *T* or *RT*.

When actually measuring the reverberation time of a room usually only the sound from –5 to –35 dB or even –25 dB is measured because background noise usually makes it difficult actually to measure anything as silent as 60 dB below the sound source measurement signal. The time this 30 or 20 dB decay takes is then multiplied by 2 or 3, respectively (most often, this multiplication is performed by the measurement device itself), assuming that the decay is linear over time so that the decay of the first some 30 dB takes the same time as the latter 30 dB. This measured reverberation time is called T_{30} (Fig. 1.7) or T_{20}. One can get a rough estimate of the reverberation time in a room at approximately 2 kHz by clapping one's hands once and counting the number of seconds it takes for that sound to die out. Low-frequency reverberation time can be estimated by producing a similar low-frequency sound. Of course, such tests are not in compliance with any standards for measurement of reverberation time where an exact measurement is made in every octave, or third octave band, for instance by firing a gun, an abrupt loud signal of noise, or measuring the decay of a sine wave sweeping through frequencies.

This reverberation time based on a long decay is also called the terminal reverberation time because it denotes a sound and the decay to more or less complete inaudibility. This seldom being the case during a concert (mostly only at the last note of a song), there is also another parameter describing sound decay called the *running reverberation time*. This takes into account only the first 10 dB of decay and to make it comparable to the terminal reverberation time, this time is multiplied by 6. If the decay slope is strictly linear from 0 to –60 dB the two reverberation times are identical, but this is seldom the case. The running reverberation time is denoted EDT—early decay time—and was suggested in 1968 by Vilhelm

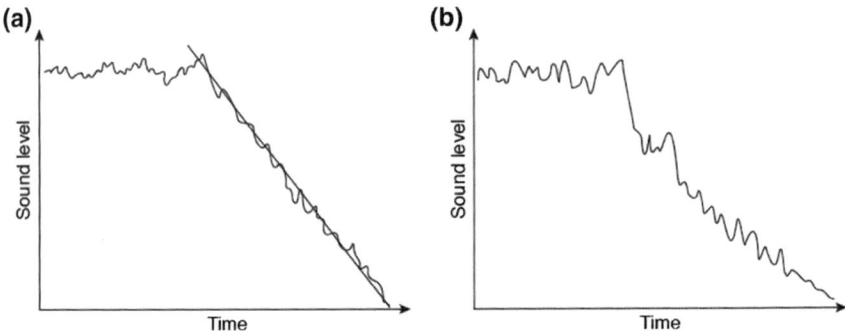

Fig. 1.8 A linear (**a**) and a nonlinear (**b**) sound decay in a room. This explains why T_{30} and EDT can be different

Lassen Jordan (1909–1982). The EDT is heavily dependent on the position in the hall where it is measured. As shown later, a given volume of hall for a given purpose has an ideal associated terminal reverberation time. Bigger halls allow, as mentioned, longer reverberation. The sound decay is often nonlinear and EDT is often a more relevant parameter to describe the reverberation actually experienced by a listener in a given position of a room. See Fig. 1.8.

The so-called Schröder frequency suggested in 1962 by Manfred R. Schröder (1926–2009) denotes the frequency below which the sound field can be regarded as being dominated by discrete standing waves

$$fs = 2,000\sqrt{\frac{T}{V}} \quad (1.4)$$

where T is the reverberation time of the room in seconds and V is the volume in cubic meters.

There is no sharp line defining above which frequency the sound field can be regarded as diffuse. Below this frequency, on the other hand, as discussed above, the modes will not be close to one another, thus standing waves at single frequencies will occur. In either case reverberation will occur at least as the decay of sound "from" the standing waves. As seen in Eq. (1.4), in bigger rooms the Schröder frequency is lower. In smaller rooms, measurements of low-frequency reverberation time can be very dependent on the position where it is measured because of the modes; therefore many measurements at different positions are averaged.

The halls investigated in this book are in almost all cases over some 1,000 m^3. With an ideal reverberation time for pop and rock music of 0.6 s for that size volume (see Chap. 5), this yields a Schröder frequency f_s below about 50 Hz, close to the lower extremity of the 63-Hz octave band. Although a lot of halls suffer from a long reverberation time at low frequencies moving f_s upwards, for the purposes of this book, sound even in the 63-Hz octave band is generally regarded as diffuse. This is important to stress, for instance, for sound engineers who often think of room acoustics solely in terms of standing waves. For studios and very small venues such as clubs and cafés with, for example, a low ceiling, it is certainly correct that distinguished annoying modes exist at lower frequencies. For mid-sized or larger halls this is not an essential point; the density of the natural frequencies is high enough even for low tones to be regarded as a diffuse sound field that decays over time as reverberation.

The relation between sound-level difference and reverberation time difference under the diffuse sound field is

$$\Delta L_p = 10 \log \frac{T1}{T2} \tag{1.5}$$

According to this formula doubling the reverberation time in a given frequency band compared to others, results in $10 \cdot \log 2 \approx 3$ dB extra response from that frequency band. A common misconception is the belief that when that frequency band has been lowered level wise, for instance on an equalizer, then the room acoustics have been corrected! This is completely incorrect because the reverberation time at that frequency is still the same. Level adjustments work in the level domain, not in the time domain. As a matter of fact the reverberation of a given room size has to be not too long and usually not too short either for a given purpose. Furthermore, musicians and sound engineers adjust their playing and their levels so that it conforms to the hall and the situation. Thereby the level difference that the possibly uneven reverberation times at different frequencies impose is evened out, still leaving the clarity of the sound, and thereby the overall sound quality, to be affected. As we show, a too-high reverberation value in a given frequency band will cause an undefined blurred sound in that frequency domain and often that is enough to damage the overall sound in a venue.

Acoustic consultants use this connection between reverberation and sound level as means to enhance, or acoustically amplify, lower frequency sound in halls for acoustical music, for example, or inversely, to reduce noise in a kindergarten or factory. In pop and rock venues, the amplification aspect of reverberation is close to uninteresting inasmuch as all levels are adjusted by the turn of a knob either by the musician or by the sound engineer. For amplified music the reverberation solely affects the sound by either washing away the core message in the music (when it is too long) or by making the performance seem dull and lacking liveliness (when too short) as discussed in Chap. 5.

Human Hearing

A sound source emits sound in many directions and the reverberation consists of thousands of reflections from many surfaces in the room (the surfaces that are not 100 % sound absorbing at all frequencies). In a concert hall for symphonic music there are approximately 8,000 reflections from one single note during the first second. Of course in smaller and less reverberant halls there are fewer reflections, but still

Fig. 1.9 Equal loudness contours at various levels. Different frequencies need different levels to be perceived equally loud for humans. The contours vary with sound level. *Lowest curve* is the hearing threshold

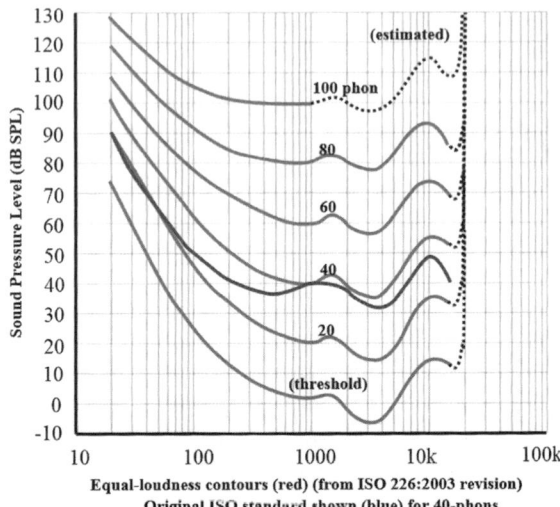

Equal-loudness contours (red) (from ISO 226:2003 revision)
Original ISO standard shown (blue) for 40-phons

each one has a delay, a direction, and a sound level associated with it. The ear is highly selective in interpreting this abundant information. The direct sound is precisely localized by the ear: even when the sum of later reflections is louder than the first direct sound, the ear is capable of detecting the direction from where it originates. The direct sound is usually not fully masked by louder later sound.

The term masking is used when the presence of a given sound *A* makes another sound *B* inaudible (fully masking) or more or less difficult to hear (partially masking). *A* masks *B* or *B* is masked by *A*: a person needs to turn down the television in order to understand what is being said on the phone. Generally a single note masks more towards the upper frequencies of other tones than to the lower. In addition, the louder the masking tone is, the broader a frequency range it masks. The term *forward masking* is used to describe the phenomenon that a loud sound signal can mask another weaker signal which is presented up to 200 ms after the first signal stops. The opposite also is possible and is called *backwards masking* because it goes back in time and the effect is restricted to 20 ms before the start of the strong signal when loud enough compared to the weak one.

Perceived strength of a sound is called loudness. The unit is son or sone. A short sound with the same sound level as a longer sound is perceived as less loud. A 5-ms sound must be 25 dB louder for the loudness to be the same as a 60-dB sound of 200-ms duration. The ear is most sensitive sounds at different frequencies but at the same sound level don't occur equally loud to the ear.

The ear is most sensitive around 3.4 kHz (the approximate resonance frequency of the typical ear canal) whereas our hearing rolls off at higher and certainly at lower frequencies below about 100 Hz although less at higher sound levels. This is one reason why at pop and rock concerts so much electric power is used to amplify the low frequencies for them to be perceived equally loud as higher frequencies. The graph showing this is referred to as the equal loudness contours (Fig. 1.9).

Human Hearing

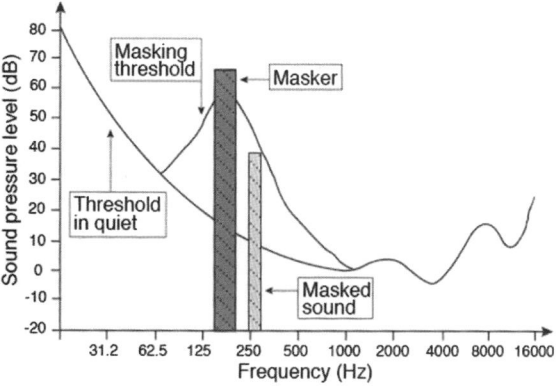

Fig. 1.10 The human ear can hear sounds that are louder than the "threshold in quiet" curve. One sound, the "masker" casts a shadow at frequencies both higher and lower than itself that prevents other sounds, "masked sounds", to be audible if they are at a lower level than the masking threshold

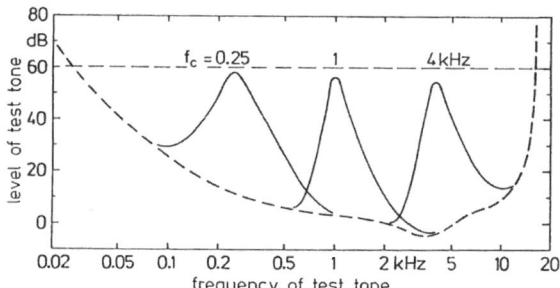

Fig. 1.11 The *curves* show the masking thresholds when the ear is exposed to about 60 dB narrowband noise at 250 Hz, 1 and 4 kHz (Zwicker and Fastl)

When regarding the figure it is seen that a sound decay from, for instance, 100 dB at 50 Hz gets inaudible quicker than a decay at, for instance, 200 Hz because the difference in dB from the perceived 100 dB curve to the hearing curve threshold is smaller at 50 Hz (given that the reverberation time of the room is close to being the same at those two frequencies). See Figs. 1.10, 1.11, 1.12 and 1.13.

Acoustic Defects: Echo

If a reflection is delayed by more than 50 ms compared to the direct sound, the ear will perceive that reflection as a distinct echo if it is louder than adjacent reflections that arrive before or after. These other reflections could otherwise mask it. An echo is always unacceptable in a music hall and must be eliminated. Echoes are more apt to occur in halls with short reverberation times because there will be few other reflections to mask them.

Flutter echoes are a series of rapidly repeated echoes mainly by parallel planar surfaces. Whereas *distinct echoes* involve long path lengths, flutter echoes are likely to occur at shorter distances such as between parallel walls on a stage, particularly if not many obstructive elements are present.

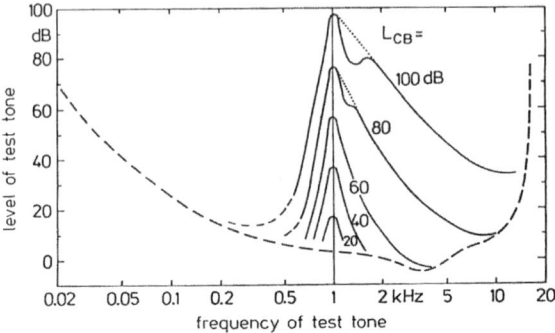

Fig. 1.12 The influence of level on the masking threshold. The masking effect gets broader with increasing level. Masking thus increases nonlinearly with level (Zwicker and Fastl)

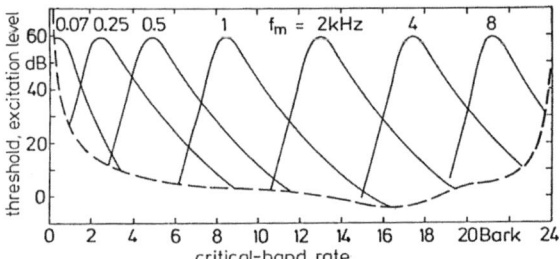

Fig. 1.13 The broadness of the masking curve changes with frequency. The slope at 70 Hz is steeper than those at higher frequencies indicating that sounds in the 63-Hz band do not mask as broadly as higher frequency sound. Here, instead of the octave band, the Bark scale is used. This scale is used in psychoacoustics instead of octave bands (Zwicker and Fastl)

Scattering

Flutter as well as distinct echoes can be avoided by applying a diffusive or angled structure on surfaces where the specular reflection originates. In (smaller) rooms with little or no obstructive elements the standing waves are not taken apart whereby the amplitude of the standing waves and thereby colorations of the sound can reach higher levels. The same goes for rooms with parallel surfaces especially, of course, if the room dimensions are low integer multiples of one another. These waves should be broken up to avoid colorations. An analogy to waves in water can be made: a pier or mole forms a breakwater that will crush the wave. But obstructive elements in a room of a size comparable to the wavelength of the wave do indeed break the wave so that a more uniform distribution of sound energy of the tone becomes present in the room. The sound field is scattered. The big peaks and dips are somewhat evened out so that one will encounter fewer level differences throughout the room. The reverberation time of a room with uneven distribution of sound absorption is greatly reduced when the room is made more diffuse.

The depth and the overall dimensions of scattering elements, also referred to as diffusers, are decisive for which frequencies are scattered. For example, many living rooms are highly diffusive (diffuse) because of the presence of lots of small and large elements randomly placed in the room. Schröder was behind a major development in diffusive elements (1979) where he proposed a series of diffusers whose properties could be calculated in advance. The pioneering work of exploiting diffusers has since been carried forward by Marshall, Cox, d'Antonio, Konnert, and others. Vorländer and Mommertz have made progress in actually measuring scattering coefficients in the laboratory.

In mid-sized and bigger halls (which this book is mainly about) additional scattering elements are not a necessity because of the high diffusivity due to the absence of isolated standing waves in large enclosures. Also an audience scatters the sound profoundly, although not as much at lower frequencies. The stage surrounding may, however, benefit from well-designed diffusive surfaces that eliminate flutter echoes and help to distribute the sound energy on stage more evenly.

Acoustic Vocal Sound

Niels W. Adelman-Larsen and Jens Jørgen Dammerud

The vocal is often a primary instrument at pop and rock performances and without clear comprehensible lyrics a lot of the message from the band to the audience is lost. Understanding what is actually being sung is related to the term "speech intelligibility". Much research is carried out on speech intelligibility. Some of the results from this research show that the mid- to high-frequency sound is the most important for enabling understanding of what is actually being said. Consonants are important and they have most of their sound energy within this frequency range. A telephone transmits almost exclusively 300–3,000-Hz sound. And yet we understand very well what's being said. It is also within this frequency range that our hearing is most sensitive. At low levels our auditory system is not very sensitive to low-frequency sound. The low-frequency sound is not critical for getting the message. It will in most cases actually be more of the opposite: low-frequency sounds will often mask mid-frequency sound, the low-frequency sound being a masker for the other sound components, and just make communication more difficult. Figure 1.14 shows this effect. The louder the low-frequency sounds are, the broader a shadow is cast.

Of course a vocal consisting only of 300–3,000-Hz sound doesn't sound very natural. A male speaker has his fundamental sound component usually within the 125-Hz octave band, whereas female speech is usually within the 250-Hz octave band. In singing, the fundamental tones are also often within these ranges.

Within the 125- and 250-Hz octave bands a speaker or singer is close to emitting sound *omnidirectionally*, meaning that the sound propagates at close to equal levels in all directions away from the person. At higher frequencies the emitted sound is much more focused towards the front. This means that the

Fig. 1.14 Low-frequency content of the voice can partially mask higher frequency content. *Vertical*: level; *horizontal*: frequency

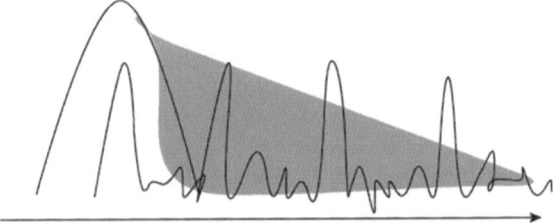

total emitted sound in all directions is usually much higher at low frequencies. Especially in an open space without reflections the directivity of the speaker is relevant with regard to the orientation of the speaker. If the person is faced away, the acoustic direct sound level within 300–3,000 Hz is significantly reduced.

What effect does the room have on speech intelligibility? For instance, in small meeting rooms the bass level due to standing waves can be so high compared to other frequencies that intelligibility is low even close to the speaker. In some cases the intelligibility is not too bad, but the low-frequency boost alone can be very disturbing. As mentioned, speech can be intelligible at a distance of 50 m in an open free field. For small rooms standing waves due to room modes will often be significant and these room modes can contribute to rather extreme acoustic gain of the speech, typically within the 125-Hz octave bands, but also within the 250-Hz octave band in very small rooms. In larger rooms where the speaker is amplified by electroacoustic means, often speech intelligibility is low if the low-frequency reverberation time is long and if the sound engineer does not lower the low-frequency content on the sound system. The long RT at low frequencies in the hall boosts the low frequency sound but it also masks the next syllable in the sentence of the speaker. The amplification is brought down to resemble the speakers' actual voice with the help of an equalizer (EQ). If the reverberant bass sound still masks the important frequencies for intelligibility (300–3,000 Hz) the sound engineer can adjust the EQ even harder and bring down the low end farther. The audience will probably not think that they did not get what they came for just because there was not a lot of low-frequency content in the voice. The important thing here is getting the message across. And that predominantly happens, for vocal sound, from 300 to 3,000 Hz. The professional sound engineer will instinctively equalize the voice so that no frequencies are masking others to the point where a transparent, "open sound" appears from the PA system in the hall. Also the sound level produced by the PA system is important for speech intelligibility; a too-high sound level introduces distortion in the listener's hearing system, which can significantly reduce intelligibility.

Reflected sound within the mid- and high-frequency range can also mask the direct sound. Furthermore, late single reflections perceived as echo are very disturbing. Such echoes can result in temporal (forward) masking where a loud reflection can make it difficult to hear new appearing direct sound. To sum up: the accumulated reverberant response provides a general "background noise" that certainly can partially mask the direct sound. Just as the sound engineer can clean out masking frequencies from a vocal to obtain a transparent sound, so he can obtain a clear mix of a whole rock band as discussed in Chap. 5.

Absorber Types

As mentioned, a certain volume of hall used for a given purpose has a recommended reverberation time associated with it. As seen from Sabine's equation, the hall volume and the area of (100 %) absorptive materials determine the reverberation time in a perfectly diffusive room. The geometry of the interior design affects the diffusivity of the enclosed space whereby the reverberation time is affected. Once the general architectural design of the interior of a hall is chosen it becomes the job of an acoustician, in accordance with the architect, to consider which building materials and acoustical products to employ, to what extent, and how to mount them because in this way they can predict which frequency bands are absorbed and to what degree. In this way a desired reverberation time as a function of frequency can be achieved in a given hall. In Table 1.1 absorption coefficients for some building materials are shown for various frequency bands. Absorption of sound can take place by three different means: porous, vibrating panel, and resonator absorption.

In porous absorption (such as mineral wool, drapes, porous concrete, etc.) the sound energy is dissipated into heat because the propagation of the sound wave is impeded. The porosity of a given material determines its flow resistance which is an important measure when calculating absorption properties of porous absorbers. The sound wave is impeded the most at the distance from the surface where the particle velocity of the wave is at its highest. This occurs 1, 3, or 5 or more times a quarter of a wavelength away from the surface; placing the porous absorber at a distance from the wall will increase the absorption effect at lower frequencies. But for the porous absorber to work at low frequencies it usually has to be of a certain thickness too in order to be obstructive for the relatively big wave. This is one reason why curtains, drapes, banners, and the like traditionally do not absorb any significant amount of sound energy below some 200 Hz. One can easily make a quick test if something is working as porous absorber: when blowing at the specimen, does the air seem to go through or is it blocked? If it goes through, and with some resistance, it will work by the porous absorption principle. See Fig. 1.15.

Vibrating panel absorption, also known as membrane absorption, occurs when the sound pressure on one side of a stiff or limb plate is significantly different from that on the other side. So if there is an airtight cavity behind the plate it will absorb sound because the plate will be forced to vibrate forth and back being pressed by the higher sound pressure on the one side and forced back by

Table 1.1 Possible acoustic absorption coefficients of different building materials (from Barron, Kuttruff, etc.)

Material	Center frequency of octave band (Hz)					
	125	250	500	1,000	2,000	4,000
Hard surfaces (brick walls, plaster, hard floors, etc.)	0.02	0.02	0.03	0.03	0.04	0.05
Slightly vibrating walls, suspended ceiling, glazing, etc.	0.10	0.07	0.05	0.04	0.04	0.05
Vibrating surfaces (single-layer gypsum board, etc.)	0.25	0.20	0.10	0.05	0.05	0.05
Carpet, 5-mm thick	0.02	0.03	0.05	0.10	0.30	0.50
Curtain (velour, draped)	0.06	0.31	0.44	0.80	0.75	0.65
Air absorption coefficient, 4 m (m^{-1})	0.00	0.00	0.00	0.00	0.01	0.03

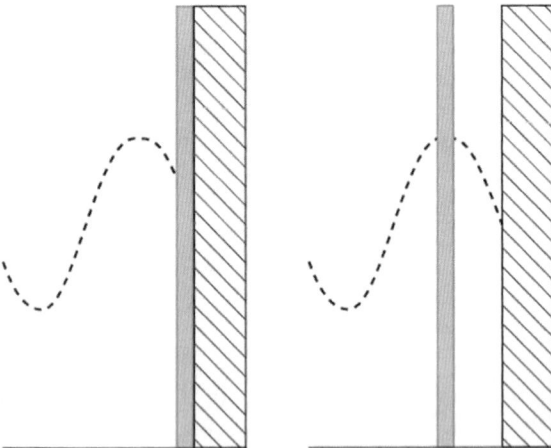

Fig. 1.15 Porous absorption works better at larger wave lengths when at a distance from the reflecting surface. The *dashed curve* represents the particle velocity in the reflected sound wave

the elasticity of the plate like a spring. Hence the system acts as does any mass-spring–damping system where the damping takes place inside the membrane as well as in the enclosed air volume of the cavity behind the plate. In order to increase the damping properties the cavity is often partly filled with porous absorption. A gypsum wall, a wooden floor on a cavity, and a window are all building elements that function by the membrane absorber principle and are certainly taken into account when designing spaces acoustically. The depth of the cavity and the mass per area of the membrane are two key factors when calculating at what frequency the absorber will achieve its maximum absorption as well as the frequency range of the absorber. Also the elasticity module of the plate material and the plate size are important attributes.

Resonating absorbers consist of an enclosed volume of air which is in open contact with the outside air through one or more openings. They are divided into three types depending on shape and number of openings: Helmholz resonators, slit resonators, and resonating panels. They possess very different absorption properties. The sound absorption that occurs in these devices is connected to the fact that the incoming sound waves meet the reflected, phase-shifted sound waves whereby they to some degree cancel each other. Where Helmholtz resonators are used to absorb single standing waves, slit resonators and resonating panels are used to absorb broader frequency ranges. Such panels are well known, such as perforated gypsum panels mounted in the ceiling.

Absorbers are used either to achieve a suitable reverberation time for a given purpose, to eliminate unwanted reflections such as echoes (diffusers or angled surfaces are maybe more obvious for this purpose), or to lower noise levels in rooms such as kindergartens, factories, and offices.

One must always remember that in order to make a significant lowering of reverberation time across frequencies in a given hall very significant areas must be occupied with absorption of a relatively high absorption coefficient across frequencies. Often the entire ceiling area or more must be used to achieve a desired effect. This can be an expensive venture and that is why it is highly recommended to get an acoustical consultant to do exact calculations. It is simply too expensive not to get it right in the first stroke.

Audience Absorption

A common comment among live sound engineers to the musicians at sound check is, "Don't worry; it will be OK once the audience is in place," usually said with an ironic smile. Well, the audience in fact does absorb a lot of sound but almost exclusively at middle and higher frequencies. A tightly packed standing audience has an absorption coefficient of above 1.0 at frequencies higher than 1 kHz. The reason why a coefficient greater than 1 is possible is both that the surface of the people is greater than the surface at which they are standing, but also because they scatter the sound whereby the probability of the sound being absorbed elsewhere heightens, leading to a greater absorption coefficient. In Fig. 1.16 absorption coefficients of a standing and seated audience are shown as a function of frequency. If the seats are heavily upholstered, higher absorption coefficients at low frequencies can be obtained.

The absorption effect of an audience is very similar to that of heavy drapery in front of a reflective surface. Also, quite few people, diversely spread over the audience area, have an impact on the reverberation time of a hall. On the other hand, the author and his colleagues have encountered at least one hall that had to be completely filled with an audience before the sound would be acceptable on stage. In Chap. 4 20 Danish halls for pop and rock are presented. Each of the T_{30} diagrams includes a calculated T_{30} curve with a packed audience. The presence of an audience does not have a big influence on the low-frequency reverberation time of a hall. This is one reason why halls for pop and rock must be designed with a low RT at low frequencies as discussed later in detail.

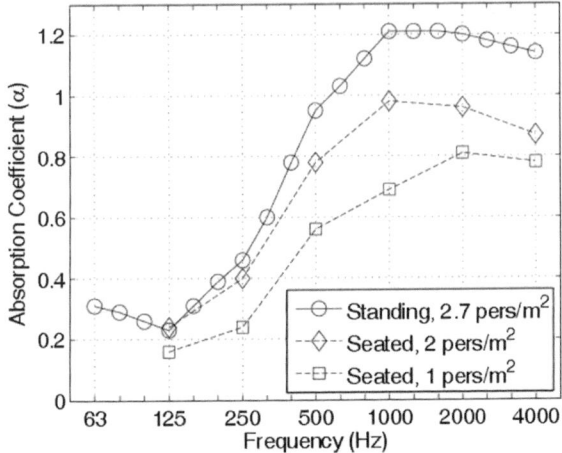

Fig. 1.16 Absorption coefficients of seated and standing audiences

Table 1.2 Attenuation constant of air at 20 °C and normal atmospheric pressure, in 10^{-3} m^{-1} (after Kuttruff)

Relative humidity (%)	Frequency (kHz)				
	0.5	1	2	4	8
40	0.6	1.1	2.6	8.4	30.0
50	0.6	1.1	2.3	6.8	24.3
60	0.6	1.1	2.2	5.9	20.1
70	0.6	1.2	2.1	5.3	17.9

Air Absorption

A shorter reverberation above 2–4 kHz is normal in halls above a certain size as in almost all halls presented in this book. This is due to the absorption of sound by air. Air absorbs a significant amount of sound at very high frequencies as seen in Table 1.2. Dry air absorbs much more sound than more humid air, so part of the high-frequency sound above 2–4 kHz which is absorbed by the audience is "given back" once they start to dance and sweat if the hall is not effectively dehumidified or the air was humid to begin with. The sound engineer will usually make up for the sound-level part of this effect on EQs. Also, at pop and rock concerts, artificial reverberation is usually added to many instruments at higher frequencies anyway, and the amount of this can be adjusted by the engineer according to the humidity changes. This effect can also be part of the reason why the above-mentioned hall only sounded good with a packed audience: the ventilation system in that hall was known to be unable to keep up with a full house. The high-frequency sound was therefore not absorbed as much, thus the low-frequency reverberation would not stand out as much but be somewhat masked by higher frequency reverberant sound, making the hall just bearable.

The air absorption is also the reason why acoustic parameters often are not given a value at the 8-kHz octave band for auditorium acoustics. In smaller size rooms such as sound studios it is in that sense a relevant frequency band to consider.

Critical Distance and Level of Reverberation

As earlier stated, the sound level of a sound source in a free field decreases by 6 dB per doubling of distance. The sound in a room under the diffuse sound field assumption consists of two parts: the direct sound from the sound source and the reverberant sound which is the sum of reflections described as the diffuse sound field. The sound level of the diffuse sound field is approximately the same in any location of the hall. In rooms with a low RT this is not completely the case though, as discussed later.

$$L_p = L_w + 10 \log \left(\frac{Q}{4\pi r^2} + \frac{4(1-\alpha')}{S\alpha'} \right) \text{dB (metric)} \quad (1.6)$$

In this equation L_p is the sound pressure level at a distance r, L_w is the sound power level of the sound source, Q is the directivity of the source, S is the surface area of the room, and α' is the area weighted average of the absorption coefficient.

The first term of the equation containing r^2 refers to the 6-dB attenuation of the direct sound field per doubling of distance, and the second term with $S\alpha'$ refers to the reverberant field. This equation is for empty rooms with diffuse sound fields with uniformly distributed sound-absorbing material.

This implies that close to the sound source, the direct sound is relatively loud and the reverberant sound has a given level which is the same anywhere in the room. Farther away from the sound source the direct sound level has decreased and the reverberant sound level is the same as closer to the source. Thereby the level of reverberant sound relative to direct sound is higher far away from the sound source compared to close to the source.

The distance from the sound source where the level of the direct sound is equal to that of the diffuse sound is called the *critical distance*, *reverberation radius*, or *room radius*.

From the above equation it can be deduced that

$$r, cr = \sqrt{\frac{QV}{100\pi T(1-\alpha')}} \text{ (metric)} \quad (1.7)$$

where Q is the directivity of the loudspeaker in a given frequency band, V is the volume of the hall, T is the reverberation time of the hall, and α' is the area weighted average of the absorption coefficient. For systems with several loudspeakers the equation yields

$$r, cr = \sqrt{\frac{QV}{100\pi T(1-\alpha')N}} \text{ (metric)} \quad (1.8)$$

where N is the number of loudspeakers or rather discrete clusters of loudspeakers. It is seen that the critical distance increases with higher values of Q and V and with smaller values of T, α', and N. This means that a larger share of the audience will enjoy a clear sound when Q, V, and α' are increased and when T and N are decreased. This is easy to understand intuitively.

This has some consequences when designing acoustics for amplified music and also in the design of loudspeakers and loudspeaker systems for halls.

The most striking information in this equation is perhaps the effect of the low Q value of any loudspeaker at low frequencies. In the 63- and 125-Hz octave band the directivity of a sound source is low due to the omnidirectional nature of sound waves emitted from loudspeakers at these frequencies. An omnidirectional source has a Q of 1. The Q of, for example, a loudspeaker with a dispersion pattern of 90° wide by 40° high has a Q of 10. This is one reason why, as we show, the reverberation time T at low frequencies must be low.

The equation that includes the number of different sound sources implies, for instance, that one should only employ extra clusters of loudspeakers such as delay speakers when no other option is possible. On the other hand, one could be led to think that as low a reverberation time as possible would be the answer for correct acoustics for amplified music. As shown later this is not correct; in fact there is a lower limit to recommended reverberation time for pop and rock music halls just as for halls for other musical genres. Achieving enough level is usually not a problem because the speaker system normally can be turned up (or extra amplification power can be assigned) so reverberation to increase sound level is not needed (although the very low tones from up to perhaps around 70 Hz might in fact benefit from a certain amount of acoustical amplification).

Reverberation Time Design

When designing halls for classical music the challenge is often to get a high enough value of reverberation time. The reverberation time is proportional to the volume of the hall, therefore the challenge is typically met by building halls with a relatively high ceiling. One strategy used is that of counting the volume of empty space in the hall per audience. For amplified music, as we show, the aim is to obtain a lower RT. Different prediction tools such as computer models are of help as well as calculations including Sabine's formula mentioned earlier. From data sheets containing the absorption coefficient of building materials in at least the octave bands 125 Hz–2 kHz, and maybe some specific absorption coefficient measurements on certain special materials used in the hall, as well as experience, the trained acoustic consultant is capable of making good estimates of RT of a hall before it is built or refurbished.

When designing halls with seats, a type of seat is often chosen so that its absorption matches that of a person. This is a way to ensure that the acoustics don't change much from rehearsal to concert or if the hall is not completely full. This is of advantage for the orchestra and thereby also beneficial to the audience. Some typical values of absorption coefficients for different building materials are found in Table 1.1. When using Sabine's formula it should be remembered that it is based on a perfectly diffuse room. This is seldom the case, especially when a large amount of absorption is present on, for instance, one entire surface. A packed audience on the floor represents such a surface.

Background Noise

Because the sound level is high at pop and rock concerts there are no real recommendations as to level of background noise within the hall. The audience is sometimes almost as loud between songs as the band playing their songs. Halls for classical performances have very strict background sound levels and this also indicates that clubs for dynamic jazz can benefit from a somewhat limited background noise level. In smaller clubs the bartender usually does not brew espresso during ballads. Ventilation and moving lights are among possible noise sources but also railways and highways can be too loud for a jazz club.

Sound Levels and Amplified Events

Niels W. Adelman-Larsen and Jacob Navne

Decibel denotes, as mentioned, the level of acoustic quantities relative to their reference values. In the case of sound pressure, the reference sound pressure, P_{ref}, is 20 μPa. One often used descriptor for absolute sound level is the sound pressure level (SPL) which is defined as

$$Lp = SPL = 10 \log \frac{p_{rms}^2}{p_{ref}^2} \tag{1.9}$$

where p_{rms} is the root mean square value of the sound pressure. RMS is the effective pressure of the time-varying sound pressure. The total SPL value is the summation of each SPL value in every octave or third octave band measured by the measurement device.

Because the human ear is less sensitive to low frequencies than to middle frequencies particularly at low levels, standardized frequency weighting filters are used to give the sound pressure a value that corresponds to the perceived hearing impression. The so-called A-weighted value, for instance, is 19.1 dB lower at 100 Hz and 10.9 dB lower at 200 Hz (third octave bands) than the sound pressure level with no filter used. The A-weighted sound pressure level is denoted L_A or can be specified by writing dB(A).

As a measure for characterizing the sound pressure level of a fluctuating noise averaged over time the equivalent sound pressure level L_{eq} is used. This measure can also be A-weighted, denoted $L_{A,eq}$. This is the basis of calculating how big a dose of sound a person is exposed to, such as during a concert or a working

Fig. 1.17 Sound level readout on a laptop computer at the mixing console of Ancienne Belgique, Bruxelles

day in a manufacturing facility. For instance, in Denmark the maximum doses during a eight- hour working day is an $L_{A,eq}$ of 85 dB. Mathematically, a doubling in terms of dB is approximately 3 dB but humans need an extra 10 dB to experience a sound level increase as a doubling.

The $L_{A,eq}$ can be measured directly on most sound-level meters. Calculations can also be made and as an example, at a concert, three tunes of respectively 3, 4, and 5 min are played; the averaged $L_{A,eq}$ level of the three songs is, respectively, 100, 95, and 105 dB at a given location. The total $L_{A,eq}$ during the 12 min of those three songs at that location is found as

$$L_{A,eq} = 10 \log \left(\frac{3}{12} 10^{10.0} + \frac{4}{12} 10^{9.5} + \frac{5}{12} 10^{10.5} \right) = 102 \, \text{dB} \quad (1.10)$$

In most countries legislation sets a maximum averaged sound pressure level for any employee to receive during a workday. These pieces of legislation are in place to ensure the health and safety of workers in noisy environments. It is the responsibility of the employer to make certain that the noise in the work environment is as little as possible. This is quite a dilemma when looking at live reinforced music, given that the noise you need to protect the workers from is the actual product that this "factory" is selling to its audience, that is, the music being performed. Two very conflicting interests are at play in this scenario. The customers (i.e., the audience) want to experience an event they cannot reproduce at home, with punchy bass and loud and clear sound and ambiance. At the same time the employees need to be exposed to as little noise as possible. Moving workplaces such as the bar, coat check, and so on outside the main concert area helps reduce the exposure by architectural means.

It has become very common to measure the actual sound level in venues during amplified music concerts. The microphone is most often placed approximately in the center of the audience area, at the sound engineer's desk, referred to as the "front of house" or FOH. The sound engineer monitors the level, for instance, on a laptop, throughout the concert (Fig. 1.17).

When measuring the sound level at a concert, the preferred measurement method is L_{eq}. As music is dynamic in nature, the L_{eq} value is a good way to monitor the averaged level such as during a whole concert. Doing SPL measurements at live events is always a compromise between measurement accuracy and realistic possibilities at the actual event. In an ideal world, a number of measurement positions would be looked at, as the SPL varies with distance to the stage and the main loudspeaker array. In reality it is, however, difficult to place expensive measurement microphones among the audience, and therefore one single fixed, protected position at the FOH is the common standard.

In some countries, such as Germany with the DIN 15905-5 standard, the measurement needs to be compensated for the difference between the loudest point in the audience, and the actual measurement point. In this norm, a test signal is played prior to allowing the audience into the venue, and using this signal and a compensation algorithm in the measurement equipment, the difference is measured and used during the concert. In this way, the values displayed by the measurement equipment are not the actual SPL at the measurement position, but a compensated calculated SPL at the loudest point in the venue. This approach is also used in Sweden and Belgium. It makes good sense inasmuch as actual hearing damage is often a result of a few minutes of extremely loud levels rather than a couple of dB higher SPL over a couple of hours or even a large number of concerts.

A number of countries have sound-level limits for live shows. These are not directly related to health and safety but mainly based on an overall compromise between many stakeholders—the audience who wants a physical experience, the organizer who wants to have a satisfied audience, neighbors who want their sleep—and an overall concern towards the comfort of the listening audience. These limits are often based on $L_{A,eq}$ values, and typically range from 99 to 103 dB and timespan averages from 15 to 60 min.

It is important to know that for a live show of any modern genre, an average $L_{A,eq}$ of 100 dB is generally needed in order to fulfill the requirement from the audience to both hear and feel the show. As mentioned, the typical maximum doses for employees in factories is $L_{A,eq} = 85$ dB over 8 h. If this same limit were to be followed at rock shows, a show at 100 dB on average could last no more than 15 min, then the audience would have received a full day of exposure due to the relationship between dB and time mentioned above (or simply by deducing that a 3-dB increase in average SPL equals a doubling of noise exposure to the ear, and thus halves the time it takes to obtain the same dose).

In that perspective, a live show will never be a "safe" event in terms of normal health and safety regulations, however, very few people attend more than 5–10 concerts a year, and when compared to the everyday exposure of MP3 music players and headphones mounted directly into the ear canal, with a much higher SPL, the risk that a live music event is the reason for a person developing a hearing impairment is low. The risk is there, though, and is biggest in small venues with a low ceiling or in other cases where the PA speakers are not elevated high above the audience. In these cases severe differences in SPL are present between audiences close to the stage and those in the rear.

In some countries, sound level limits at live shows are fairly gentle. Denmark, Norway, and the Netherlands, for instance, are using an $L_{A,eq}$ of 103 dB over

15 min. This makes for loud shows, and only limits very few acts in being as loud as they want to: in fact even with the right to be at 103 dB, most shows are played at 99–101 dB on average, as this often proves to be an adequate SPL. Of course if this measurement is made at the sound engineer's position, without being corrected with regard to the potentially shorter distance from the speakers to some audience members it may, depending on the layout of the hall, mean that other audience members receive considerable larger doses. Other countries including Sweden and Switzerland have a stricter approach, where shows are to be at 99 dB(A), and for Sweden, the limit is lowered if any audience member is below the age of 13. Then it is set at $L_{A,eq}$ 93 dB over 60 min. A number of exceptions exist.

Especially for outdoor events in closely populated areas, the limits are very strict, and sometimes even based on instant values instead of an average. This makes it very difficult to produce a concert at a sound level expected by the audience, and complaints from ticke tholders are often the end result.

One of the main limitations with the current legislation is that the maximum levels set forth are almost always based on A-weighted values. Part of the reason for that is also that the majority of noise regulations and guidance literature was written in the mid-twentieth century. At that time loudspeaker design was in its infancy, and not comparable to today's powerful line arrays and huge sub bass cabinets. The development in loudspeaker design has enabled sound engineers to play with full range systems, frequencies from around 30 Hz all the way to 20 kHz.

Bass frequencies are the most difficult to control and sound insulate. This fact often results in neighbors mainly being bothered by bass frequency sound, and not the sound stemming from other instruments. As mentioned, A-weighting removes a lot of bass content from the measurement to better mimic the behavior of the ear, and thus there is very little correlation between the measured SPL at the event and the nuisance experienced by the neighbor. An example that describes the inadequacy of A-weighting for this purpose could be to look at the difference between a British guitar rock band, and an electronic dance act. Both orchestras may play at the same $L_{A,eq}$ of 100 dB measured inside the venue, but due to the very heavy bass that is often associated with the electronic music genre, the neighbor of such an event will be bothered much more by the electronic act than by the rock band, as the bass content is much, much louder, but not reflected on the A-weighted measurement. Switching to more frequency flat C-weighted measurements would provide a much better correlation between the value measured inside the venue and the nuisance to the neighbors. Switching to C-weighted measurements would at the same time require a significant increase in the value of the maximum allowed L_{eq}. When looking at measurements of A- and C-weighted values performed simultaneously, there is often a 10–20 dB difference between the two quantities, so a limit of $L_{A,eq}$ 15 min = 100 dB, would have to be $L_{C,eq}$ 15 min = 110 dB or maybe even more.

Regardless of the important safety issue of sound-level control at amplified concerts, it should be remembered that our hearing system distorts more at higher levels. The masking curves in Fig. 1.12 show that at higher levels the sound engineer has a more difficult job creating a clear sound.

Chapter 2
Auditorium Acoustics: Terms, Language, and Concepts

There have been primarily three methods for performing subjective studies of the acoustics in concert halls for classical music, each of which has its advantages and disadvantages. One method has been to create a virtual concert hall in a laboratory, where either recordings from a given hall or simulations from room-acoustics software were presented to the listeners. The acoustics could either be simulated with an array of loudspeakers in an anechoic chamber or auralized and presented over headphones. With this method, listeners can quickly rate many halls without having to travel long distances. Some other benefits are that halls can be presented anonymously and blindly so that there is no bias based on a hall's reputation or visual appeal, and the exact same performance of a piece of music can be evaluated in all halls and positions. Despite these advantages, it can be difficult to get truly qualified listeners, such as professional musicians with their busy schedules, to participate in a laboratory experiment. More important, the actual perceived sound in such experiments is very far from the actual listening impressions in the real halls. It is the author's opinion that it is questionable whether the approach has a good enough connection to reality.

In search of a research model that is manageable and practical, decisive information may be lost. Fortunately great care is taken before making recommendations for the building of concert halls based on results originating from this approach. An elaborated test method, used for instance by Lokki and Pätynen, is to place numerous loudspeakers, each with an anechoic recording of a particular instrument in the relevant position on stage in real concert halls. The total "loudspeaker orchestra" is then recorded in different locations among the audiences. Each recording is then later played back over several loudspeakers in a quite anechoic room for test persons to evaluate. This gives quite surprisingly detailed auditory information. Testing isolated acoustical attributes and how they affect humans may be tested in such set-ups. For truly evaluating a hall the method may not prove adequate.

As an alternative, listening tests can be performed in an existing hall where there is a possibility of changing the acoustics. However, the acoustical changes that have been possible to make have formerly been somewhat limited and typically have not included the low frequencies. The variations have been based on changes of the amount and placement of absorption in the room leading to different reverberation times and decay characteristics. When investigating, for instance, ideal reverberation time or for a specific type of music, the advantage of this method is that the basic system, hall geometry, overall diffusivity, and the like remain constant whereby the one parameter, the reverberation time, is somewhat isolated and thereby easier to judge. Evidently such experiments do not include different size halls and thus recommendations of reverberation time do not stretch across hall volume.

Surveys have also been done of existing halls through interviews of people who have experience with the acoustics in many halls. These subjective "measurements," often in the shape of a questionnaire the participants are asked to fill out, are then correlated with objective, acoustic measurements of the halls. One drawback of this method is that acoustic memory for some people is short and can be colored by many nonacoustic factors. This can make the comparison of halls imprecise. Therefore great care must be taken when choosing the test persons and when interviewing them. Those who don't feel completely capable of performing the interview, and with certainty be able to remember the acoustics of certain halls, must be told not to participate in the investigation regarding those halls. Such extremely professional and experienced individuals who end up participating in the survey have trained their sense of hearing as well as their acoustic memory to a very high level, in many cases without even knowing it. This method of subjective responses to actual halls from memory was selected in the study leading to the recommendations later in this book, primarily because it was deemed to be important to avoid the lack of acoustical information connected with the two other methods mentioned above but also because the author had an extensive network among musicians and sound engineers and thus the investigation was relatively easily set up.

During the twentieth century acousticians have introduced a number of words that characterize a set of acoustical attributes that have proven to be of importance when seeking to describe halls for symphonic music, opera, and other classical music genres. This has enabled acousticians and musicians to share a common vocabulary and thereby to communicate in a somewhat unambiguous way. Doelle (1972), Barron (1993), and Beranek (1996) have provided a list of musical/acoustical terms listed, among others, in *Architectural Acoustics* by Marshall Long (2006). Please see Table 2.1. Each of the terms is associated with at least one measureable acoustical property of a hall. A few of these terms are used in the chapter on the design of halls for amplified music and certainly some of the terminologies are useful when discussing auditorium acoustics in general.

Likewise, in classical music concert halls at least five independent acoustic qualities have been found. These are acoustic properties that the human ear is able to identify isolated from one another. This work was primarily done by Hawkes and Douglas (1971) and as articulated by Michael Barron (2010) discussing the five qualities: "The major concerns are that the *clarity* should be adequate to

Table 2.1 Commonly employed musical and acoustical terms and their definitions

Term	Definition
Balance	Equal loudness among the various orchestral and vocal participants
Blend	A harmonious mixture of orchestral sounds
Brilliance	A bright, clear, ringing sound, rich in harmonics, with slowly decaying high-frequency components
Clarity	The degree to which rapidly occurring individual sounds are distinguishable
Definition	Same as clarity
Dry or dead	Lacking reverberation (little reverberation[a])
Dynamic range	The range of sound levels heard in the hall (or recording); dependent on the difference between the loudest level and the lowest background level in a space
Echo	A long delayed reflection of sufficient loudness returned to the listener
Ensemble	The perception that musicians can easily play together[b]
Envelopment	The impression that sound is arriving from all directions and surrounding the listener
Glare	High-frequency harshness, due to reflection from flat surfaces
Immediacy	The sense that a hall responds quickly to a note. This depends on early reflections returned to the musician
Liveness	The same as reverberation above 350 Hz
Presence/intimacy[b]	The sense that we are close to the source, based on a high direct-to-reverberant level
Reverberation	The sound that remains in the room after the source has been turned off. It is characterized by the reverberation time
Spaciousness	The perceived widening of the source beyond its visible limits. The apparent source width is another descriptor
Texture	The subjective impression that a listener receives from the sequence of reflections returned by the hall
Timbre	The quality of sound that distinguishes one instrument from another
Tonal color	The contents of harmonics or overtones and their strength relative to the fundamental of a tone[b]
Tonal quality	The beauty or fullness of tone in a space. It can be marred by unwanted noises or resonances in the hall
Uniformity	The evenness of sound distribution
Warmth	Low-frequency reverberation, between 75 Hz and 350 Hz

[a]Added by the author. For instance, a quite dead room at low frequencies is favorable for amplified music
[b]Added/altered by Jens Holger Rindel

enable musical detail to be appreciated, that the *reverberant response* of the room should be suitable, that the sound should provide the listener with an *impression of space*, that the listener should sense the acoustic experience as *intimate* and that he/she should judge it as having adequate *loudness*." In this short summary a part of the challenge of auditorium acoustics becomes apparent: although clarity and intimacy, to a large extent, call for a low reverberation time, the factors of reverberant response and impression of space, as well as (acoustic) loudness call for a longer reverberation time. Acoustical consultants operate within a narrow window in order to get all parameters right in a given space.

These principal acoustical subjective qualities have been determined from the questionnaire investigations mentioned earlier. Such investigations have led to an understanding of basic important properties of reflections of sound from boundaries and the like in a room. The late, most quiet reflections are inaudible whereas strong and delayed single reflections are perceived as disturbing echoes. One of the pioneers in these investigations was Helmuth Haas who in 1951 found that the human ear uses the first arriving sound to locate its position and later reflections, even up to approximately 10 dB louder still will appear to stem from the origin, if they are delayed up to some 50 ms compared to the direct sound. This is used advantageously in connection with PA speaker coverage in order to make the emitted sound seem to originate from the orchestra.

Harold Marshall proposed in the late 1960s that early reflections are important and that lateral reflections, meaning reflections from walls or other somewhat vertical surfaces, are the most important. Later Barron and Marshall found that the louder the sound is, and the higher the proportion of lateral sound, the greater is the perceived source broadening. Later investigations by Morimoto and Maekwa and by Souldore and Bradley showed that there are two such spatial effects occurring, both stemming from lateral reflections: the early sound gives a sense of source broadening and the later reflections create a sense by the listener of being enveloped in sound (Barron). This sense of being inside the sound is believed to have an importance in pop and rock venues too, at least for the musicians and probably also the audience. This is discussed further in Chap. 5. Both source broadening and envelopment are connected to the term "spaciousness" listed in Table 2.1.

Marshall Long lists in his book eight parameters concerning the listening environment itself in halls for music. Later Barron investigated different acoustic spaces in detail, but halls for amplified music do not occupy a specific chapter as do acoustics for synagogues or music practice rooms, for example. He notes that in halls for music:

- The audience should feel *enveloped* or surrounded by sound. This requires strong lateral reflections with a significant fraction arriving from the side.
- The room should support instrumental sound by providing a *reverberant* field whose duration depends on the type of music being played. *A reverberation time that rises with deceasing frequency below* 500 Hz *yields a pleasant sense of musical warmth.*[1]
- There must be *clarity* and definition in the rapid musical passages so that they can be appreciated in detail. This requires reflections from supporting surfaces

---

[1] In order to avoid any misunderstanding it is pointed out that the higher value of reverberation time at low frequencies proposed here (and in many other textbooks on room acoustics) which some people find beneficial for classical music is the worst enemy of the acoustics for pop and rock music. This is derived in Chap. 5. Furthermore, the PA speaker system is responsible for carrying out the sound at amplified music concerts, not the hall itself, therefore the loudness factor is generally not important in halls for pop and rock music.

located close to the source or the receiver so that the initial time delay gap is short.
- Sound must have adequate *loudness* that is evenly distributed throughout the hall. When the number of seats becomes excessive (above 2,600 seats), loudness and definition are reduced. In small auditoria the loudness must not be overbearing.
- A wide *bandwidth* must be supported. Musical instruments generate sounds from 30 to 12,000 Hz which is much broader than the speech spectrum. The room must not colour the natural spectrum of the music.
- The detailed *reverberation characteristics* of the space should be well controlled with a smooth reverberant tail and no echoes, shadowing coloration or other defects.
- The *performers* should have the ability to hear each other clearly and to receive from the space a reverberant return that is close to that experienced by the audience.
- *Noise* from exterior sources and mechanical equipment must be controlled so that the quietest instrumental sound can be heard.

The above-mentioned subjective parameters are at any given listener position affected by the general design of the classical music hall. Some of these general perspectives are, of course, hall size, hall shape, interior geometry, hall volume, surface materials, balconies and overhangs, seating, and platforms. This is discussed with respect to the design of pop and rock venues in Chap. 6.

Objective Parameters

Various objective parameters, other than reverberation time and EDT, have been defined in an attempt to better describe sound in rooms. The following objective parameters are commonly used and have been shown to be in good agreement with some of the subjective parameters mentioned above.

EDT, Reverberation Time, Liveliness, and Reverberance

Ever since it was defined by Sabine, reverberation time has been the single most important objective acoustic parameter. Whether targeting a reverberation time for an empty hall or a hall with an audience, the reverberation time is the parameter that rooms and halls are designed from based on knowledge of absorption coefficients of different building materials, the volume of the hall, and its purpose. Furthermore, in room acoustics with computer modeling programs including Odeon and Catt among others, more precise calculations can be undertaken by including the modeling of how the sound waves behave in their meeting with surfaces in a room. The reverberation time is defined by a 60-dB decay and the measurement of it is (as earlier mentioned) based on a 30-dB decay from −5 to −35 dB or even a 20-dB decay.

The early decay time is an expression of reverberation time but based on the decay from 0 to −10 dB and has proven to be better correlated with a test person's judgment of reverberation than T_{30}. A 60-dB, or even a 30-dB, decay is rarely encountered in music because new notes will be played before one gets a chance to hear the full decay of the first note. Although pop and rock music can be highly percussive and syncopated with at least a loud and clear backbeat on the snare drum and a usually somewhat less pronounced downbeat on bass and bass drum, a full 60-dB decay will be seldom encountered. Even an isolated 30-dB decay from −5 to −35 dB is not often heard because it will be masked by other sounds, and therefore EDT is a good descriptor of *reverberance* which is the term used to describe the subjective experience of reverberation. The term *running reverberation* is used for EDT and describes the liveness, or liveliness, of the room.

If the full decay in a hall is perfectly linear, the EDT and T_{30} are the same. EDT is heavily dependent on position in the room where T_{30} is more stable. It can be very short, for instance, under balconies and the like, where the strongest reflections come from large surfaces close to the microphone that do not form the actual room. In such places the decay tail will be steep in its first stage and thereafter flatten out and attain a value in accordance with the T_{30} of the room. This effect is also called the *coupled rooms effect* inasmuch as the first part of the decay takes the shape of the closest "room" with a short reverberation time but dies out in an adjacent room that also has been acoustically evoked and has a longer reverberation. It's easily comprehended by clapping one's hands once in an acoustically quite dead room close to an open door that leads to a room with more lively acoustics.

EDT is the best objective parameter describing a listener's judgment of reverberance while the music is playing. Therefore a room with a seemingly too long T_{30} may actually be acceptable if the EDT is shorter. This is one reason why the ratio EDT:T_{30} is relevant. Furthermore the ratio is a measure for the diffuseness of a room. If the sound in a room is very well diffused the ratio should be close to 1 because a diffuse room is characterized by a linear decay. As a matter of fact it has been shown that a longer reverberation time than expected can be accepted for a room for music if the room is well diffused. In a room design where early energy is directed towards the rear of the hall the ratio is often smaller than 1.

C_{80}, D_{50}, Early Reflections, Clarity, and Intimacy

The early reflections have been found to have an important influence on our impression of reflected sound as containing the qualities of clarity and intimacy. Therefore a parameter describing spaces with early reflections compared to later reflections is a measure of these qualities. Objective clarity is simply the ratio of early to late sound energy at a given position. The most common measure is C80 for which, as the name indicates, the cut between early and late energy is made at 80 ms; sound arriving before this is considered as early whereas energy that arrives later is defined as late energy. But other definitions are used such as D_{50}, where 50 refers to the cut being applied at 50 ms and D is called definition (in German: *deutlichkeit*). Pop and rock music are often highly syncopated musical genres where even 50 ms seems to be too high a value when noting that sound travels 17 m in 50 ms or 8 m to a reflective surface and 8 m back. It has therefore been suggested by the author to investigate shorter time spans.

According to Barron there is evidence that the ear's response to low tones in the 125- and 250-Hz octave bands is slow, wherefore objective clarity is usually calculated as an average of the 500-, 1-k, and 2-kHz bands. As shown in Chap. 5, bass clarity is indeed of importance in pop and rock music although the objective characterization can be of another type than the C_{80} or D_{50} parameters at low tones. Objective clarity is a fraction and is expressed in dB. Clarity is inversely proportional to reverberation.

LF, Envelopment, and Lateral Reflections

Reflections from vertical boundaries such as walls are of importance in order for listeners to obtain a sensation of being enveloped in sound. It has been found that particularly the later-arriving lateral reflections are the ones most responsible for this sensation. It is believed that also in pop and rock music especially the performers but probably the audience as well want this acoustic attribute although there has been no specific study made to witness this. It is a fact though that at least some sound engineers prefer clarity to a degree where almost no envelopment can be apparent. The lateral energy fraction in dB is measured with two microphones, one being a figure-eight microphone directed with its zero towards the source and the other being an omnidirectional microphone. The fraction of these two gives an idea of how much sound energy comes from the sides. Of course another way of more rapidly and less precisely finding out about these attributes is simply to take notice of whether the side walls of a room are heavily absorptive or whether the ceiling is relatively low compared to the width of the room. Lateral fraction (LF) is a ratio and is unitless. Envelopment is proportional to reverberation.

G, Strength, and Room Gain

At unamplified concerts the hall alone must bring forward the sound to a degree where the audience even in the back of the hall experiences an appropriate sound level of the acoustic information brought forward from the stage. On the contrary, in smaller recital halls a large orchestra playing *fortissimo* (very loud) will sound too loud and the loudness of the room becomes overwhelming. The objective measure for the acoustic gain of a room is called strength with the symbol G. G is also known as the room gain.

G of a hall equals the ratio between the SPL of an omnidirectional sound source at a calibrated level at 10 m distance from the source, and the same source, at the same level, at the same distance in an anechoic room.

According to theory, the value of G decreases by 3 dB by doubling the total absorption in the space. The absorption in concert halls is not measured directly and therefore the room gain is computed from the ratio of the reverberation time and the volume of the hall. This makes it clear that increasing the reverberation time of a hall also makes it louder and vice versa. The strength of the reverberant field is often of interest and is denoted G_{late}.

As mentioned earlier, G is not of importance in halls for amplified music because the speaker system is dimensioned to provide enough level although a certain room gain may be beneficial in the 63-Hz band. In overly dampened halls for amplified music, delay speakers must be applied supplementary to the main PA system in order to make up for the sound energy being lost rapidly as it propagates through the hall. This seems like an undesirable design, wasting resources. Of course delay speakers may advantageously be used in very long halls. G is proportional to reverberation.

Bass Ratio, Warmth, and Bass Response

Some people prefer a rise in low-frequency (LF) reverberation time in halls for symphonic music so that at 125 Hz a value of a factor of up to 1.4 times that at mid-frequencies is attained. Evidently this will increase the level of the bass or at least guarantee that the musicians playing the low tones do not have to play unnaturally loud to achieve an acceptable level and blend. Indeed a large number of double-bass players are needed in very large orchestras in order to achieve a sufficient sound level. The rise should in any case bring a sense of "warmth" to the music preferred by some. The "warmth" parameter was suggested by Leo L. Beranek and used by John O'Keafe, for instance.

Beranek has proposed the parameter *Bass Ratio* computed as BR = (T125 + T250)/(T500 + T1 k).

In pop and rock music there is a much higher level in the 63-Hz band compared to classical music. But as it turns out, the reverberation time in the 63-Hz band can be justified longer here than in the 125-Hz octave band, also in halls for amplified music. In the appendix of this book two different BR are calculated, one exclusively including the 125-Hz band versus mid-frequencies and one also including the 63-Hz band versus middle frequencies. Other BRs can be calculated depending on what is being investigated. In pop and rock music the 250-Hz band is not really perceived as bass but rather as a mid-low range.

Similarly, the BR can also be defined from the strength G instead of reverberation time, thus referring to the sound in the steady-state condition instead of the decaying sound. Some researchers find this BR more relevant for a concert hall.

ST, Support, and Ensemble

The stage parameter "support" was suggested by Gade in the 1980s. The support parameter ST describes the sound energy returning to the musicians on stage. Skålevik has gathered information about acoustic parameters on his website:

ST_{early} The early, reflected 20–100-ms sound energy level relative to the initial 0–10 ms direct sound, measured at 1.0 m from an omnidirectional source.

ST_{late} The late, reflected 100–1,000-ms sound energy level relative to the initial 0–10-ms direct sound, measured at 1.0 m from an omnidirectional source.

The early support, ST_{early}, is now commonly used to describe the degree of *mutual hearing*, also referred to as *ensemble*, on stage. On most stages the early reflected energy is expected to contain sound from the whole ensemble as well as the musician's own instrument.

The late support describes the degree to which the musician hears the late reverberant sound. ST late is suggested as a descriptor of the performer's subjective reverberance. Singers often appreciate hearing their own voices filling the auditorium and they often prefer high ST_{late} values. ST_{late} is almost solely determined by the ratio between RT and volume of the hall.

However, one should take the balance ST_{late}–ST_{early} into account, because if ST_{late} is high compared to other halls, the late reverberant sound may still appear weak if ST_{early} is also very high. If this balance is too low (say ≪ −3 dB) musicians may feel that the stage is acoustically decoupled from the auditorium. This may be the result when introducing a canopy that is too low and too dense. Likewise, it seems evident that the ST_{early} value should be considered in relation to ST_{late}. If there is not enough early sound compared to late, the musician may feel not in touch with his or her instrument and this can harm his or her timing. Mutual hearing conditions on a stage can't be fully measured or predicted with an omnidirectional source on an empty stage because instruments do not radiate sound omnidirectionally and because musicians have an impact on the acoustics on stage both in terms of absorption but also because they block the sound propagation. For symphonic music Christopher Blair notes: "The art of designing good on-stage acoustics boils down to providing just enough early energy to help with coordination, but not so much as to mask audibility of the late-energy room response."

This parameter indeed also has relevance for amplified music. More support on stage from the immediate surroundings in a given hall lowers the need for loud monitor speaker levels and gives the musicians a feeling of being more enveloped in sound from their own and their colleagues' instruments, rather than that stemming from the PA system and reflecting surfaces farther away. This is of major importance for the musicians to enjoy the hall in general. The author finds that the acoustics must be similar on stage to that of the audience area, and that these two spaces must not be acoustically separated. This will automatically lead to louder early reflections than later reflections as perceived by the musicians because of the distance law (sound level decays over distance). In other terms, ST_{early} must be stronger than ST_{late}. This is sometimes a challenge to achieve inasmuch as sound engineers especially (definitely no musicians) want the stage dead to easily handle feedback and the like. This is dealt with in more detail in later chapters.

Late support is proportional to reverberation.

Chapter 3
Reinforcement of Sound Sources

One of the things that characterize pop and rock music is that most sound sources have to be reinforced (amplified). This either because the initial balance between the different instruments' acoustic sound is problematic (especially in small venues), or because a too-low sound level leads to the sound not being engaging for the public (larger venues). An acoustic drum set produces very high sound levels in excess of 100-dB SPL several meters away from it in a small venue. Therefore other more quiet sources, such as the vocal, need to be amplified. In bigger venues, as with most of the halls described in this book, the sound level for the audience is too low altogether without amplification of all instruments. A sound engineer takes care of creating a mix of the various instruments that is amplified and made audible to the audience and him- or herself through the PA (public address) loudspeakers. The mixing board position is sometimes referred to as the front of house (FOH). The layout of a typical amplified concert is seen in Fig. 3.1. The art of mixing reinforced performances begins with finding a balance between instruments that makes the entire band sound as much as possible like one instrument. If the band produces a beautiful sound and its musicians play well together, as do most professional orchestras, mixing is made easier. This chapter discusses sound system design mainly to the extent where it has an impact on the room acoustical design. The more precisely the PA sound can be aimed at the audience and not onto reflecting wall and ceiling surfaces, the less will reverberation be evoked.

On stage, the musicians also need reinforcement to be able to hear themselves and each other well. Earlier, monitor loudspeakers were relied on exclusively to create the extra sound needed on stage, but since the 1990s in-ear monitoring has become popular. With in-ear monitoring musicians can move around on stage without getting away from the sound of directional monitors and thereby losing track of the music. An attempt to achieve the same effect with open monitor speakers has been to provide the stage with so-called side-fill monitors, placed on the sides of the stage. Another just as important feature of in-ear monitoring is though, that the excess of sound sources is limited greatly, reducing the

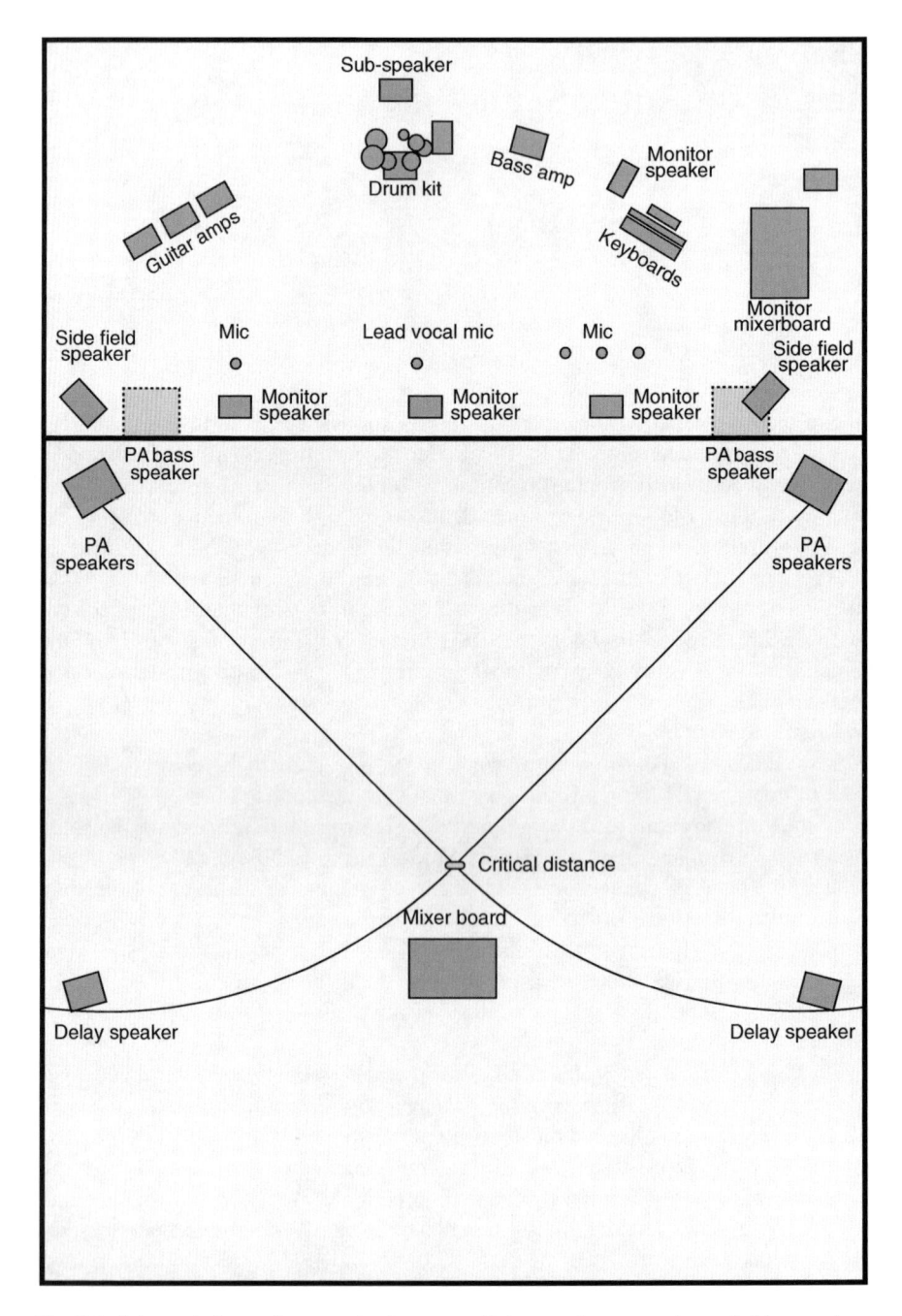

Fig. 3.1 Schematic layout for a pop/rock concert. Delay speakers are only needed in very long halls, under balconies, and so on

amount of unwanted sound in open microphones. This helps the sound engineer in producing a "clean" mix. Each musician usually needs an individual mix and an empathic monitor sound engineer is crucial for the band to feel good during a concert.

The sound system by itself (the whole chain from microphones, mixing consoles, and amplifiers to the loudspeakers) usually by itself amplifies some frequencies more than others, and this is also true for a room if it does not have a flat reverberation time curve across all frequency bands. The combined uneven response of this complete system (sound system and room) is often equalized levelwise in, for instance, third octaves with the use of a parametric called "house-EQ" or by use of, for instance, combinations of IIR and FIR filters. (As with any equalization there are consequences for the phase response of the loudspeaker sound and therefore the more untouched the house-EQ can be left the better). The FOH engineer will further equalize each instrument to sound right according to his and the band's taste in order to create a mix where no instrument heavily masks other instruments and also to avoid even some frequency bands of one instrument masking other frequencies of that same instrument. Every frequency band of each instrument has been allocated an appropriate "shelf" in the mix. A transparent mix is created. Sound engineering can truly be an art form.

It is important to note that this equalization has no impact on the actual room acoustics. The room acoustic properties are a product of the surface materials of the room and the way they are assembled, and the geometry of the room as well as the interior including chairs, audience, and so on. The room acoustics cannot be changed by electronic means. But if the whole sound-chain, as described above, is evened out levelwise and all instruments have found their appropriate place in the mix, then why do we need to worry about room acoustics? That is because of the duration of the reflected sound. The reverberation time of the hall is evidently not altered by the electronic equalization. The challenge lies in the time-domain, not the level-domain. And when some frequencies are dying out too slowly, then this reflected remaining sound will (1) partially mask the direct sound of all instruments in that frequency domain, making them appear unarticulated and thereby destroying the core message that those instruments are trying to get across; and (2) partially mask wanted reverberant sound in other frequency ranges. The typical example, of course, here being rumbling, late bass reflections. Then why don't we just create halls with as little reverberation as possible and boost the levels accordingly by amplification? The answer to this is briefly, that concerts in such acoustics are a dull experience. In Chap. 5 it is shown, that a too long or a too short reverberation time spoils the sound experience for musicians, sound engineers, and most important, the audience. There is a narrow window, the *recommended reverberation time*, which depends on hall size and frequency, that allows for the greatest satisfaction for musicians and audiences attending the concert.

The Sound of a Rock Band

Typically pop and rock music is characterized, among many ingredients, by its percussive nature. In pop and rock the low frequencies are also percussive. An active bass line supported by a syncopated bass drum is the basic recipe of almost any pop song. Professional rhythm sections have a very well-articulated beat. A few hundredths of a second of difference in bass tone duration or point of attack can easily make or break the grooving, swinging sensation of a bass line. That's one very basic message that rock and pop musicians are trying to get across to the listener. The entire band respects this structure and supports it in their every phrasing. The audience's as well as the musicians' perception of this core message is made difficult when reverberation causes a tone not to stop when it is supposed to, thereby blurring the percussive articulation of the music.

What we want at pop/rock concerts, on the other hand, is among other things the excitement of having the loud bass sound vibrating our body. In fact it has been shown by psychologists that bass sounds louder than some 90-dB SPL trigger a sensation of pleasure in human beings (Todd et al. 2000). And pleasure is a desirable feeling. Bass levels have been brought up significantly over the last decades along with the development of speaker systems that in a practical way can produce such levels. Our sensation of low-frequency tones is as much a physical experience (body parts vibrate and the tones also transmit vibrational low-frequency energy direct to our middle and inner ear) as it is actual hearing airborne sound through our outer ears. Loud bass sound simply sounds and feels right in many musical subgenres within the amplified music idiom. And without it, it would hardly be regarded a pop/rock live concert experience any more. Certainly the sound engineer will EQ each instrument to obtain an open transparent mix where one instrument does not cast shadows over the others spectrally. Each instrument gets its own "shelf" in the complete mix as mentioned. But the total mix must incorporate quite a loud level of bass sound, primarily stemming from the bass guitar and bass drum, but also keyboards, guitars, male voice, and so on. So unlike the situation of mixing a male speaker, where rolling off the bass on the EQ is an obvious option, a rock concert simply has to contain loud bass levels for it to be perceived a real rock concert.

Evidently, if a hall has a long RT at low frequencies the loud levels of bass sound will lead to loud reverberant bass sound. This reverberant low-frequency sound has, as mentioned, a tendency to partially mask direct sound (both at low and mid frequency), and hereby the core message of the music, the beat and the text, is lost. In fact, in Chap. 5, this is proven scientifically: halls with a low RT at low frequencies are considered the best. A high reverberation time at mid frequencies can also be problematic but because the loudspeakers are much more directional at these frequencies, the sound can be directed onto the audience who absorb the mid–hi-frequency sound to a large extent. In Chap. 1 it was shown how the audience does not absorb low frequencies, even if we do try to direct the low-frequency sound towards them. In Fig. 3.2 some typical L_{eq} values in the octave bands from 31.5 Hz–8 kHz from concerts of amplified music are shown. These are average values

The Sound of a Rock Band

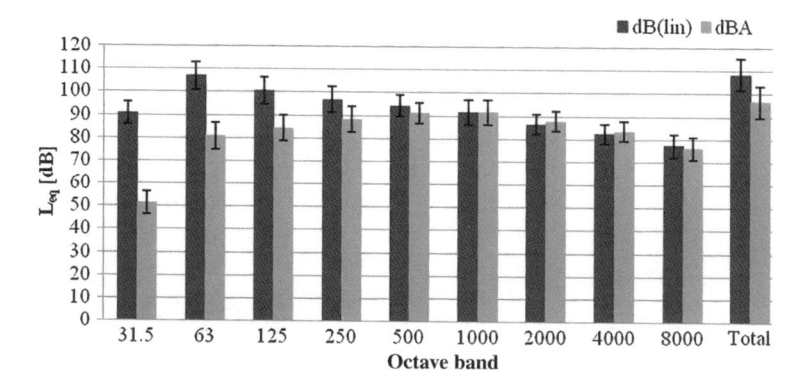

Fig. 3.2 Averaged L_{eq} values of 15 rock and pop acts measured in one hall

of 15 different pop and rock bands each playing about three songs in the same indoor venue at a band contest, as measured by Rosenberg in Denmark 2013.

Providing Amplified Direct Sound from Loudspeakers

Jens Jørgen Dammerud

The intention with a sound system is to provide sufficiently loud and clear sound for all audience members. The total sound includes both the direct sound from the loudspeakers and the room's reflections of this direct sound. The level balance between direct sound and reflected sound affects the sound quality perceived by the audience. This level balance depends on four major factors:

- The directivity of the direct sound from the loudspeakers
- Interference of direct sound waves from several loudspeakers or between loudspeakers and sound reflected by surfaces close to loudspeakers or listeners
- The sound-absorbing properties of the audience
- The acoustic properties of the room

At mid and high frequencies significant loudspeaker directivity, audience absorption, and favorable acoustic properties of the room are easily achieved. When using more than one loudspeaker (which is almost always the case) there will often be interference between the loudspeakers that will result in low direct sound at certain locations within the audience area. The locations of "dark spots" of direct sound (low levels) vary with frequency (Fig. 3.7). Low direct sound levels within the frequency area of 500–4000 Hz are considered crucial regarding perceived definition/intelligibility (Davis and Patronis 2006). At low frequencies there are significant challenges regarding all four factors. This section concerning amplified direct sound focuses on the two top factors in the above listing: achieving directivity and avoiding significant interference.

Single Point Sources

If using only one loudspeaker the most common way to achieve directivity is to install a horn in front of the radiating loudspeaker membrane. The loudspeaker will function here as one point source. A horn also improves the efficiency of the loudspeaker, which means louder sound for the same effect provided to the loudspeaker. A horn will provide better directivity as long as the horn is sufficiently long compared to the wavelength. Some subwoofer models have a coiled horn inside the loudspeaker enclosure. At low frequencies the horn is often not long enough to provide significant directivity, but the horn improves the efficiency. Loudspeakers that are not extremely large have problems generating low-frequency sound at high levels without a horn construction attached.

Some loudspeakers are made very directional at higher frequencies, much like a megaphone, with Q values being typically 10–20 ($Q = 1$ being omnidirectional). The high Q value helps keep the level of reflected sound at a minimum, whereby a good balance between direct and reflected sound can be maintained at larger distances. Such loudspeakers are often referred to as *long-throwing*. The directivity also helps reduce direct sound-level variations at different distances; the loudspeaker is normally oriented in such a way that on-axis is pointed towards the farthest listener. The direct sound level is normally highest on-axis allowing the directivity to counteract the inverse-square law dictating the direct sound level of a single point sound to drop 6 dB if the distance to the loudspeaker is doubled (or correspondingly a 20-dB drop with 10 times longer distance). For listeners far away from the stage an additional speaker can be introduced that is close to these listeners. This will help avoid a too-low direct sound level for these listeners. Due to the speed of sound being 343 ms (at 20 °C) this new loudspeaker will play ahead of the original loudspeaker close to the stage. Approximately 3 ms delay is introduced for each meter of separation between the speakers. To get the loudspeakers to act closer to each other in time, a time delay is normally introduced to the loudspeaker close to listeners at the far distance. Hence the name *delay-speaker* is used. For large arenas it is a great challenge to synchronize all sources, to avoid some speakers acting more like reflected sound rather than direct sound.

Virtual Point Sources

To achieve the wanted directivity at mid and high frequencies, desired sound levels, and some directivity at lower frequencies, it is normal to put several loudspeakers close to each other, called loudspeaker *clusters*. The intention of such a constellation of loudspeakers is that the loudspeakers together operate as one single source, referred to as a *virtual point source* (see Fig. 3.8a). Such clusters also make it practical to adjust the total direct sound coverage by varying the number of loudspeakers in the cluster. The loudspeaker is arranged here in such a way that the on-axis lines of the speakers meet in one point in space behind the speakers.

This creates *coherent* sound originating from a virtual single point in space; the loudspeakers are in-phase and related to each other and are said to *couple*. A virtual point source corresponds to using several torches (directional point light sources) to light up several spots in the dark. Due to the physical separation between the speakers and the speed of sound, it is difficult to achieve coherent sound at all frequencies. At high and mid frequencies the difference in distance to the different loudspeakers is significant compared to the wavelength. The total direct sound will be more or less out of phase (with a phase difference of 180° the waves will cancel each other out if the waves have equal levels). Such interference, so-called *comb filtering*, will result in constructive and destructive summation of the sound waves that create large variations of the total direct sound level provided by the virtual point source. The interference results in both spatial-level variations at a single frequency and spectral variations at a single location. The spatial variations lead to what are often referred to as *side lobes* (see more details below). Such side lobes can appear in both the vertical or horizontal plane depending on the arrangement of the loudspeaker cluster. Horizontally stacked speakers will result in interference in the horizontal plane and vice versa. At a single position the comb filtering can result in large variations of sound levels at different frequencies. The problem with such constructive and destructive interference, depending on location and frequency, is that the direct sound level within the audience does not show a consistent level and spectral balance. Additionally, the side lobes will often point in directions away from the audience and will contribute to increasing the level of reflected sound. With good directivity control for the individual speakers there will be only small overlapping regions of significant interference because the individual speakers will provide direct sound within their own angle sectors (solid angle). But the directivity pattern will normally vary with frequency, making it virtually impossible to find one angle between the loudspeaker cabinets that leads to maximum separation at all frequencies. This frequency-dependent overlap and unavoidable interference is one problem associated with virtual point sources.

Directional Subwoofer Arrays

At low frequencies interference can be employed to create directivity by use of several omnidirectional speakers, for example, the *cardioid subs* configuration, named after the directivity the subwoofers are intended to produce when placed together. Sound waves oscillate with both positive and negative amplitude within their cycle of repetition (one period). If two sources interfere with equal amplitude value but opposite polarity they can in principle fully cancel each other out. This can be used to obtain little sound emission at the rear of the speakers, thus achieving significant directivity. This directivity is achieved at the expense of reduced efficiency and total sound level, because portions of the radiated sound are ancelled out.

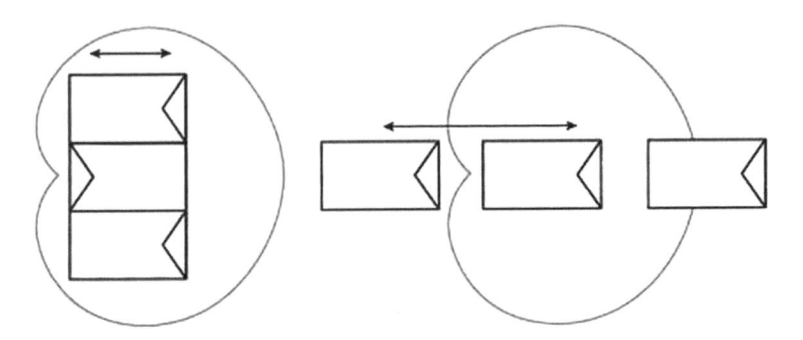

Fig. 3.3 Directional subwoofer configurations, with vertically stacked cardioid subs shown on the *left* and the horizontally stacked end fire configuration shown on the *right*. The cardioid pattern shown for the end fire configuration is idealized; in practice there will be some low-frequency radiation towards the rear

Figure 3.3 shows two ways to accomplish directivity with three subwoofers stacked or spread out horizontally (the prior called cardioid sub and the latter called end fire). Due to the relatively low speed of sound waves the physical separation between the acoustic centers of the subwoofers introduces some sort of phase-shift (increasing with separation distance and frequency). By adding individual delays to the speakers the added phase-shift can be compensated for to result in coherent sound waves either towards the front (audience) or towards the back. The left approach in Fig. 3.3 aims at having fully synchronized sound waves towards the front (right side in the figure), with one subwoofer having reversed polarity resulting in maximum cancellation in this direction. The approach to the right in Fig. 3.3 (end fire) aims at fully synchronized waves towards the front (audience; right side in the figure) with significant delays between the subwoofers towards the rear. An important distinction between the cardioid and end fire configurations is that the end fire configuration does not lead to a full cancellation towards the back. The cancellation towards the back is more frequency dependent with end fire. The benefit with end fire is that efficiency in the front direction is higher, because the sound waves here are fully coherent. In addition to higher efficiency towards the audience, the fully synchronized sound with the end fire configuration will often lead to a more distinct bass sound experience, which can be crucial for certain musical genres. The physical horizontal separation between the acoustic centers of the subwoofers is indicated with arrows in Fig. 3.3.

As an example, a 1.5-m separation will lead to an approximately 4.5-ms delay between the two emitted sound waves. For the end fire case 4.5-ms delay on the mid subwoofer and 9-ms delay on the front subwoofer will provide synchronized in-phase sound towards the right. The signals sent to the loudspeakers are normally exact copies, which means the signals are in principle perfectly correlated and in-phase (coherent) relative to each other. The signals/waves match each other perfectly and result in high levels for the audience. Towards the rear there will now be a 4.5 and 9-ms delay. Acoustically a 9-ms delay will lead to the two emitted waves being 180° out of phase if the cycle of the oscillating signal is 18-ms. Using the formula $f = 1/T$, one cycle, or period, of 6-ms is found to correspond to 56 Hz. This will result in significantly reduced levels towards the back in the lower bass. For the vertically stacked case the cancellation towards the rear will in principle exist at all frequencies due to the polarity reversal. A 1-m separation will result in an approximately 3-ms delay between the sources in the front direction. This will lead to a wave cancellation at 167 Hz which is normally above the frequency band for the subwoofer and hence does not represent a problem.

A challenge with the cardioid configuration can be to match the compensating delay exactly with the physical separation of the subwoofers. The location of the acoustic center of the subwoofer can vary with frequency, making it difficult, if not impossible, to find one delay setting that suits all frequencies. Failure to match delay and physical separation fully can result in significant radiation backwards within certain frequency bands, that can represent a problem, for instance, for the musicians on stage. All pass filters can be used here to make phase adjustments at specific frequencies. Another challenge is the phase relation between the subwoofers and the main system (e.g., a line array) in the crossover frequency region (typically around 125–160 Hz). This challenge applies to all subwoofer systems.

Split Mono Systems

Instead of one central cluster above the middle of the stage front it is common to operate with speakers at each side of the stage. A true stereo experience will exist only close to the center line with close to equal distance to the speaker clusters, and in reality sound engineers use only very little panning of instruments between left and right speakers. For listeners close to one side the sound from the closest speaker cluster will dominate leading to the impression that all loudspeaker sound is coming from that side of the stage. It is therefore more representative to look at the stereo system as a *split mono system*, where the left cluster provides mono sound for the listeners at the left side of the audience center line. Close to the center line the audience will experience some interference because the levels from the two sides are very similar whereas there is some delay between the direct sound from the different sides. Such interference along the center line is inevitable and can exist in a large area if the loudspeakers do not have significant directivity. The upper picture in Fig. 3.4 shows the horizontal variation of the direct sound level at 120 Hz for a split mono configuration based on an omnidirectional source. In the dark regions for the split mono case the direct sound level is very low. This type of interference also creates a directivity pattern with side lobes; in certain directions *off-axis* (not directly in front of the sound system) there are high levels (with low levels between). At the center line the two sources are coherent or close to being coherent (in-phase). At the sides of the center line the delay between the two sound waves is significant, leading to cancellation. Due to the long wavelength at 120 Hz, the lobes occupy large angle sectors; at higher frequencies the

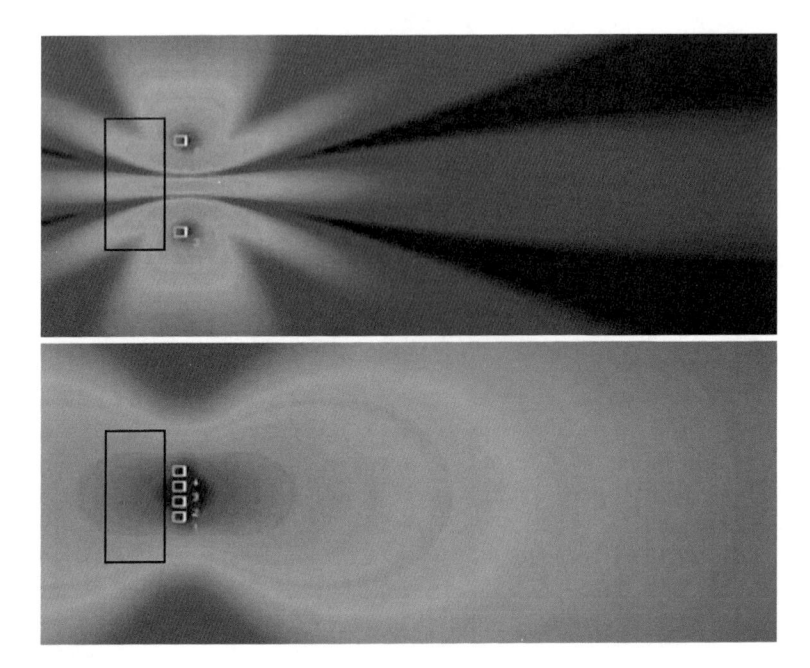

Fig. 3.4 Split mono (*top*) and line subwoofer (*bottom*) configuration. The stage is indicated with a black rectangle

distances between dark and hot spots will be reduced. Significant directivity at higher frequencies also helps isolate the left from the right side. A more spatially dense interference pattern, directivity, and psychoacoustic effects make such interference less of a problem at mid and high frequencies.

A strategy to reduce the level variation for close to omnidirectional subwoofers is to introduce more sources on a line across the stage front. If the separation distance between the subwoofers is less than $\lambda/2$ for all relevant frequencies, the line of subwoofers will operate as a line source, leading to a reduced loss of sound level versus distance (see below). The increased number of sources with different mutual delays helps avoid all sources being out of phase at a single position. This line configuration of subwoofers under the stage is shown in Fig. 3.4 (bottom). By adjusting the delay between the subwoofers, the subwoofers will instead appear as a curved line, leading to a directivity that better suits the audience area (avoiding high levels and fully coherent sound right in front of the subwoofers and significant level drop towards the sides). By applying a frequency-dependent level reduction on the outmost subwoofers, negative side lobes can be avoided. Such a strategy of delaying and level-adjusting the line of speakers to create a wanted directivity is called *delay-shading*.

Some reflections (preferably early) can help avoid very low total sound levels (dark spots) while maintaining sufficient clarity of sound, which is an example of the beneficial effects of reflected sound. At low frequencies standing wave patterns instead of discrete reflections will be most relevant, particularly for small rooms. When using reflected sound to reduce level variations, the level balance between early and late reflected sound may be a crucial property of the reflected sound.

Line Arrays

Line array systems, shown in Fig. 3.8b can cover larger source-to-receiver distances than point source clusters. They also maintain a significant vertical directivity with relatively little destructive comb filtering between the speakers in the array. The frequency area of significant directivity will depend on the array length; longer arrays result in a more broadband and more low-frequency directivity control. All the loudspeakers constituting the array will be very close to being perfectly in-phase and coherent directly in front of the array. To achieve an effective line array the separation between the loudspeakers must be least than $\lambda/2$ for a point source array. At high frequencies wave guides are often employed to achieve a continuous line source effectively without significant separation between the sources. This means that most conventional line arrays function as close to real line sources at high frequencies but only as coupled point sources at mid and low frequencies. Good and even coverage of audience segments at various distances from the stage especially in large venues or at outdoor concerts is achievable with line array systems. Avoiding too many speakers in the array while ensuring a very precise suspension will ensure that sound does not propagate, for instance, directly onto the rear wall of the hall. The horizontal directivity is in rough terms approximately the same as for each single loudspeaker in the array, although not exactly the same due to the vertical separation between the sources. The high degree of coherence makes it easier to achieve high total sound levels without having to drive each individual loudspeaker at the maximum tolerable level (speaker overheating or distorting). For a typical line array the desired directivity is achieved through both the construction of the loudspeaker and the array itself. A second very advantageous bonus is that the direct sound-level reduction can be 3 dB instead of 6 dB per doubling of distance (Fig. 3.5). This helps maintain a high ratio of direct to reflected sound for listeners at far distances. The key factors that control 3 dB instead of 6 dB reduction per distance doubling are the distance to the line array and the length of the line array. Line arrays are typically 1–10 m long leading to the array not being long compared to the wavelength at low frequencies, leading to the array functioning more like a point source. At far distances the array will appear more as a single point source. Line arrays are therefore most efficient at producing directional direct sound with a 3-dB drop per doubling distance mainly at mid and high frequencies in the vicinity of the array. The longer arrays are consequently effective line sources in a larger frequency and distance range. Equation (3.1) from Davis and Patronis (2006) can be used to calculate the limiting frequency f for an effective line source depending on the listener's

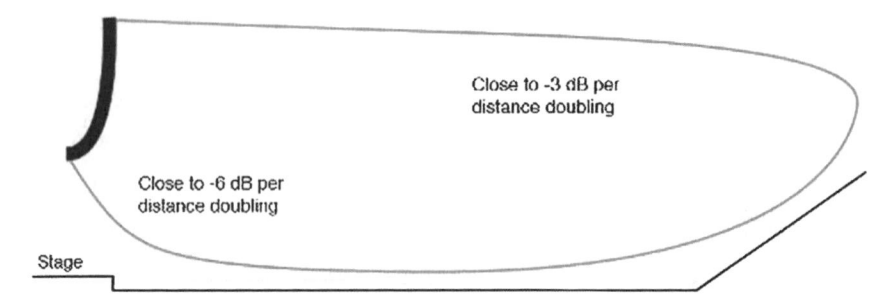

Fig. 3.5 Line array for providing direct sound to the audience

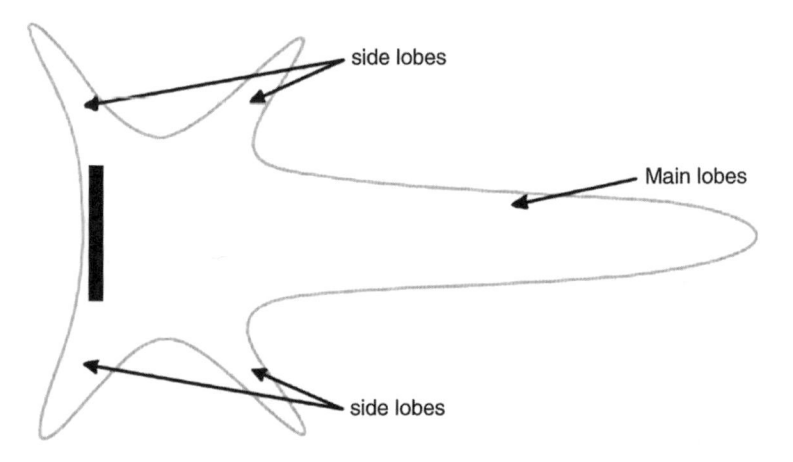

Fig. 3.6 Main lobe and side lobes for uniform line arrays seen from the side. The side lobes here are vertical

distance to the array d and the length of the array (L). The constant c in Eq. (3.1) is the speed of sound (343 ms at 20 °C).

$$f \geq \frac{2 \cdot c \cdot d}{L^2} \qquad (3.1)$$

Bending the array converts it more to a virtual point source with a 6-dB drop per distance doubling. This is useful to reduce the direct sound levels for the audience close to the stage; see Fig. 3.5. With long line arrays constituting many loudspeakers, the problematic interference associated with virtual point sources is often less significant due to the large number of sources (comparable to the situations in Fig. 3.4). But off-axis line arrays based on feeding the same signal to all the loudspeakers within the array (*uniform arrays*) will often show significant side lobes in their vertical directivity in mid and high frequencies, as illustrated a bit exaggerated in Fig. 3.6. By feeding individual signals to each loudspeaker,

Line Arrays

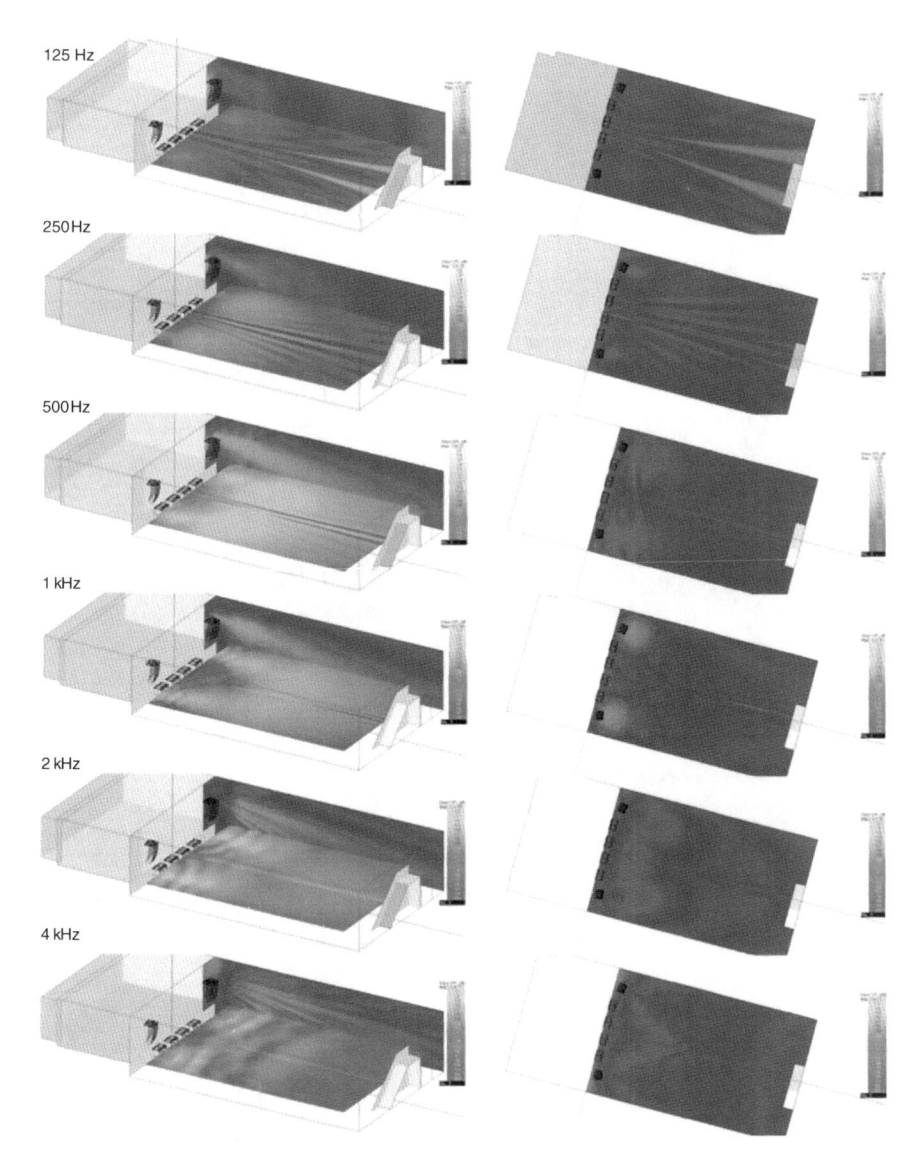

Fig. 3.7 Direct sound coverage plots for different octave bands for a hypothetical venue. Audience area plus right side wall (*left*) and ceiling (*right*). (Plots courtesy of Bård Støfringsdal)

with customized levels and delays, side lobes can be reduced, but at some expense of how the line array behaves as an effective line source (3-, not 6-dB–level drop per distance doubling). One example of such a strategy is *CBT* systems (*constant beamwidth transducers*; Keele 2002).

Fig. 3.8 Main loudspeaker system design principles: **a** point source clusters and **b** line arrays

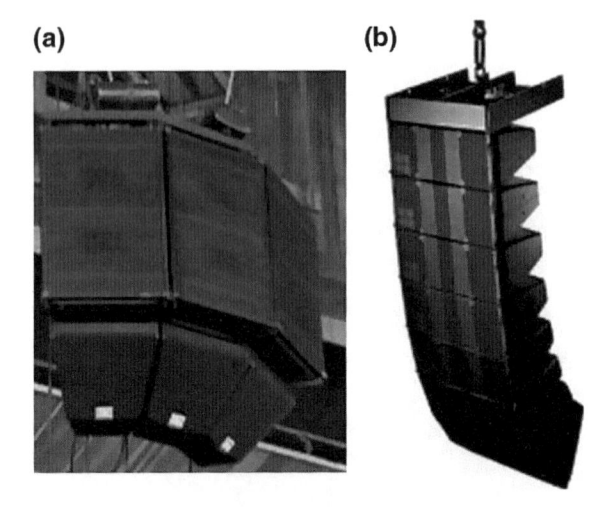

It is worth noting that such side lobes will exist only in the vertical plane for normal line array configurations. The array consists of vertically stacked speakers and seen from above the array will be close to behaving like a point source horizontally. In the horizontal plane there is normally less spatial interference although the line source cannot fully be seen as one coherent source horizontally. For line arrays close to the side walls there can be a problem with wall reflections providing significant interference on the direct sound as for a virtual point source.

Examples of calculated direct sound coverage and examples of point source and line source loudspeaker cabinets are shown in Figs. 3.7 and 3.8, respectively.

Chapter 4
Assessments of 20 Halls

In this chapter the 20 halls that were assessed for the recommendations in the paper "Suitable Reverberation Times for Halls for Rock and Pop Music" are presented. All measurements were made in 2005 and since then many of the halls have been refurbished or rebuilt. For each hall there is a photo, a scaled drawing, and the subjective ratings and objective measurement results are presented as well. There is the T_{30} curve as a function of frequency of the measurement in the empty hall with an omnidirectional loudspeaker. In the halls where there was a preinstalled PA system available, measurements were also made with that as a source and the T_{30} curve deriving from these measurements is presented in the same diagram. An estimated T_{30} including a densely packed audience is also presented in the diagram. The estimate is based on the omnisource measurements, the floor area available for the audience, and the coefficients from Fig. 1.16. SD stands for standard deviation. St. stands for stage, and ratings relates to the musicians whereas Aud. stands for audience and relates to sound engineers' ratings and the audience area. The T_{30} diagrams stretch down to only the 63-Hz third octave band whereas the EDT and D_{50} diagrams include the entire 63-Hz octave band. The bass ratio is calculated as the average T_{30} in the 63- and 125-Hz octave bands to the average T_{30} in the 0.5–2 kHz octave bands. Other acoustical data are self-explanatory.

In the drawings, the symbol for PA speaker is ⬭

And for the omnisource the symbol is ◆
Microphone positions are indicated with the symbol ● and the sound engineer's position is denoted RSG.

The subjective ratings are shown for both musicians and sound engineers. The ratings are commented on by the author inasmuch as in some incidents they do not agree with what would be expected from the objective data of the hall. These comments are only to be regarded as speculation. Nevertheless it can be valuable information when designing a hall for pop and rock music.

Some information on constructive details such as building materials used is included for each hall. The capacity of the hall refers to the number of standing audience members that the hall houses.

Amager Bio

BUILT: in Copenhagen in 1941 as a cinema. Renovated 1997 as a music venue.
   CAPACITY: 1,000
   NUMBER OF CONCERTS PER YEAR: 60–100

   BUILDING MATERIALS USED: Ceiling: Concrete, about half is covered by suspended mineral wool slabs. Floor: Wood on joists. Walls: Perforated, wave-shaped metal plates, with mineral wool behind. Backdrop. Balcony level in the rear of the hall.

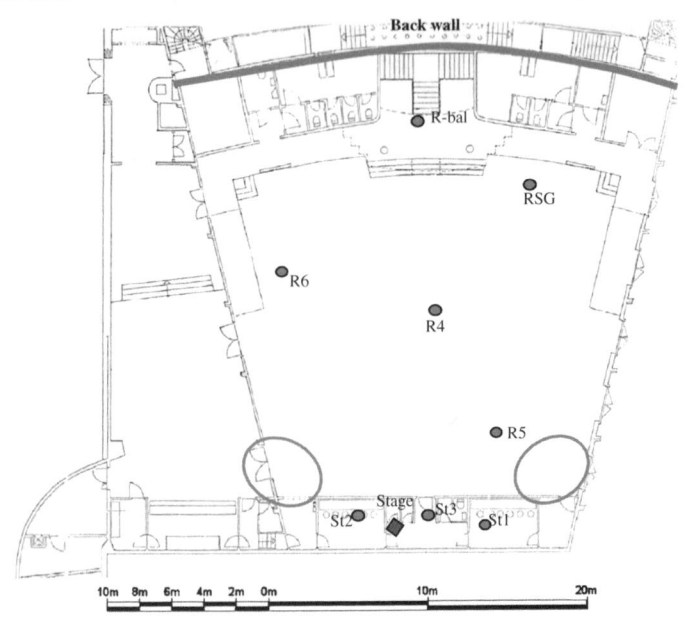

Amager Bio

Geometrical data	
Volume (m^3)	4,500
Height (m)	8.4–9.1
Surface area stage (m^2)	70
Surface area audience (m^2)	480
Total surface area of hall (m^2)	2,000
Capacity of standing persons	1,000
Acoustical data—omnisource	
Audience area	
T_{tom}—(500–1k) (s)	1.0
T_{fuld}—(500–1k) (s)	0.7
D50—(250–2k)	0.56
EDT—(250–2k) (s)	1.1
$BR_{rock,a}$	1.1
Stage area	
T_{tom}—(500–1k) (s)	1.0
D50—(250–2k)	0.83
EDT—(250–2k) (s)	0.8
Subjective data	
Audience area	
Rating out of 20	9
Stage area	
Rating out of 20	4

The sound engineer's position is in the rear corner of the hall. This may negatively influence the sound engineer's rating. The venue sounds great; the fan shape may be ideal for amplified music.

52　　　　　　　　　　　　　　　　　　　　　　　　4　Assessments of 20 Halls

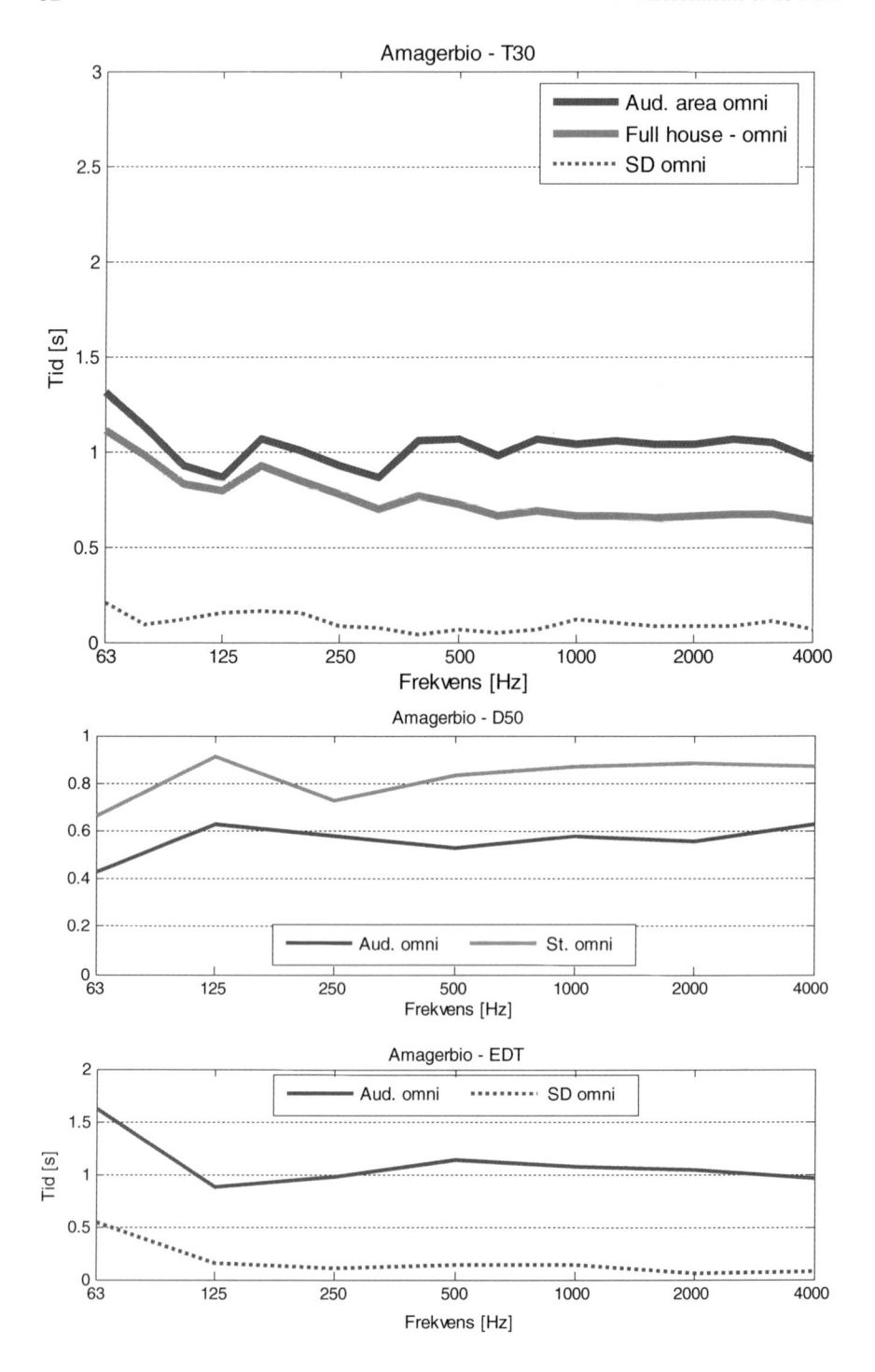

Forbrændingen

BUILT: in Albertslund, 1996.
　CAPACITY: 450
　NUMBER OF CONCERTS PER YEAR: 70

BUILDING MATERIALS USED: Ceiling: Suspended wood fiber slabs. Floor: Wood on joists. Walls: Painted concrete and drywall. Two balcony levels.

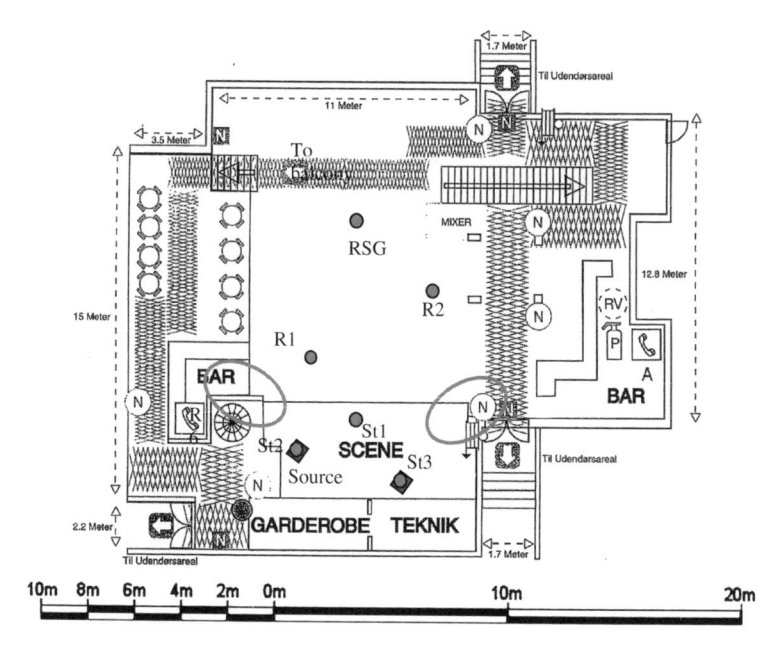

Geometrical data	
Volume (m^3)	3,050
Height (m)	8.1
Surface area stage (m^2)	36
Surface area audience (m^2)	170
Total surface area of hall (m^2)	1,400
Capacity of standing persons	450
Acoustical data—omnisource	
Audience area	
T_{tom}—(500–1k) (s)	0.9
T_{fuld}—(500–1k) (s)	0.5
D50—(250–2k)	0.65
EDT—(250–2k) (s)	0.8
$BR_{rock,a}$	1.2
Stage area	
T_{tom}—(500–1k) (s)	1.0
D50—(250–2k)	0.89
EDT—(250–2k) (s)	0.5
Subjective data	
Audience area	
Rating out of 20	11
Stage area	
Rating out of 20	19

These ratings do not reflect the almost acceptable acoustics of the room. Earlier this venue was not well liked by musicians due to lack of audience. Stage may lack some early reflections.

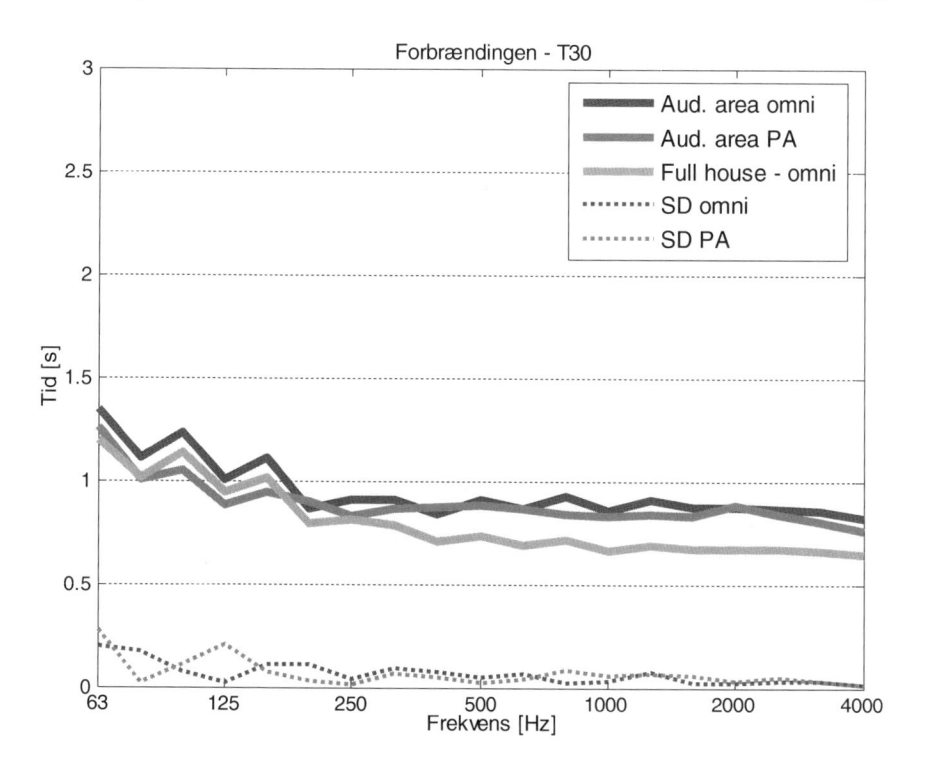

Forbrændingen

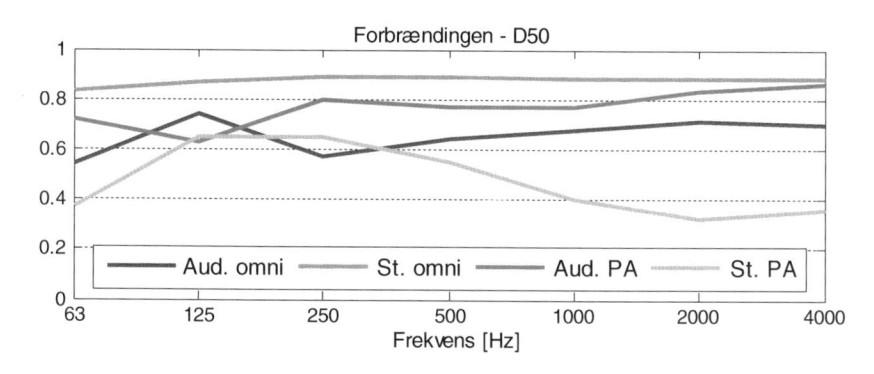

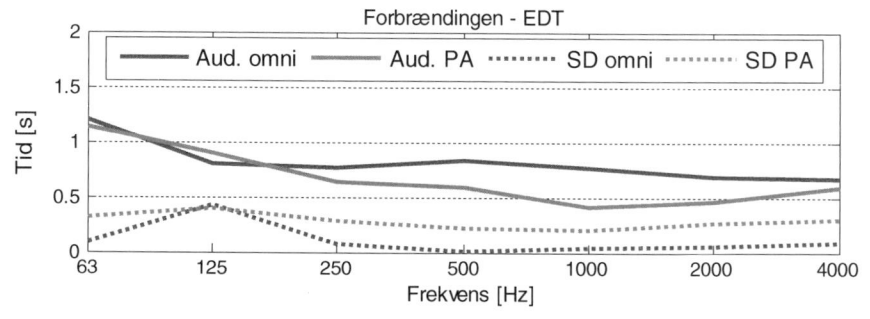

Forbrændingen - T30

Forbrændingen - D50

Forbrændingen - EDT

Godset

BUILT: in Kolding in 1920 as railway premises. Rebuilt 2001 as a music venue.
CAPACITY: 700
NUMBER OF CONCERTS PER YEAR: about 70

BUILDING MATERIALS USED: Floor: Wood on joists. Ceiling: Wood fiber slabs. Walls: Painted brick.
End walls: Painted drywall. A backdrop can be mounted behind the stage.

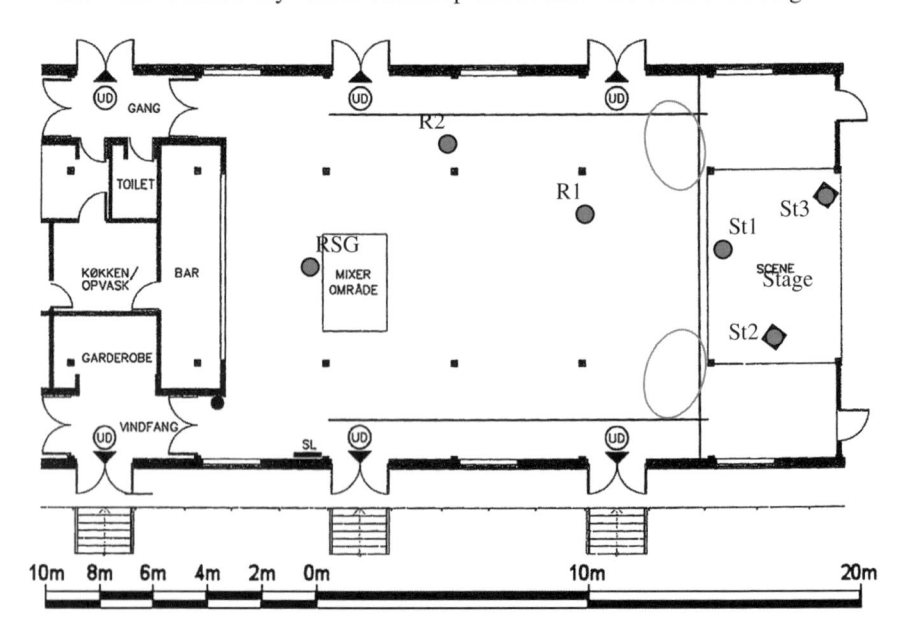

Godset

Geometrical data	
Volume (m^3)	2,150
Height (m)	3.8–9.2
Surface area stage (m^2)	35
Surface area audience (m^2)	260
Total surface area of hall (m^2)	1,100
Capacity of standing persons	700

Acoustical data—omnisource	
Audience area	
T_{tom}—(500–1k) (s)	0.8
T_{fuld}—(500–1k) (s)	0.5
D50—(250–2k)	0.38
EDT—(250–2k) (s)	0.9
$BR_{rock,a}$	0.8
Stage area	
T_{tom}—(500–1k) (s)	0.8
D50—(250–2k)	0.85
EDT—(250–2k) (s)	0.5

Subjective data	
Audience area	
Rating out of 20	4
Stage area	
Rating out of 20	6

58   4   Assessments of 20 Halls

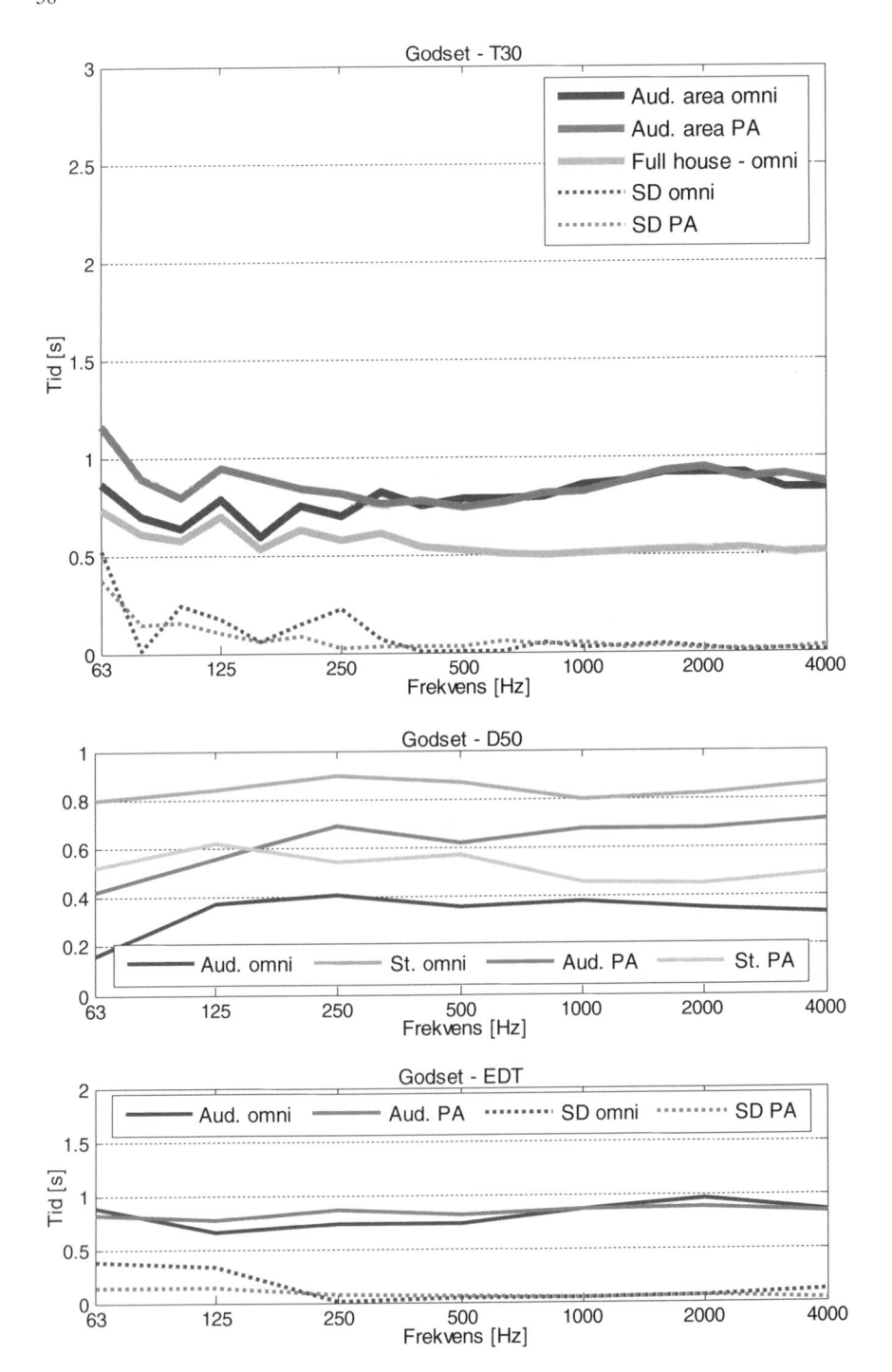

Godset - T30

Godset - D50

Godset - EDT

Lille VEGA

BUILT: in Copenhagen in 1956. Rebuilt in 1996 as a music venue.
CAPACITY: 500
NUMBER OF CONCERTS PER YEAR: 150

BUILDING MATERIALS USED: Floor and walls: Wood on joists. Ceiling: Concrete, suspended fabric.

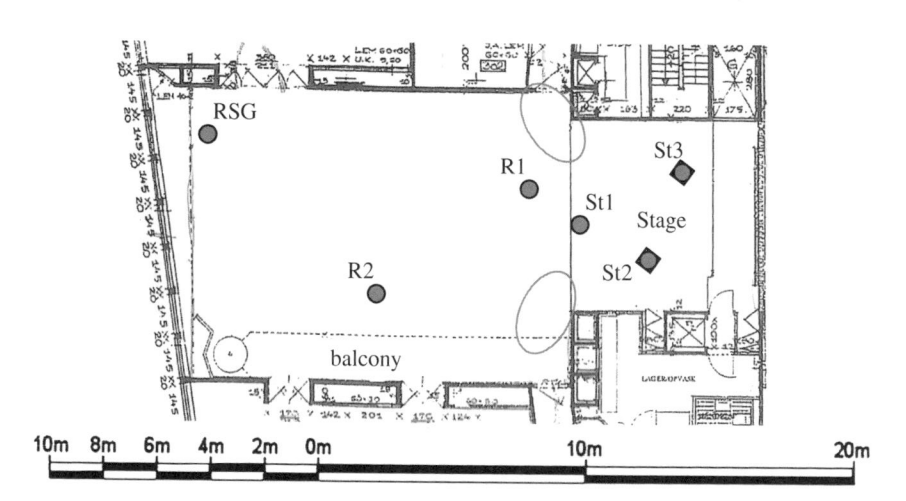

Geometrical data	
Volume (m^3)	785
Height (m)	4.4
Surface area stage (m^2)	47
Surface area audience (m^2)	165
Total surface area of hall (m^2)	625
Capacity of standing persons	500

Acoustical data—omnisource	
Audience area	
T_{tom}—(500–1k) (s)	0.7
T_{fuld}—(500–1k) (s)	0.3
D50—(250–2k)	0.65
EDT—(250–2k) (s)	0.7
$BR_{rock,a}$	0.7
Stage area	
T_{tom}—(500–1k) (s)	0.6
D50—(250–2k)	0.86
EDT—(250–2k) (s)	0.4

Subjective data	
Audience area	
Rating out of 20	6
Stage area	
Rating out of 20	1

The sound engineer's position is in the back of the hall in a corner.

Lille VEGA

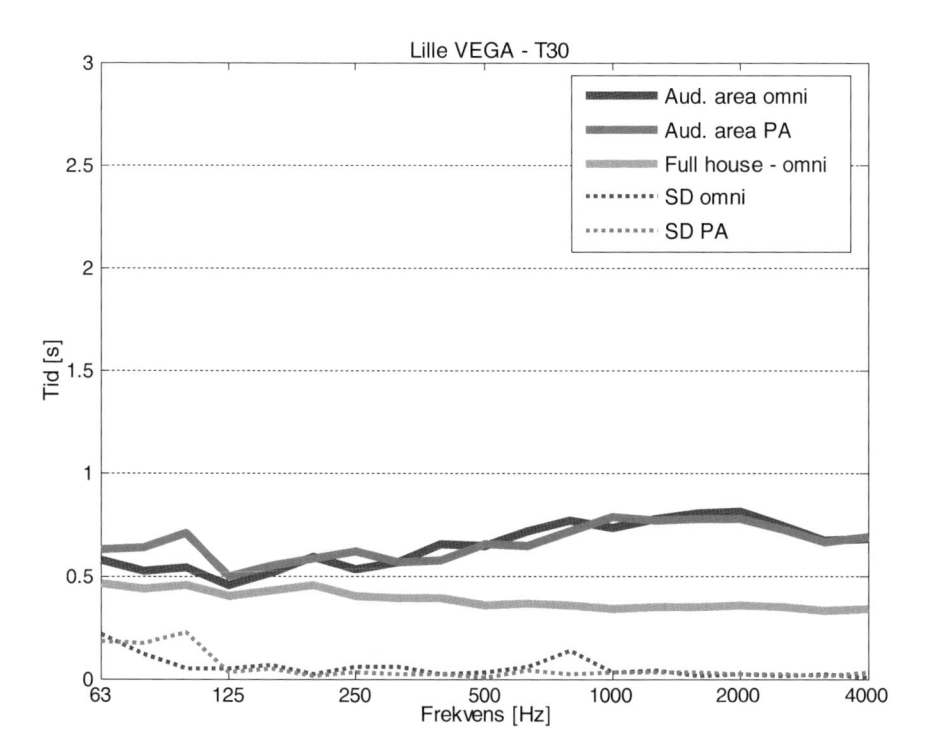

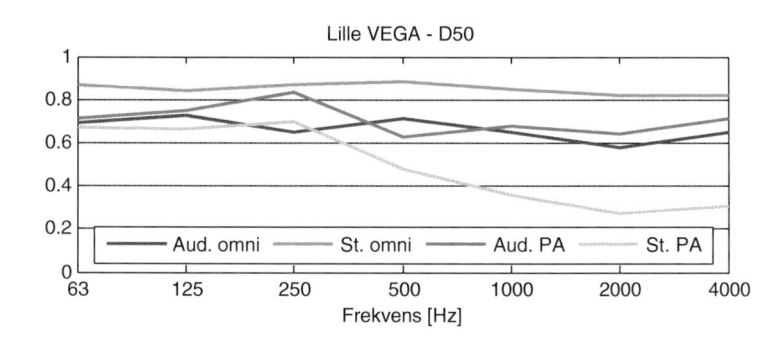

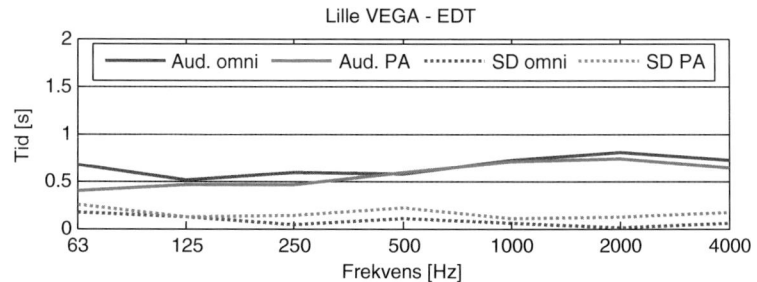

Loppen

BUILT: in Copenhagen in 1880. Rebuilt as a music venue in 1973.
  CAPACITY: 350
  NUMBER OF CONCERTS PER YEAR: 180

BUILDING MATERIALS USED: Ceiling: Wood. Floor: Wood on joists. Walls: Painted bricks. No backdrop.

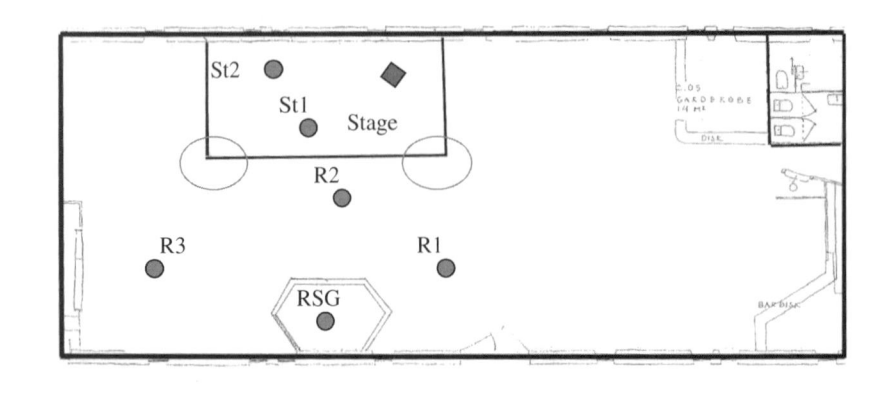

Loppen

Geometrical data	
Volume (m^3)	890
Height (m)	2.7
Surface area stage (m^2)	34
Surface area audience (m^2)	150
Total surface area of hall (m^2)	870
Capacity of standing persons	350

Acoustical data—omnisource	
Audience area	
T_{tom}—(500–1k) (s)	0.8
T_{fuld}—(500–1k) (s)	0.5
D50—(250–2k)	0.69
EDT—(250–2k) (s)	0.7
BR$_{rock,a}$	1.2
Stage area	
T_{tom}—(500–1k) (s)	0.7
D50—(250–2k)	0.82
EDT—(250–2k) (s)	0.6

Subjective data	
Audience area	
Rating out of 20	12
Stage area	
Rating out of 20	5

Difficult peak in T_{30} around 200 Hz due to low ceiling. Very intimate venue; musicians meet their audience one-to-one.

4 Assessments of 20 Halls

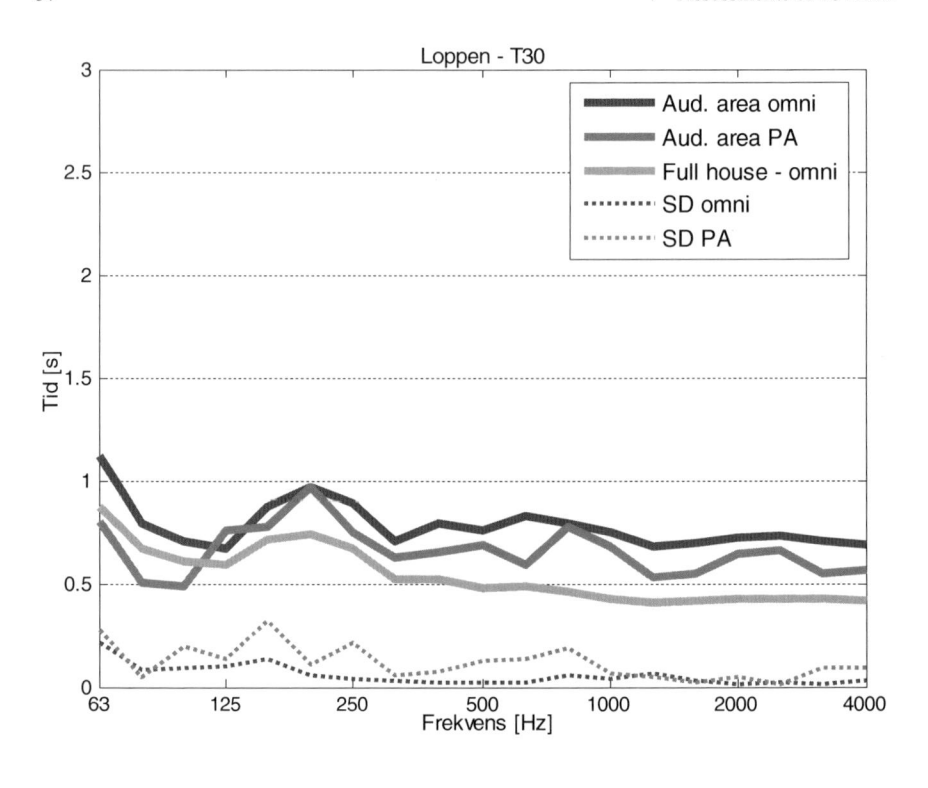

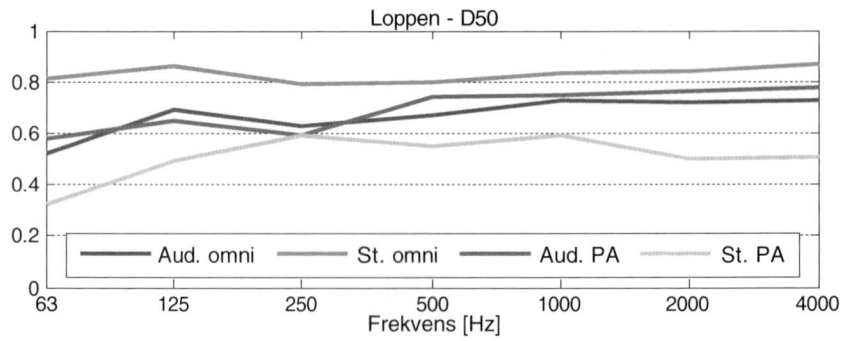

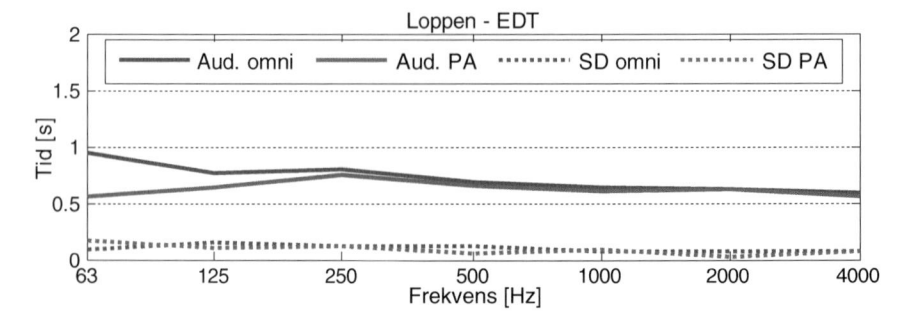

Magasinet

BUILT: in Odense 1933. Rebuilt as a music venue 1988.
CAPACITY: 525
NUMBER OF CONCERTS PER YEAR: 100

BUILDING MATERIALS USED: Ceiling and walls: Painted concrete. Floor: Rubber on concrete. Backdrop.

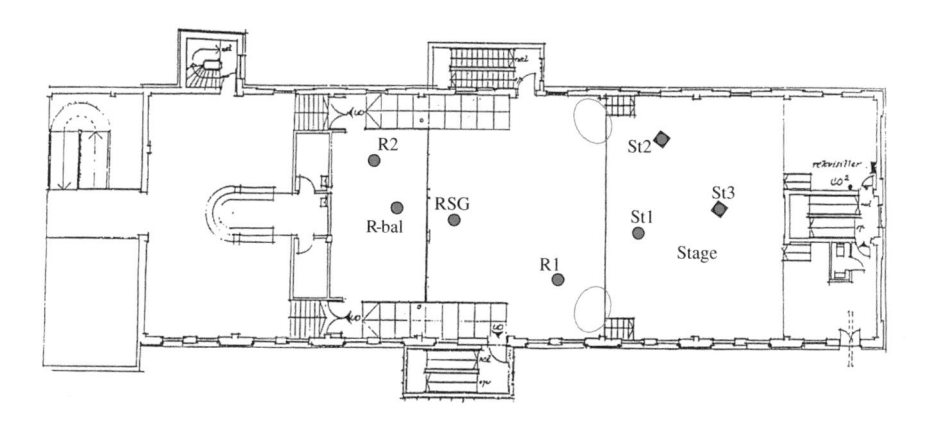

Geometrical data	
Volume (m^3)	2,540
Height (m)	7.8
Surface area stage (m^2)	120
Surface area audience (m^2)	230
Total surface area of hall (m^2)	1,400
Capacity of standing persons	525
Acoustical data—omnisource	
Audience area	
T_{tom}—(500–1k) (s)	1.3
T_{fuld}—(500–1k) (s)	0.8
D50—(250–2k)	0.33
EDT—(250–2k) (s)	1.5
BR$_{rock,a}$	1.4
Stage area	
T_{tom}—(500–1k) (s)	1.3
D50—(250–2k)	0.85
EDT—(250–2k) (s)	1.2
Subjective data	
Audience area	
Rating out of 20	19
Stage area	
Rating out of 20	12

The low rating reflects the peak in T_{30} at 125 Hz. This venue was renewed in 2008 (see Fig. 5.5) and the peak was brought down to the recommended value for this size hall. It is now one of the best venues in Denmark according to musicians and audience, but due to the deliberate absence of porous absorption, sound engineers can have a hard time controlling the sound, especially at sound check, if the band does not have a good sound by itself.

Magasinet

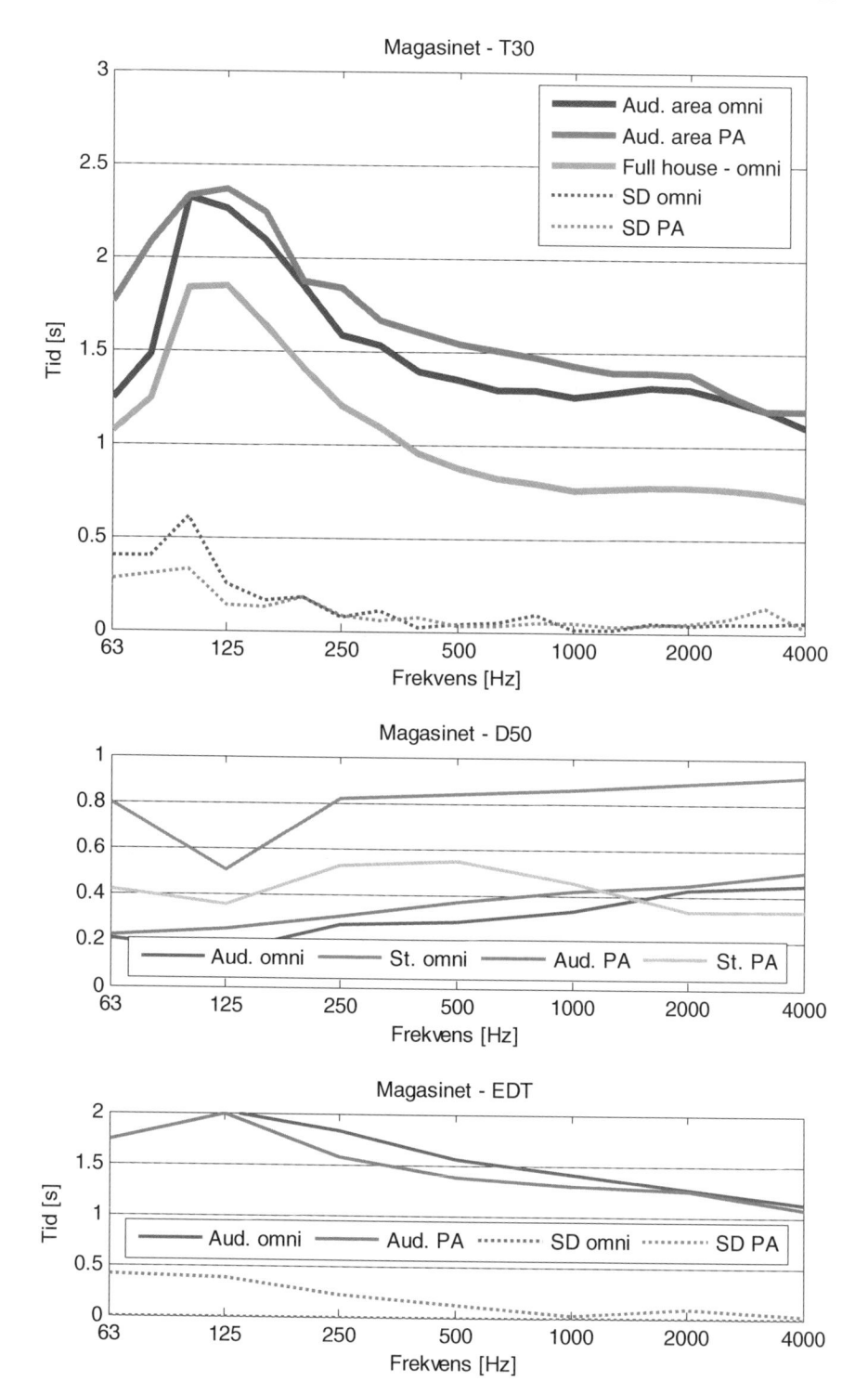

Palletten

BUILT: in Viborg, as a cinema. Rebuilt as a music venue 1993.
 CAPACITY: 375
 NUMBER OF CONCERTS PER YEAR: 75

BUILDING MATERIALS USED: Ceiling: Painted mineral wool.
Floor: Wood on joists.
Walls: Painted concrete. Backdrop.

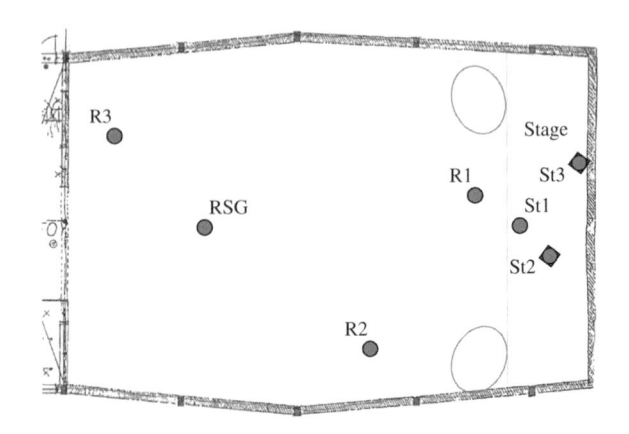

Palletten

Geometrical data	
Volume (m^3)	1,420
Height (m)	2.9–5.5
Surface area stage (m^2)	43
Surface area audience (m^2)	185
Total surface area of hall (m^2)	890
Capacity of standing persons	375
Acoustical data—omnisource	
Audience area	
T_{tom}—(500–1k) (s)	0.8
T_{fuld}—(500–1k) (s)	0.6
D50—(250–2k)	0.53
EDT—(250–2k) (s)	0.7
$BR_{rock,a}$	1.0
Stage area	
T_{tom}—(500–1k) (s)	0.7
D50—(250–2k)	0.84
EDT—(250–2k) (s)	0.54
Subjective data	
Audience area	
Rating out of 20	8
Gennemsnit	3.63
Median	3.5
Spredning	1.30
Varians	1.70
Antal besvarelser	8
Stage area	
Rating out of 20	7
Gennemsnit	3.23
Median	3
Spredning	1.01
Varians	1.03
Antal besvarelser	13

4 Assessments of 20 Halls

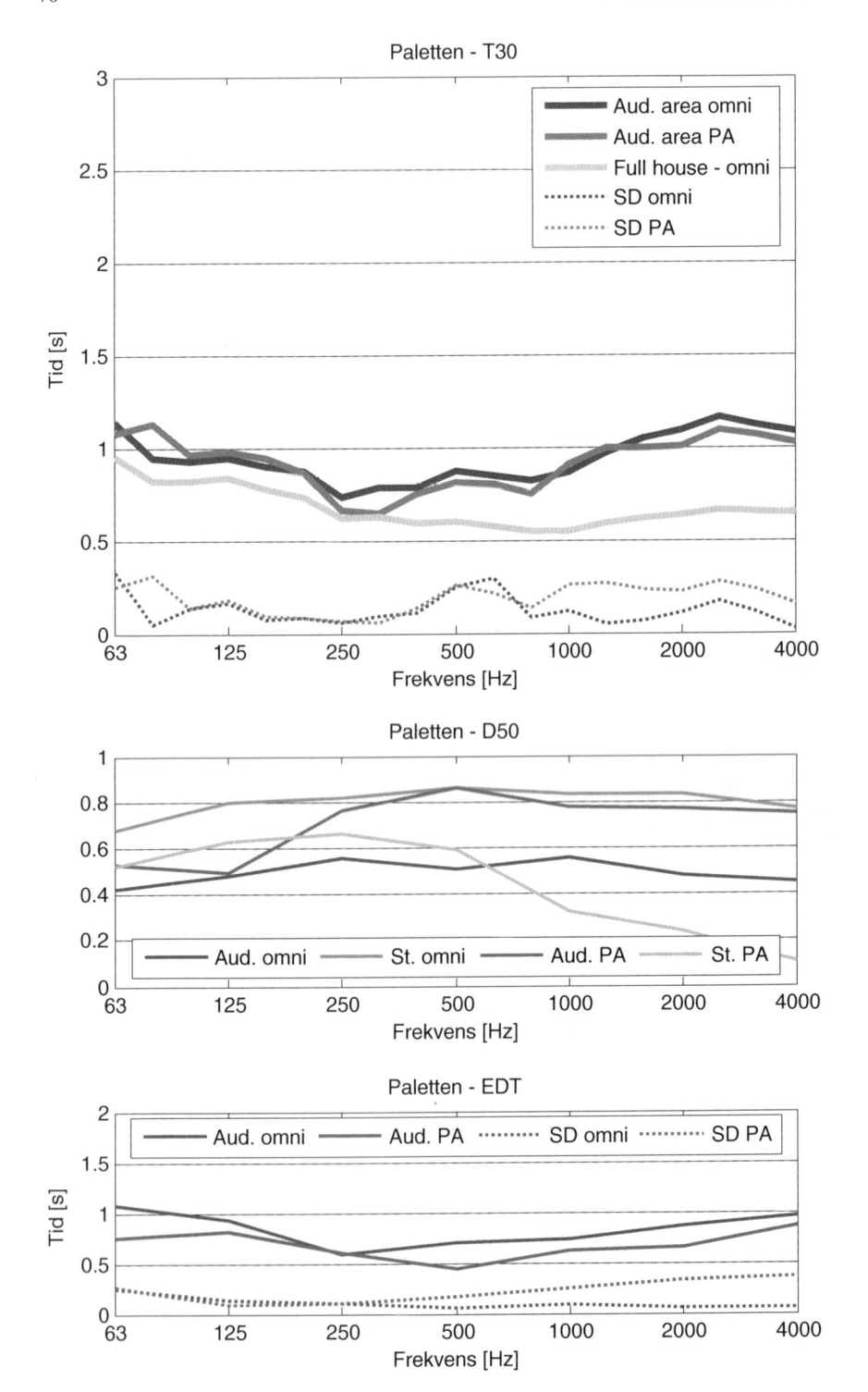

Paletten - T30

Paletten - D50

Paletten - EDT

Pumpehuset

BUILT: in Copenhagen, 1852 as a freshwater pumping station. Rebuilt as a music venue in 1986.
CAPACITY: 600
NUMBER OF CONCERTS PER YEAR: 125

BUILDING MATERIALS USED: Ceiling: Wood. Floor: Rubber on concrete. Side walls: Painted brick. Upper triangle on each end wall: Wood fiber slabs direct on brick. End wall opposite stage: Brick wall with perforated bricks (hole width: 25 mm, hole length: 100 mm and a 25-cm cavity with mineral wool behind). Resonator absorbers on side walls. Small thin baffles suspended near ceiling.

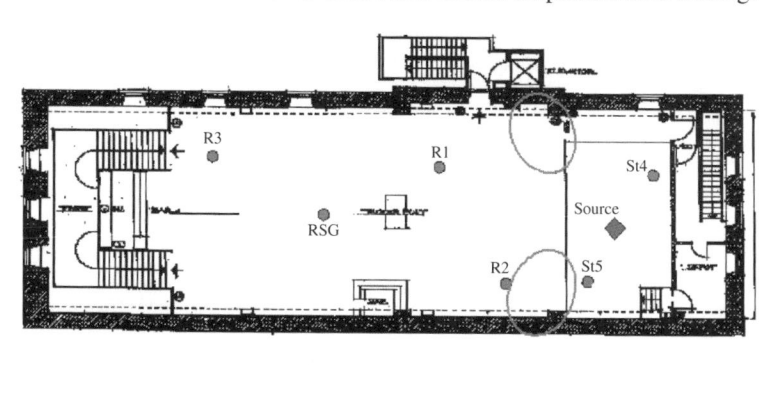

Geometrical data	
Volume (m^3)	3,000
Height (m)	6.7–9.4
Surface area stage (m^2)	55
Surface area audience (m^2)	225
Total surface area of hall (m^2)	1,400
Capacity of standing persons	600
Acoustical data—omnisource	
Audience area	
T_{tom}—(500–1k) (s)	1.2
T_{fuld}—(500–1k) (s)	0.8
D50—(250–2k)	0.60
EDT—(250–2k) (s)	1.0
$BR_{rock,a}$	0.9
Stage area	
T_{tom}—(500–1k) (s)	1.2
D50—(250–2k)	0.62
EDT—(250–2k) (s)	1.0
Subjective data	
Audience area	
Rating out of 20	15
Stage area	
Rating out of 20	16

Everybody has an opinion about this hall. The poor ratings are to some degree justified. There is too much reverberation altogether (except at 250 Hz where probably the perforated brick wall has an impact). The hall is very long and narrow and from the stage it seems that the sound decay is indeed too long and it's difficult to get loud enough levels of early sound in order to be able to mask the loud level of late reflections. On the other hand, when the venue is completely packed it is actually ok and a certain sound atmosphere arises. It seems that the air conditioning cannot keep up with the many active audience members whereby the high level of moisture in the room creates a new frequency balance that makes the room exciting to create music in.

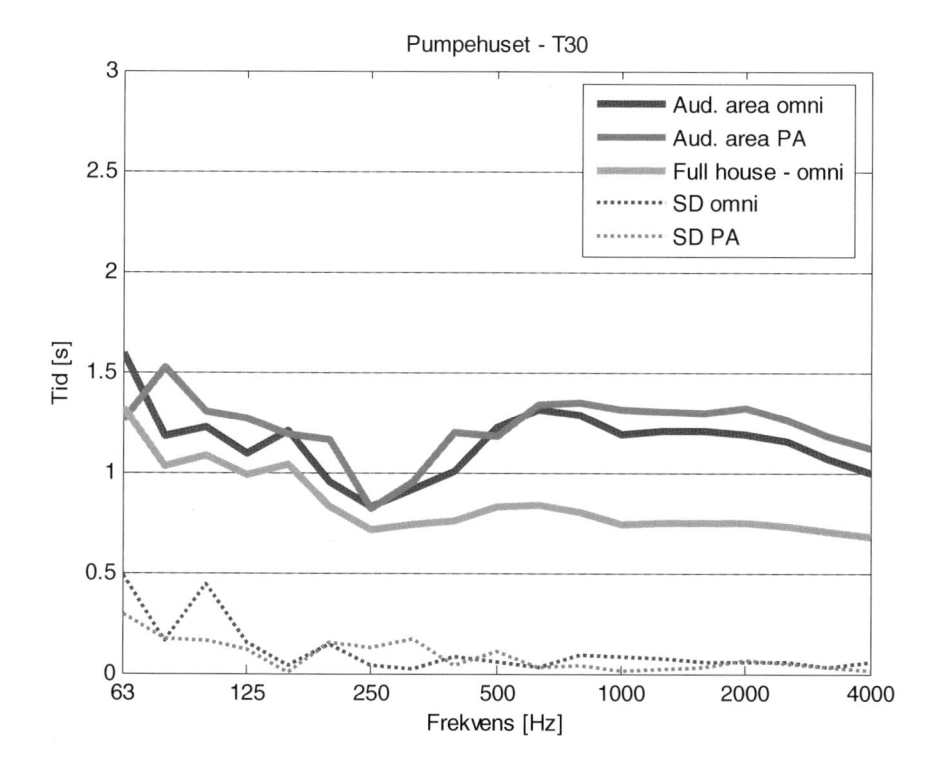

Pumpehuset

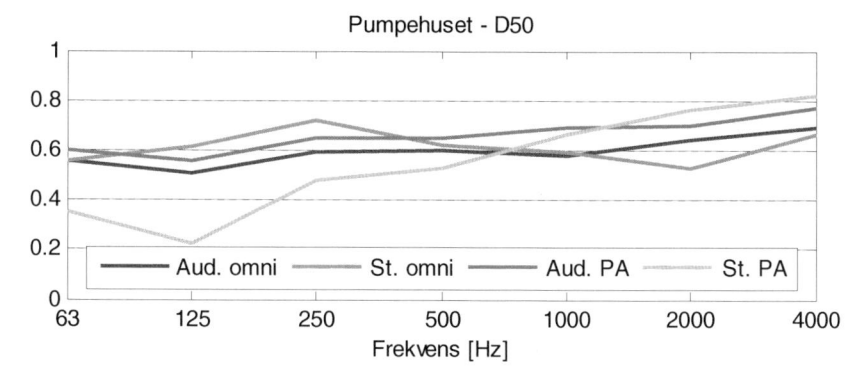

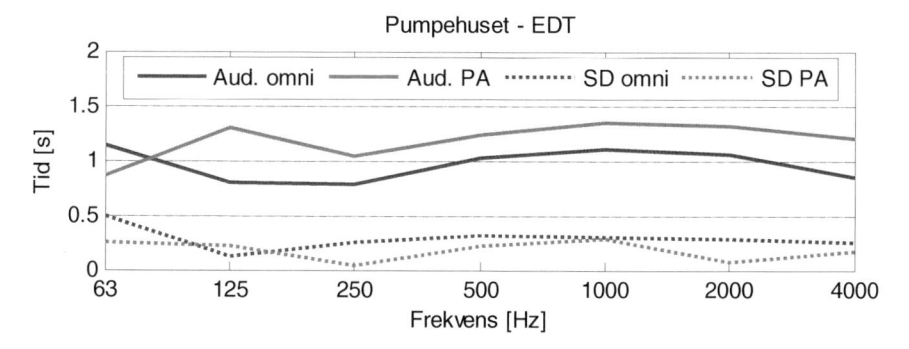

Pumpehuset - T30

Pumpehuset - D50

Pumpehuset - EDT

Rytmeposten

BUILT: in Odense, 1915 as a post office; rebuilt 1988 as a music venue. Rebuilt again in 2007.
CAPACITY: 300
NUMBER OF CONCERTS PER YEAR: 130

BUILDING MATERIALS USED: Ceiling. Gypsum board on cavity. Suspended mineral wool slabs on half the ceiling area (the half above and close to the stage). Floor: Wood on joists. Walls: Painted bricks. Some mineral wool slabs on side walls. Stage walls are completely covered by mineral wool slabs, carpet on stage floor.

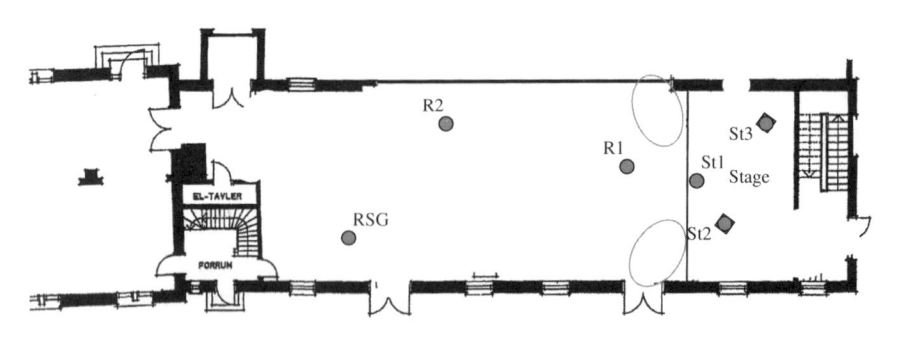

Geometrical data	
Volume (m^3)	655
Height (m)	3.3
Surface area stage (m^2)	41
Surface area audience (m^2)	130
Total surface area of hall (m^2)	605
Capacity of standing persons	300

Acoustical data—omnisource	
Audience area	
T_{tom}—(500–1k) (s)	0.8
T_{fuld}—(500–1k) (s)	0.5
D50—(250–2k)	0.71
EDT—(250–2k) (s)	0.7
$BR_{rock,a}$	1.0
Stage area	
T_{tom}—(500–1k) (s)	0.6
D50—(250–2k)	0.92
EDT—(250–2k) (s)	0.3

Subjective data	
Audience area	
Rating out of 20	10
Stage area	
Rating out of 20	14

The stage had been dampened too much compared to the hall itself, mostly at mid/high frequencies (the BR on stage was 1.6). The low ceiling makes it difficult to obtain proper PA coverage. Today a completely new hall has been built and this former stage is merely a secondary stage in the building.

76   4   Assessments of 20 Halls

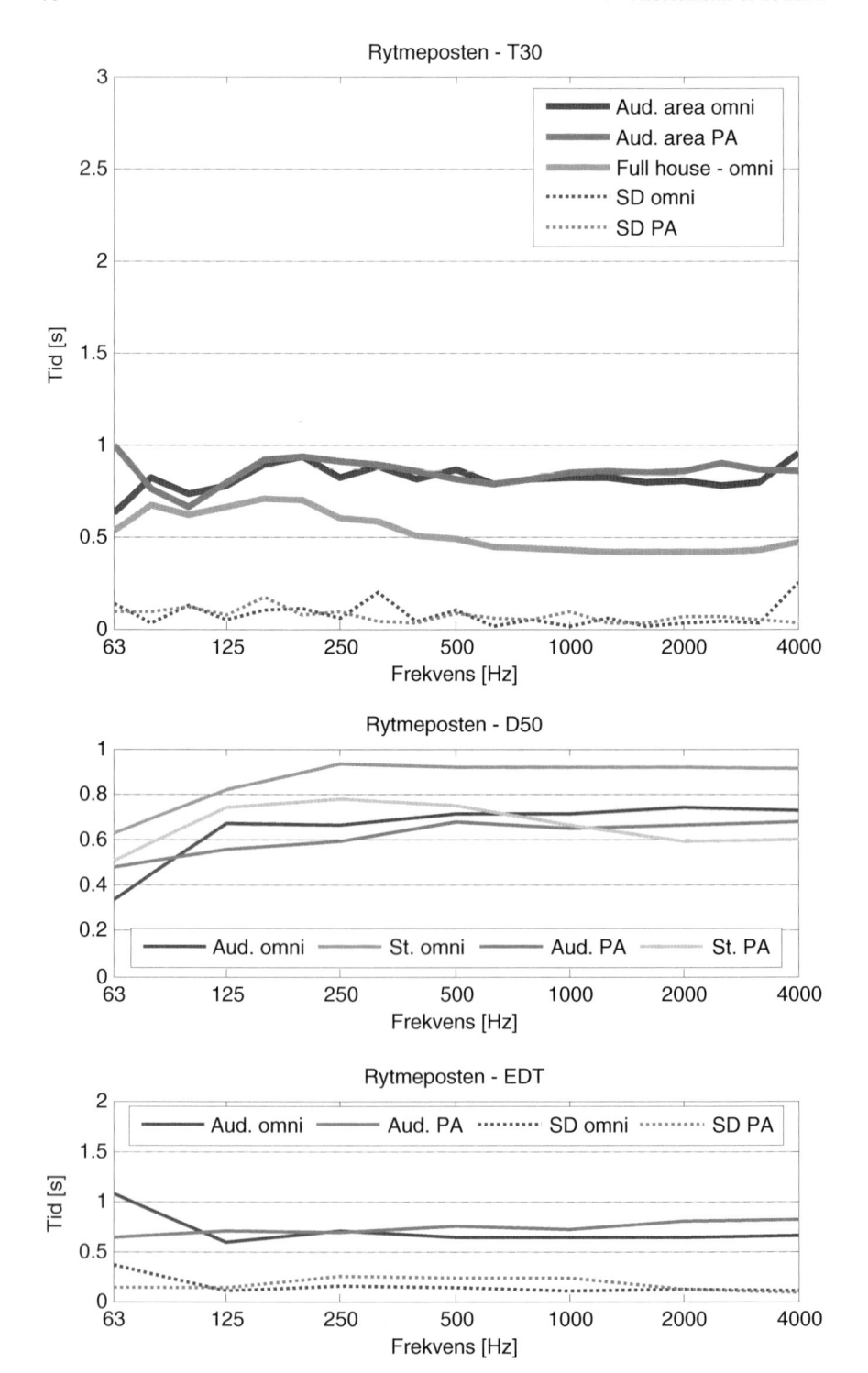

Musikhuzet

BUILT: in Rønne, 1930s as a cinema. Rebuilt as a music venue in 1992. Refurbished again in 2004.
   CAPACITY: 640
   NUMBER OF CONCERTS PER YEAR: 50

BUILDING MATERIALS USED: Ceiling: Gypsum board on cavity. In 2004 partial coverage with suspended mineral wool slabs. Full coverage above stage. Floor: Wood on cavity. Walls: Wooden panels. Backdrop.

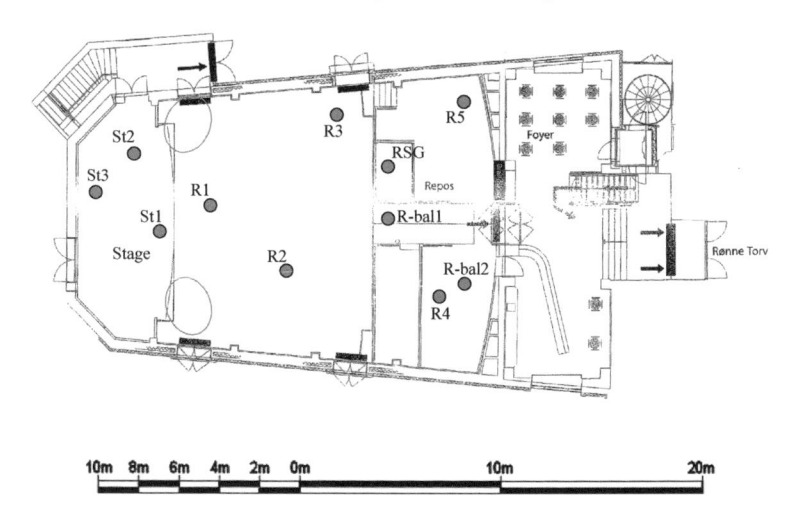

Geometrical data	
Volume (m^3)	2,080
Height (m)	6.0–8.0
Surface area stage (m^2)	48
Surface area audience (m^2)	275
Total surface area of hall (m^2)	1,100
Capacity of standing persons	700
Acoustical data—PA	
Audience area	
T_{tom}—(500–1k) (s)	0.9
T_{fuld}—(500–1k) (s)	0.5
D50—(250–2k)	0.58
EDT—(250–2k) (s)	0.9
$BR_{rock,a}$	1.2
Stage area	
T_{tom}—(500–1k) (s)	0.8
D50—(250–2k)	0.51
EDT—(250–2k) (s)	1.0
Subjective data	
Audience area	
Rating out of 20	17
Stage area	
Rating out of 20	9

The ratings do not correspond to the objective measurements because the hall was refurbished shortly before these measurements were taken.

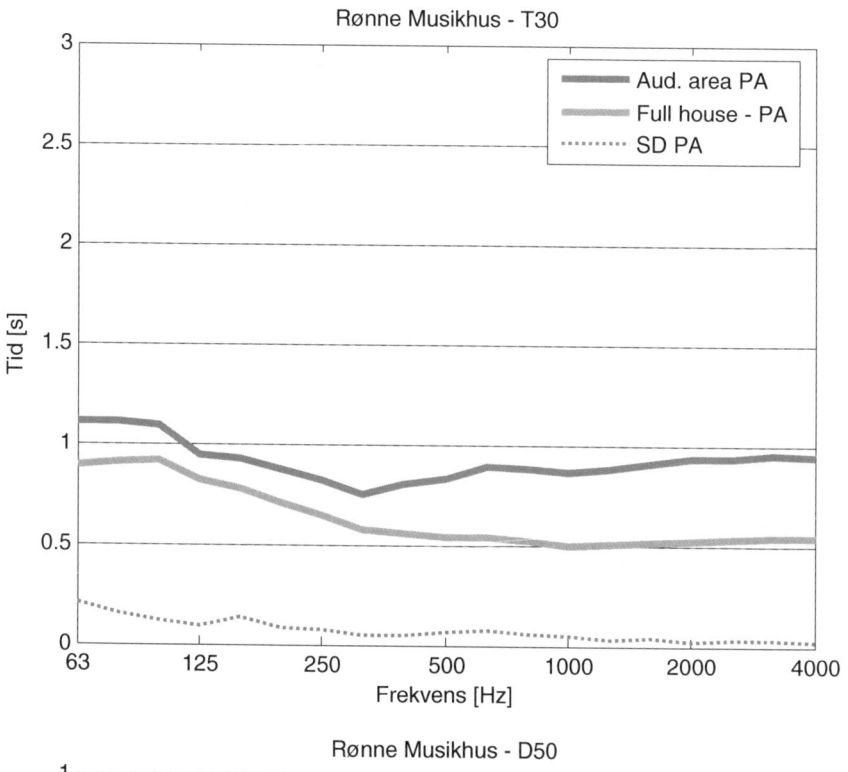

Musikhuzet

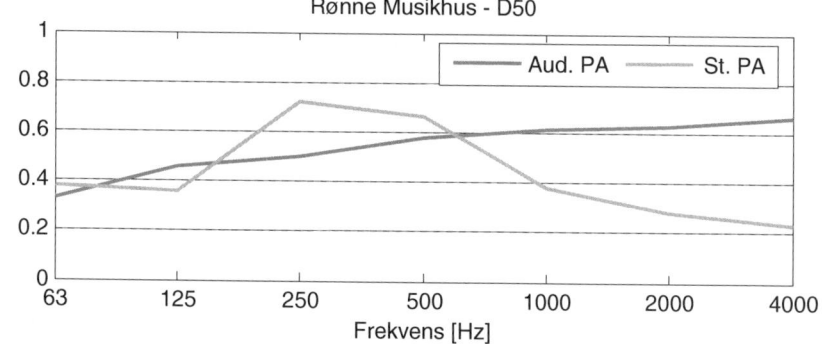

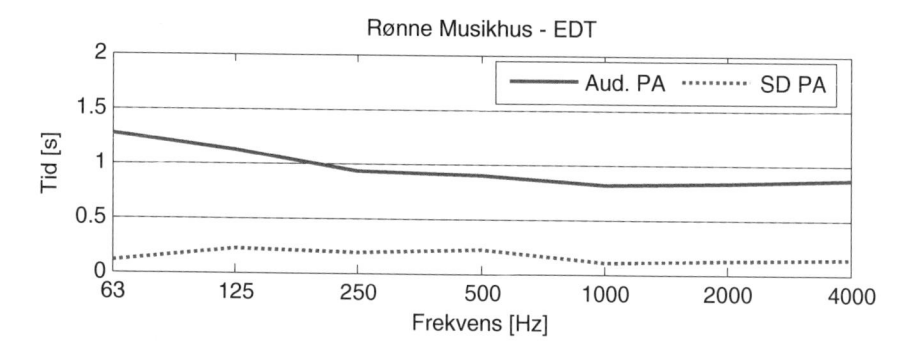

Rønne Musikhus - T30

Rønne Musikhus - D50

Rønne Musikhus - EDT

Skråen

BUILT: in Ålborg, 1978.
   CAPACITY: 375
   NUMBER OF CONCERTS PER YEAR: 100

BUILDING MATERIALS USED: Ceiling: Concrete partially covered with thin suspended mineral wool slabs. Floor: Rubber on concrete. Walls: Painted concrete. Backdrop.

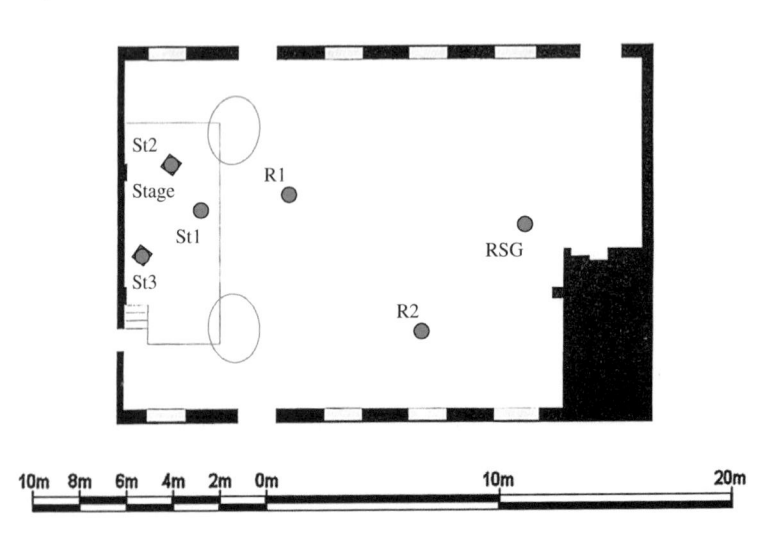

Geometrical data	
Volume (m^3)	1,100
Height (m)	3.9
Surface area stage (m^2)	37
Surface area audience (m^2)	180
Total surface area of hall (m^2)	870
Capacity of standing persons	375

Acoustical data—omnisource	
Audience area	
T_{tom}—(500–1k) (s)	0.7
T_{fuld}—(500–1k) (s)	0.4
D50—(250–2k)	0.53
EDT—(250–2k) (s)	0.7
Stage area	
T_{tom}—(500–1k) (s)	0.7
D50—(250–2k)	0.82
EDT—(250–2k) (s)	0.7
$BR_{rock,s}$	1.8

Subjective data	
Audience area	
Rating out of 20	14
Stage area	
Rating out of 20	13

The T_{30} in the 50–100-Hz third octave bands dominate this venue. However, when the packed the venue is almost bearable. This fact leaves a hint that a longer T_{30} in the 63 Hz octave band is acceptable. A fine new Skråen was built in 2007.

4 Assessments of 20 Halls

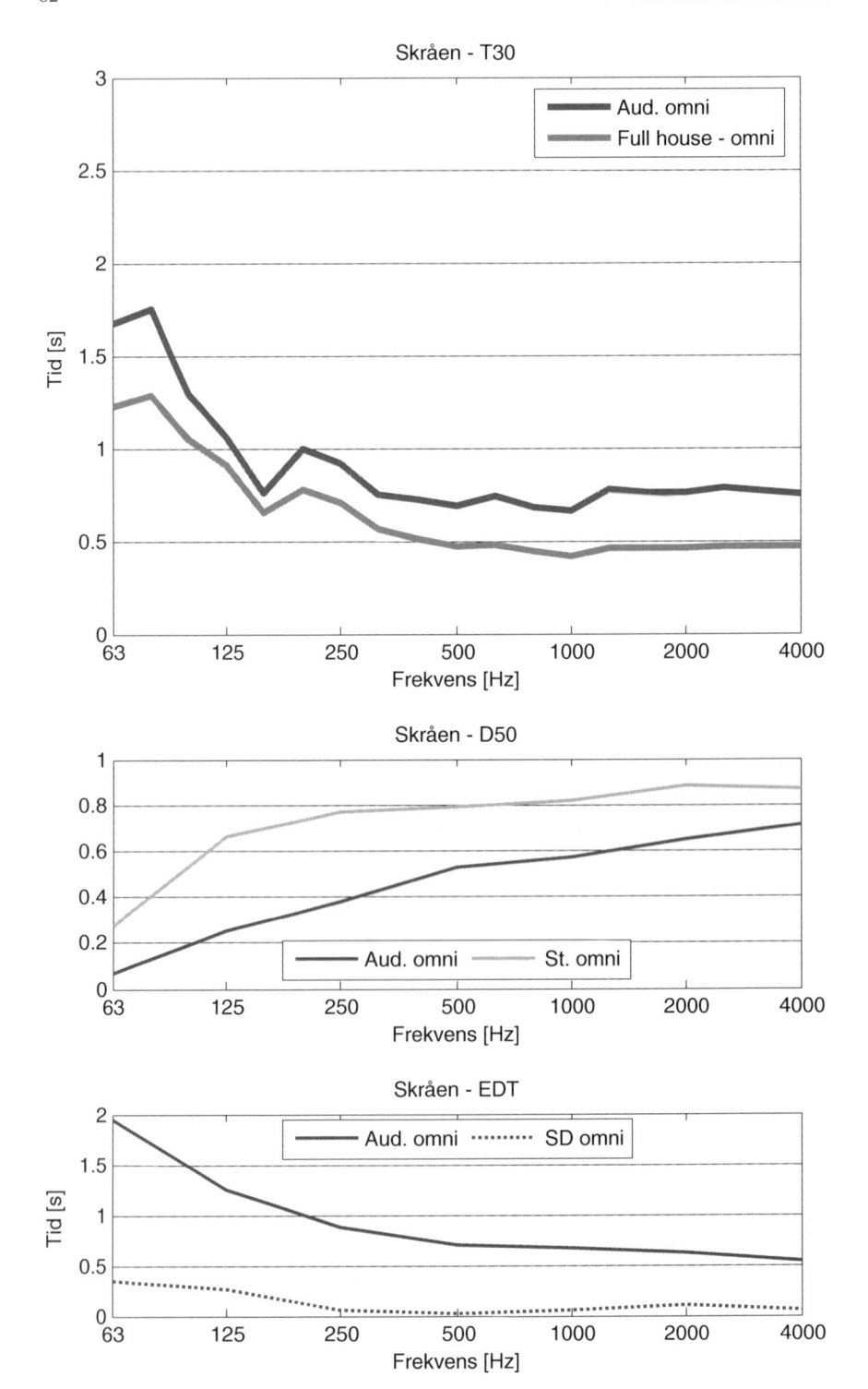

Skråen - T30

Skråen - D50

Skråen - EDT

Slagelse Musikhus

BUILT: in Aalborg, 1909 as a power plant. Rebuilt in 1994 as a music venue.
CAPACITY: 700
NUMBER OF CONCERTS PER YEAR: 30

BUILDING MATERIALS USED: Ceiling: Concrete. Floor: Wood on concrete. Side walls: Bricks. End walls large windows. Perforated gypsum boards above windows. Upholstered chairs on balcony. Upholstered, retractable seats stuffed in rear end of hall.

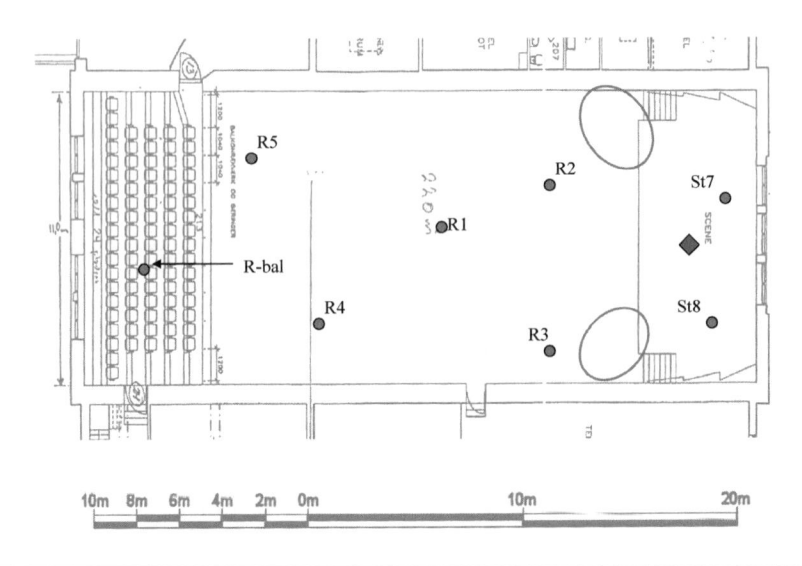

Geometrical data	
Volume (m^3)	3,800
Height (m)	10.6
Surface area stage (m^2)	60
Surface area audience (m^2)	285
Total surface area of hall (m^2)	1,650
Capacity of standing persons	700
Acoustical data—omnisource	
Audience area	
T_{tom}—(500–1k) (s)	1.7
T_{fuld}—(500–1k) (s)	1.0
D50—(250–2k)	0.32
EDT—(250–2k) (s)	1.5
Stage area	
T_{tom}—(500–1k) (s)	1.6
D50—(250–2k)	0.77
EDT—(250–2k) (s)	1.1
$BR_{rock,s}$	1.1
Subjective data	
Audience area	
Rating out of 20	20
Stage area	
Rating out of 20	17

Evidently much too long T_{30} across all frequencies. Smaller venues at the same address host most of the amplified concerts here.

Tobakken

BUILT: in Esbjerg, 1900 as a tobacco factory. Rebuilt 1993 as a music venue.
CAPACITY: 1,150
NUMBER OF CONCERTS PER YEAR: 75

BUILDING MATERIALS USED: The hall is a covered courtyard situated between two buildings from which the facades have been partially removed. Ceiling: Nonperforated steel plates on cavity. Floor: Hard wood on concrete. Walls: Bricks, large window at one end.

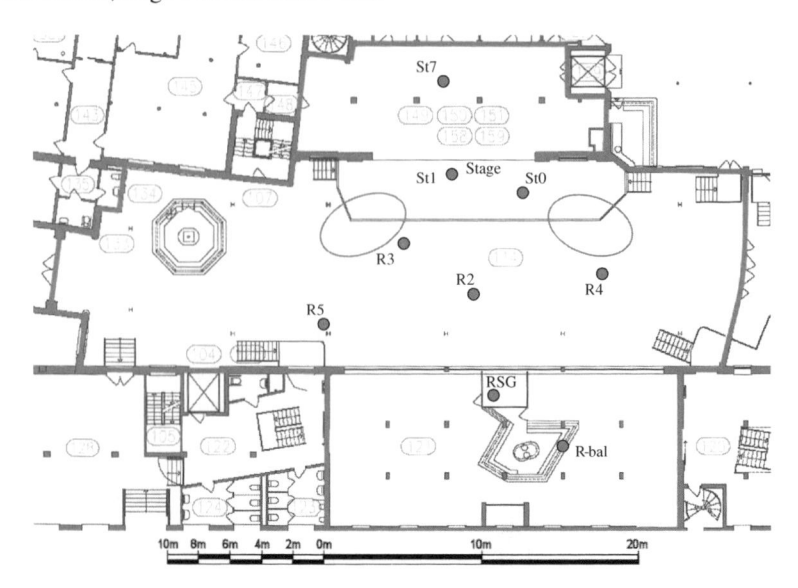

Slagelse Musikhus

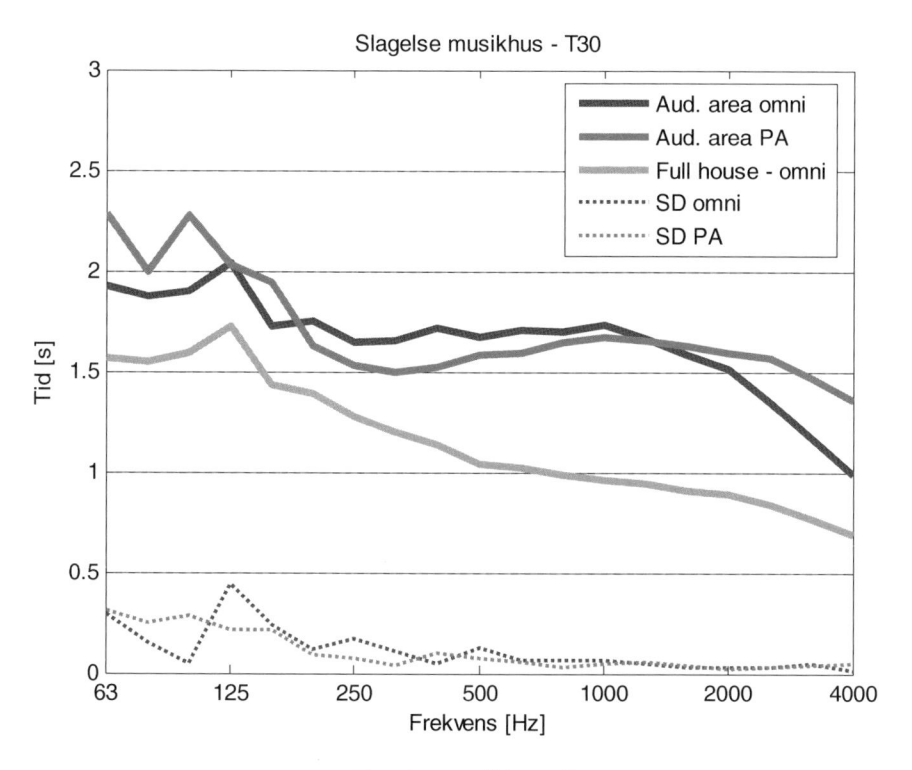

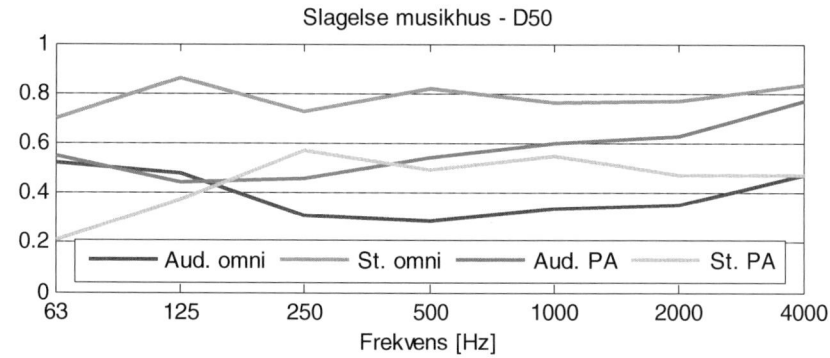

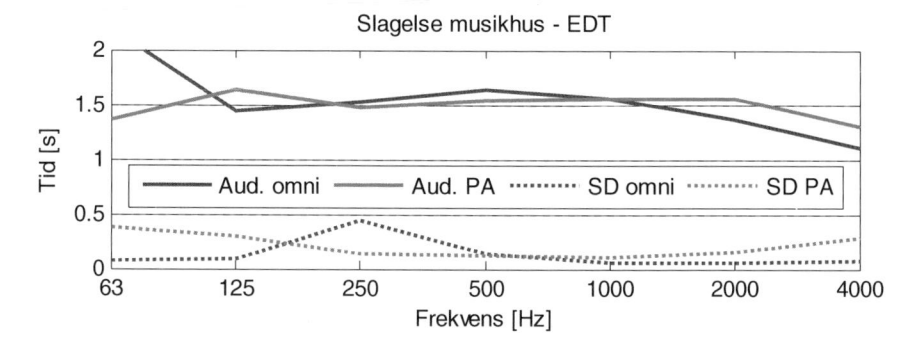

Stars

BUILT: in Vordingborg around 1900. Rebuilt as a rock venue in 1997.
CAPACITY: 460
NUMBER OF CONCERTS PER YEAR: 100

BUILDING MATERIALS USED: Ceiling: Mineral wool slabs on cavity. Floor: Wood on joists. Walls: Large areas are covered by mineral wool on large cavity. Carpet on stage floor, backdrop.

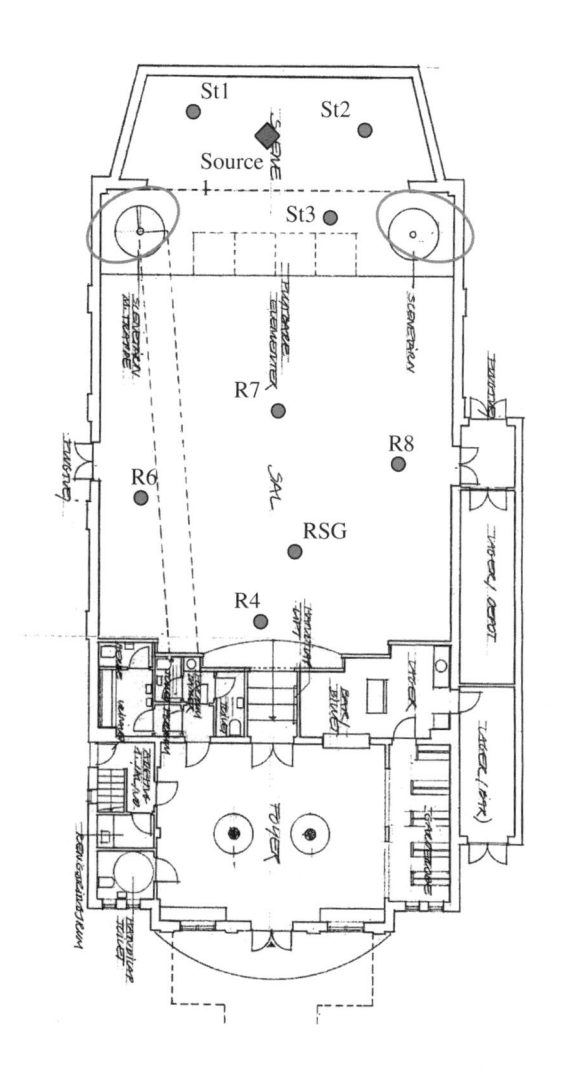

Stars

St1
St2
Source
St3
R7
R8
R6
RSG
R4

10m 8m 6m 4m 2m 0m 10m 20m

Geometrical data	
Volume (m^3)	1,440
Height (m)	4.7–6.7
Surface area stage (m^2)	42
Surface area audience (m^2)	220
Total surface area of hall (m^2)	970
Capacity of standing persons	400
Acoustical data—omnisource	
Audience area	
T_{tom}—(500–1k) (s)	0.6
T_{fuld}—(500–1k) (s)	0.4
D50—(250–2k)	0.73
EDT—(250–2k) (s)	0.5
$BR_{rock,a}$	0.9
Stage area	
T_{tom}—(500–1k) (s)	0.5
D50—(250–2k)	0.94
EDT—(250–2k) (s)	0.3
Subjective data	
Audience area	
Rating out of 20	1
Stage area	
Rating out of 20	10

This is a very interesting hall inasmuch as it is the only one among the 20 that was rated too dry by the musicians. This hall is responsible for our knowledge that halls for pop and rock can in fact have a too low reverberation time, not enough vivacity, and a lack of communication between stage and hall. The musicians' rating may increase by simply removing sound absorption such as the carpet from the stage. But the sound level of the audience will still occur too low on stage. Sound engineers enjoy the possibility of full control of outboard sound-processing devices such as artificial reverberation here.

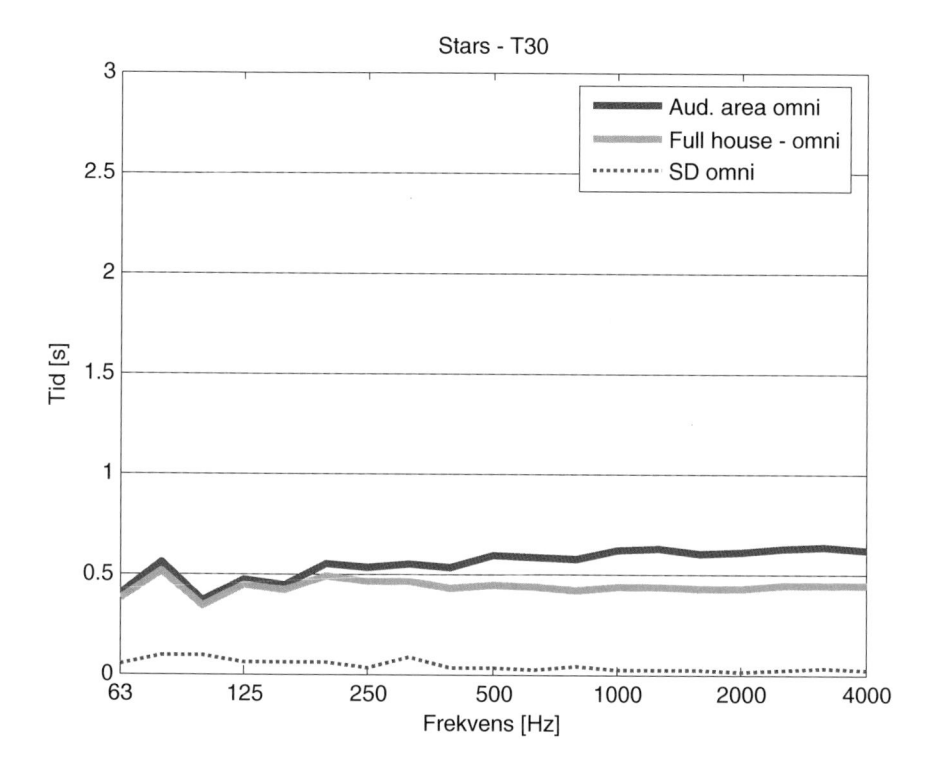

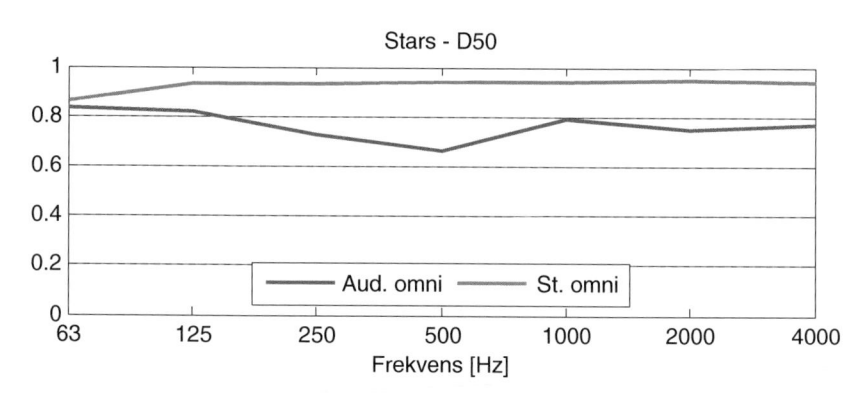

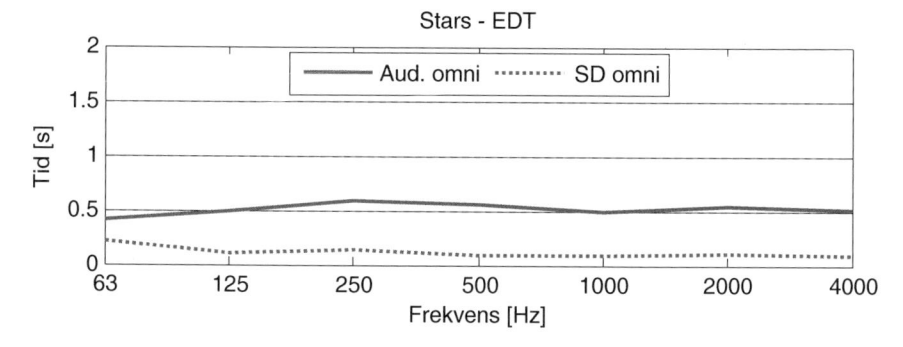

Stars

Stars - T30

Legend: Aud. area omni, Full house - omni, SD omni

Y-axis: Tid [s], X-axis: Frekvens [Hz]

Stars - D50

Legend: Aud. omni, St. omni

X-axis: Frekvens [Hz]

Stars - EDT

Legend: Aud. omni, SD omni

Y-axis: Tid [s], X-axis: Frekvens [Hz]

Store VEGA

BUILT: in Copenhagen,1956. Rebuilt 1996 as a music venue.
 CAPACITY: 1,500
 NUMBER OF CONCERTS PER YEAR: 100

BUILDING MATERIALS USED: Ceiling: Suspended mineral wool slabs on large cavity. Perforated gypsum boards on ceilings under balconies. Floor: Wooden. Walls: Wooden panels on cavity. Backdrop. Stage tower above stage.

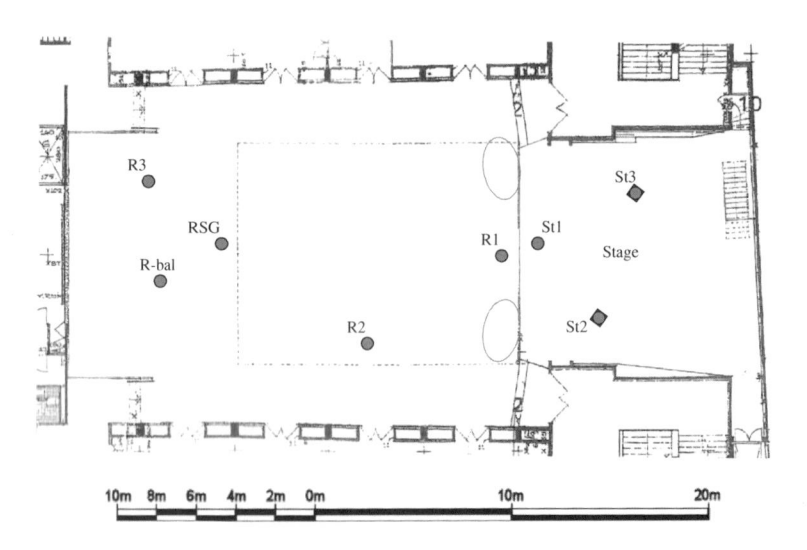

Store VEGA

Geometrical data	
Volume (m^3)	5,800
Height (m)	10.1
Surface area stage (m^2)	124
Surface area audience (m^2)	460
Total surface area of hall (m^2)	2,200
Capacity of standing persons	1,430

Acoustical data—omnisource	
Audience area	
T_{tom}—(500–1k) (s)	1.2
T_{fuld}—(500–1k) (s)	0.7
D50—(250–2k)	0.63
EDT—(250–2k) (s)	1.0
Stage area	
T_{tom}—(500–1k) (s)	0.9
D50—(250–2k)	0.91
EDT—(250–2k) (s)	0.44
$BR_{rock,s}$	1.1

Subjective data	
Audience area	
Rating out of 20	3
Stage area	
Rating out of 20	2

Store Vega is the unofficial national stage for pop and rock in Denmark and it's always a pleasure to perform there as well as to be a member of the audience. There are multiple bars in the adjacent rooms. Being placed far back on the large stage one may find that late low-frequency sound gets a little too dominant. This is a result of a rise in the T_{30} curve at very low frequencies together with the fact that the tall stage tower does not create early sound energy to mask the late LF response (and maybe the tower itself also creates a long LF reverberation).

4 Assessments of 20 Halls

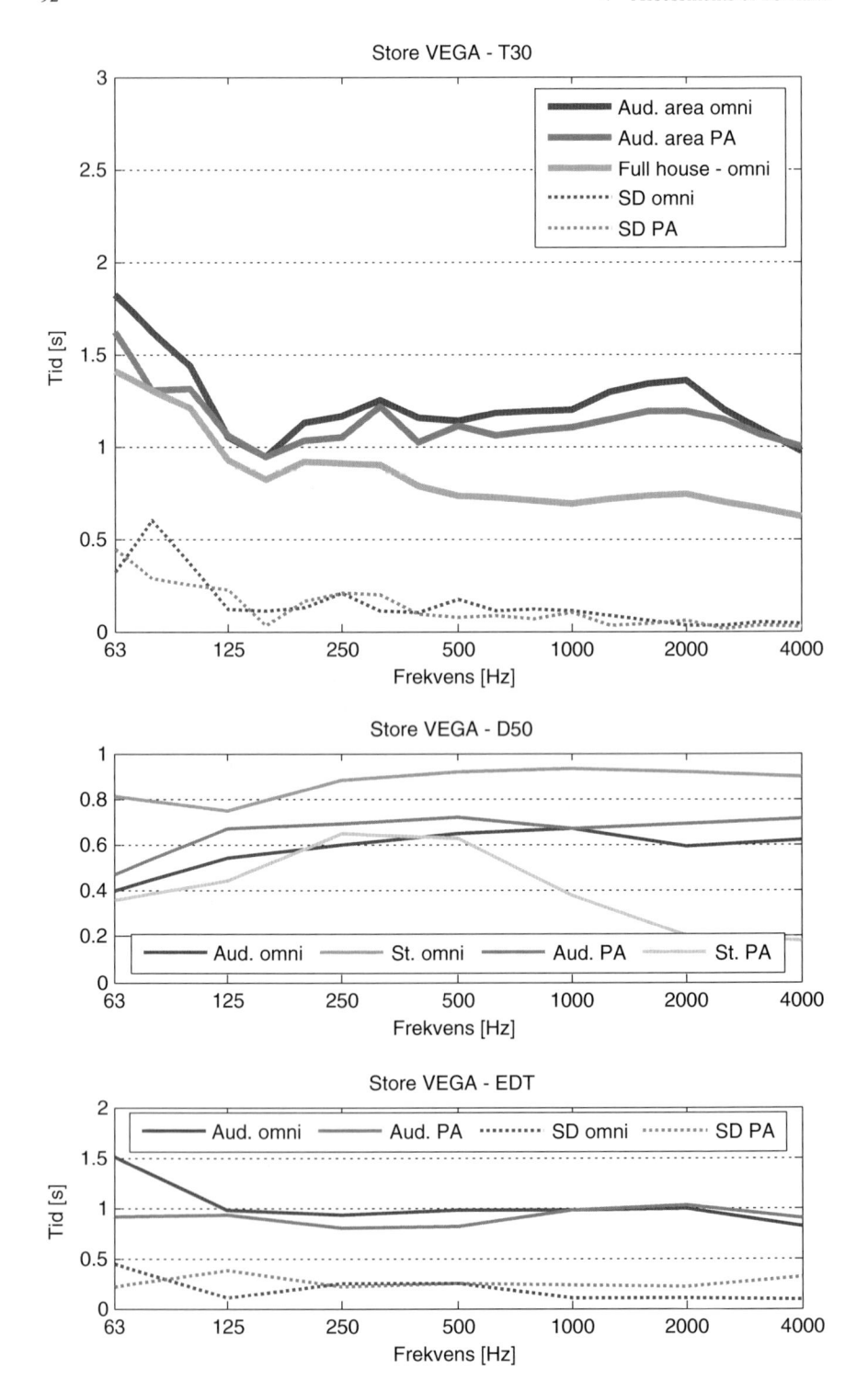

Sønderborghus

BUILT: in Sønderborg, 1913 as a community house for the Danish minority. Rebuilt 1976.
CAPACITY: 350
NUMBER OF CONCERTS PER YEAR: 150–180

BUILDING MATERIALS USED: Ceiling: Masonry. Some mineral wool slabs mounted in ceiling and on wall areas (installed 2004). Floor: Wood on cavity. Walls: Painted bricks, large windows, backdrop.

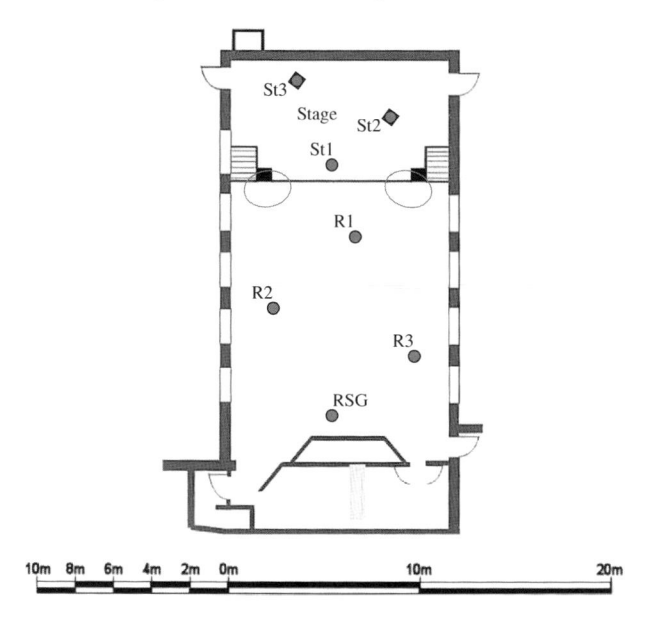

Geometrical data	
Volume (m^3)	1,600
Height (m)	6.0
Surface area stage (m^2)	55
Surface area audience (m^2)	185
Total surface area of hall (m^2)	1,000
Capacity of standing persons	420

Acoustical data—omnisource	
Audience area	
T_{tom}—(500–1k) (s)	1.0
T_{fuld}—(500–1k) (s)	0.6
D50—(250–2k)	0.51
EDT—(250–2k) (s)	1.0
Stage area	
T_{tom}—(500–1k) (s)	0.8
D50—(250–2k)	0.78
EDT—(250–2k) (s)	0.6
$BR_{rock,s}$	1.3

Subjective data	
Audience area	
Rating out of 20	18
Stage area	
Rating out of 20	20

Just prior to the measurement of this hall some sound absorptive materials were installed as well as a new PA system. It is therefore uncertain how well the ratings reflect these measurements. However, the T_{30} is still too long for the relatively modest volume of the hall, there is a bass ratio of 1.3, and the musicians dislike being put away in a smaller enclosure detached from the hall which is even, as mentioned, too reverberant.

Sønderborghus

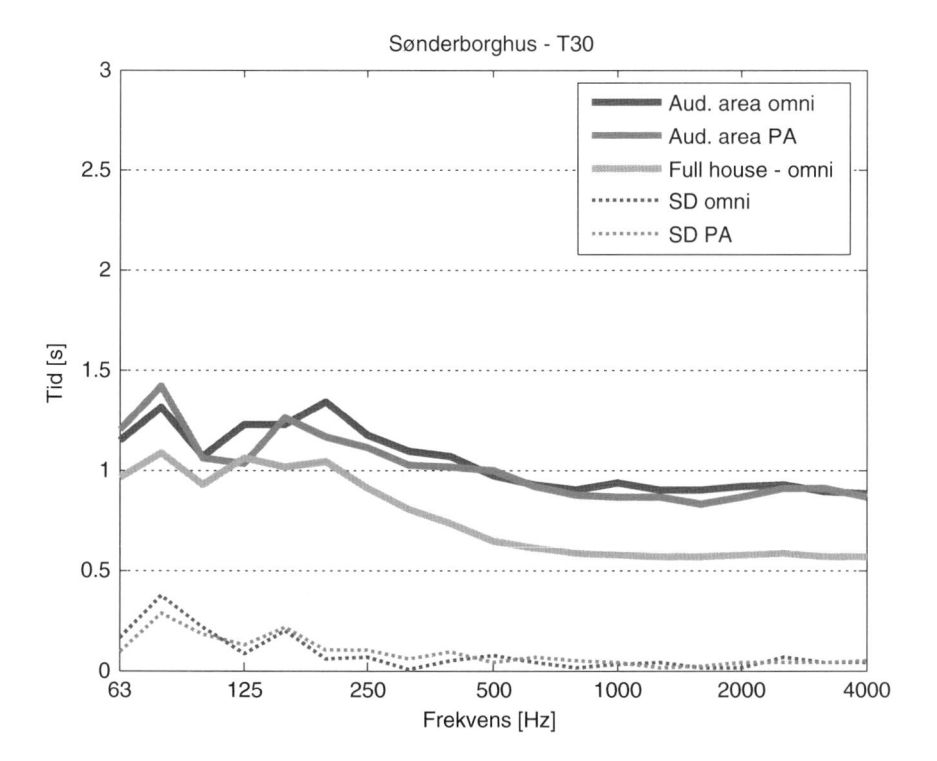

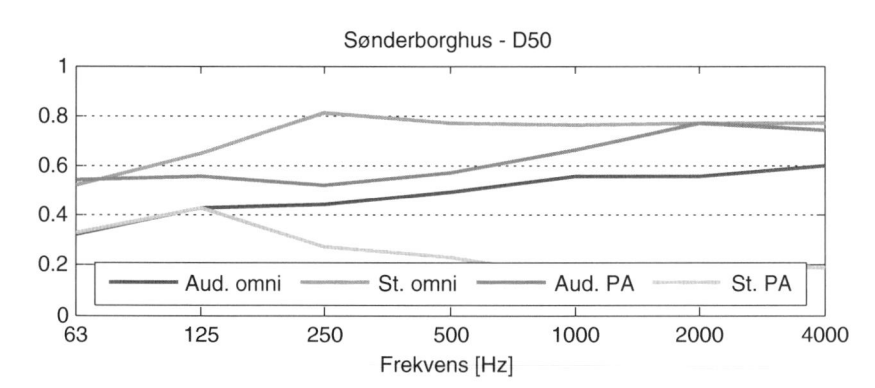

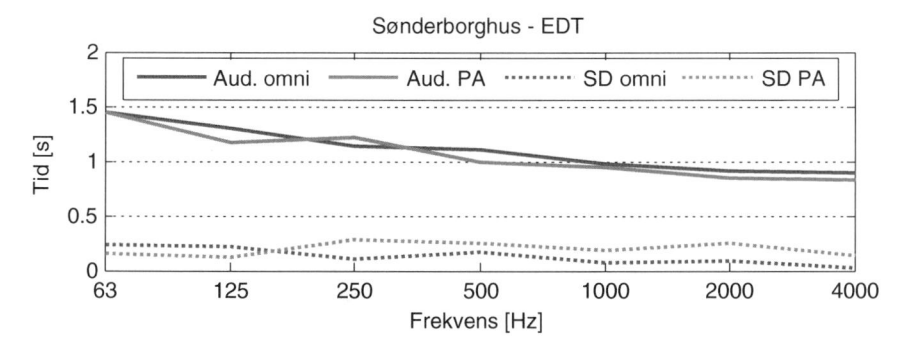

Geometrical data	
Volume (m^3)	6,500
Height (m)	9.7
Surface area stage (m^2)	100
Surface area audience (m^2)	600
Total surface area of hall (m^2)	3,300
Capacity of standing persons	1,200
Acoustical data—PA	
Audience area	
T_{tom}—(500–1k) (s)	1.0
T_{fuld}—(500–1k) (s)	0.7
D50—(250–2k)	0.67
EDT—(250–2k) (s)	1.0
Stage area	
T_{tom}—(500–1k) (s)	1.0
D50—(250–2k)	0.83
EDT—(250–2k) (s)	0.6
$BR_{rock,s}$	1.73
Subjective data	
Audience area	
Rating out of 20	7
Stage area	
Rating out of 20	11

Musicians suffer from a too-long RT at low frequencies in combination with too few early reflections due to lack of reflective surfaces behind the stage area.

98 4 Assessments of 20 Halls

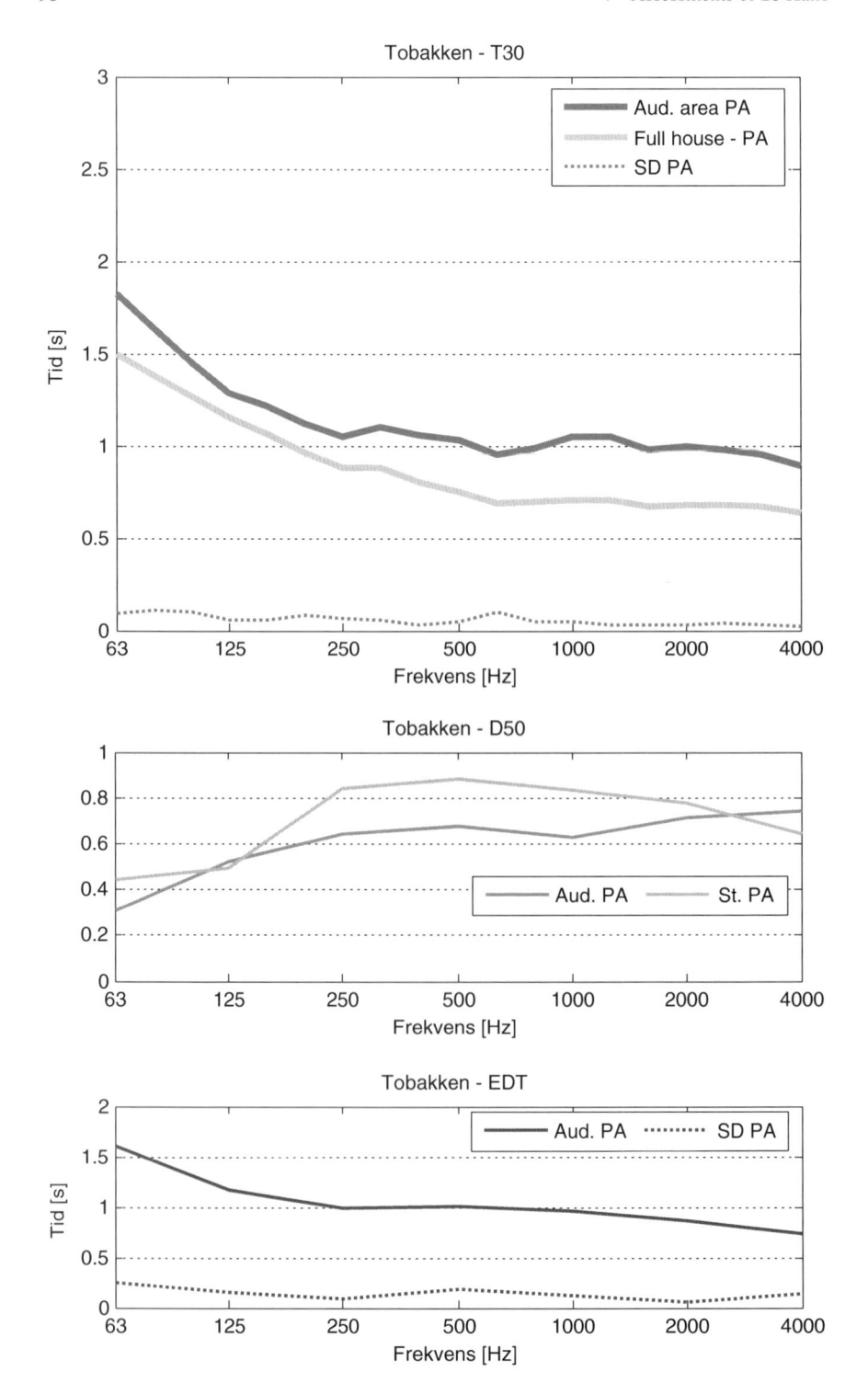

Torvehallerne

BUILT: in Vejle, 1992.
　CAPACITY: 700
　NUMBER OF CONCERTS PER YEAR: 40

BUILDING MATERIALS USED: Ceiling: Painted masonry, gypsum boards on cavity. Floor: Wood on cavity. Walls: Perforated gypsum boards, painted concrete. Backdrop.

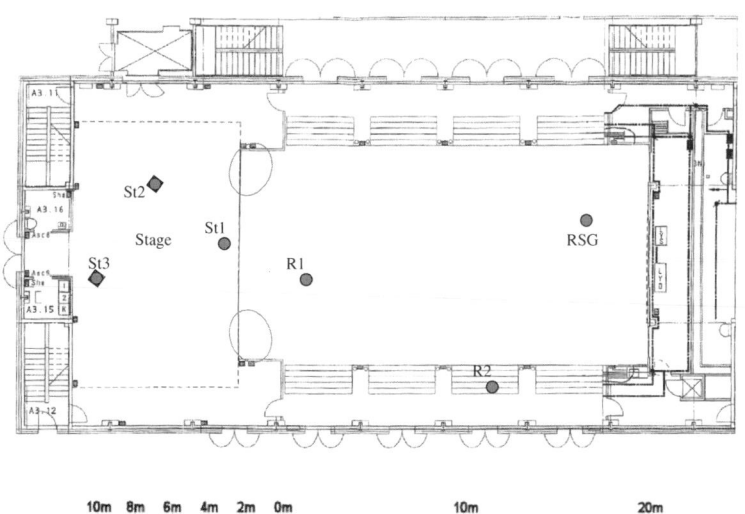

Geometrical data	
Volume (m^3)	5,400
Height (m)	9.8
Surface area stage (m^2)	98
Surface area audience (m^2)	280
Total surface area of hall (m^2)	2,100
Capacity of standing persons	700
Acoustical data—omnisource	
Audience area	
T_{tom}—(500–1k) (s)	1.6
T_{fuld}—(500–1k) (s)	1.1
D50—(250–2k)	0.26
EDT—(250–2k) (s)	1.6
Stage area	
T_{tom}—(500–1k) (s)	1.4
D50—(250–2k)	0.60
EDT—(250–2k) (s)	0.89
$BR_{rock,s}$	0.8
Subjective data	
Audience area	
Rating out of 20	16
Stage area	
Rating out of 20	15

The hall is too reverberant at mid and higher frequencies. This could be controlled by a better loudspeaker coverage, but the upper speakers in the two arrays point directly towards the big reflective back wall. Musicians lack early reflections on the vast stage to mask the loud level of late sound.

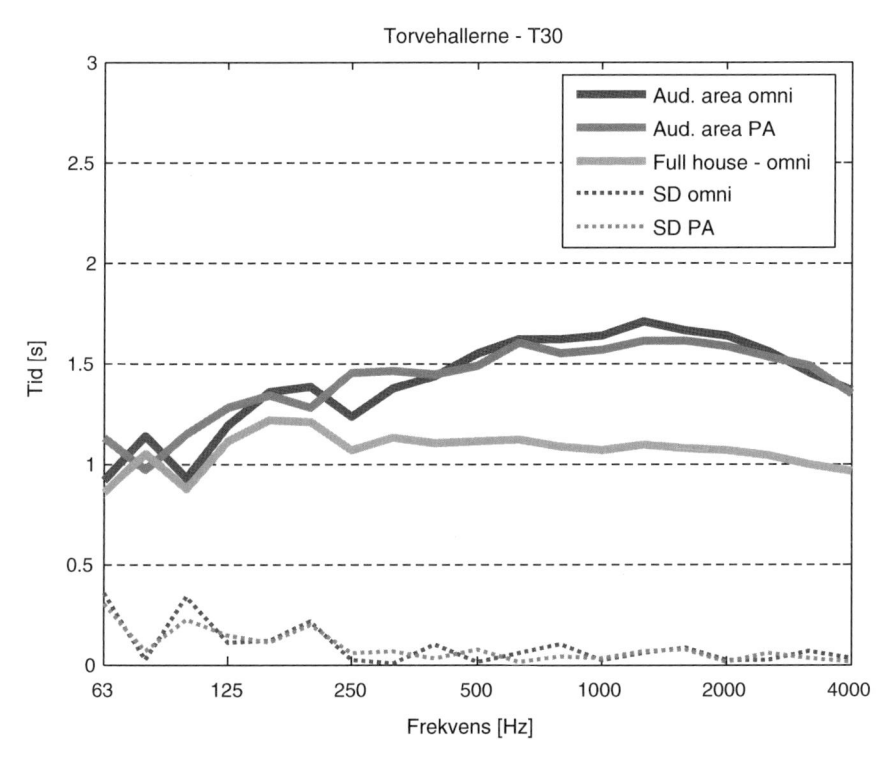

Torvehallerne 101

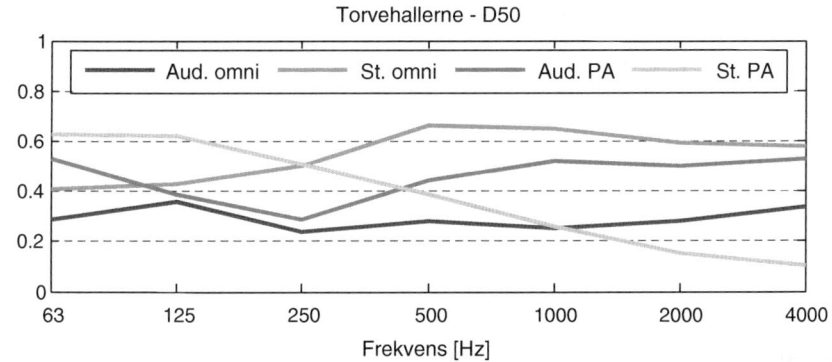

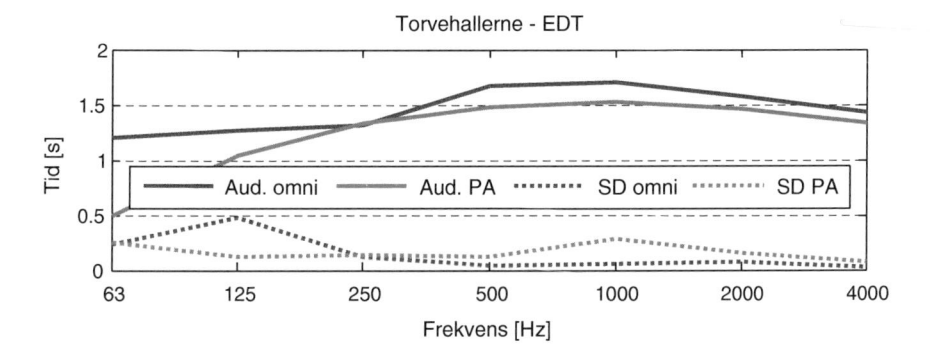

Torvehallerne - T30

Torvehallerne - D50

Torvehallerne - EDT

Train

BUILT: in Århus for storage. Rebuilt in the 1990s first as a discothèque, then as a music venue.
CAPACITY: 900
NUMBER OF CONCERTS PER YEAR: 100

BUILDING MATERIALS USED: Ceiling: Gypsum board on cavity, suspended wood fiber slabs. Floor: Linoleum on concrete. Wood on cavity on stage. Walls: Gypsum board on cavity.

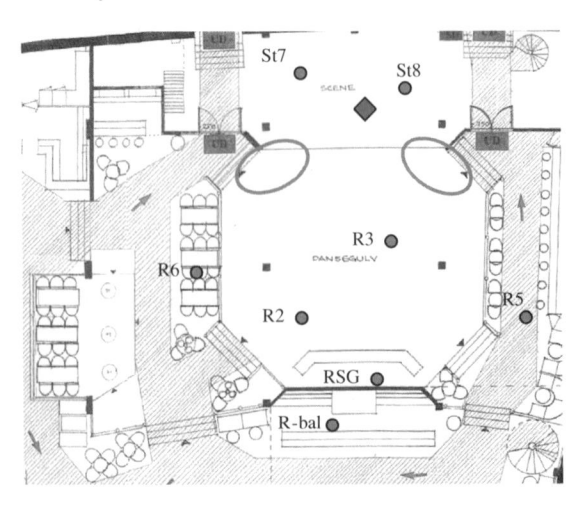

Train

Geometrical data	
Volume (m^3)	3,300
Height (m)	average: 4, 9
Surface area stage (m^2)	60
Surface area audience (m^2)	390
Total surface area of hall (m^2)	2,000
Capacity of standing persons	900

Acoustical data—omnisource	
Audience area	
T$_{tom}$—(500–1k) (s)	1.0
T$_{fuld}$—(500–1k) (s)	0.6
D50—(250–2k)	0.60
EDT—(250–2k) (s)	0.85
Stage area	
T$_{tom}$—(500–1k) (s)	0.8
D50—(250–2k)	0.86
EDT—(250–2k) (s)	0.4
BR$_{rock,s}$	0.9

Subjective data	
Audience area	
Rating out of 20	2
Stage area	
Rating out of 20	3

This is a great hall for pop and rock music. Not only are the acoustics close to ideal, there is also a great view from the stage onto the different bar areas and floor levels for the audience. The sound engineer's position is also ideal. The height of the hall just allows for one balcony level and for acceptable PA coverage. A few more side reflections on stage would be good.

104   4   Assessments of 20 Halls

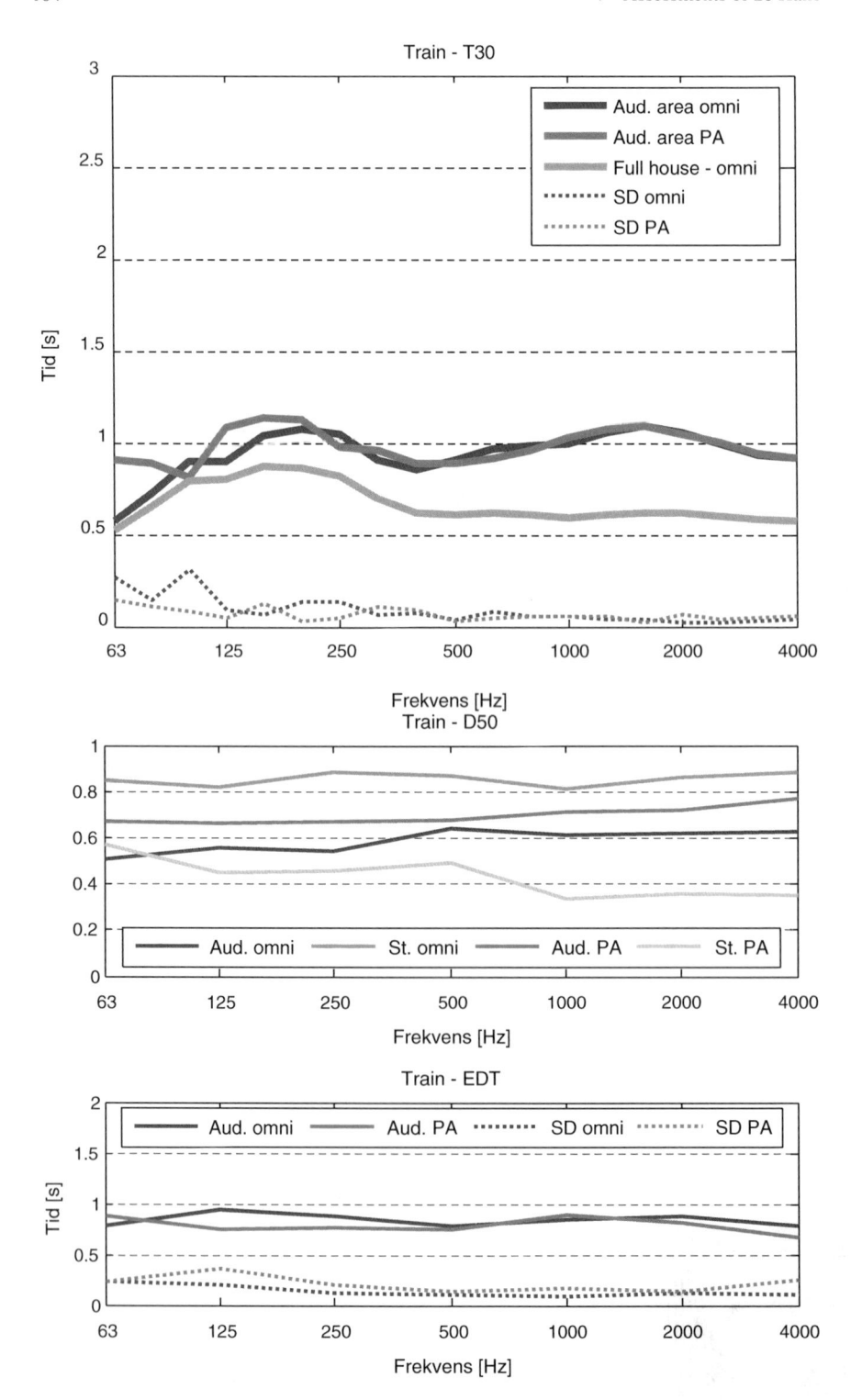

Viften

BUILT: in Rødovre, 1989 as an inexpensive performing arts center/multipurpose hall.
  CAPACITY: 700
  NUMBER OF CONCERTS PER YEAR: 55

BUILDING MATERIALS USED: Ceiling: Concrete. Floor: Cork on concrete. Walls: Concrete. Banners on all walls. Backdrop.

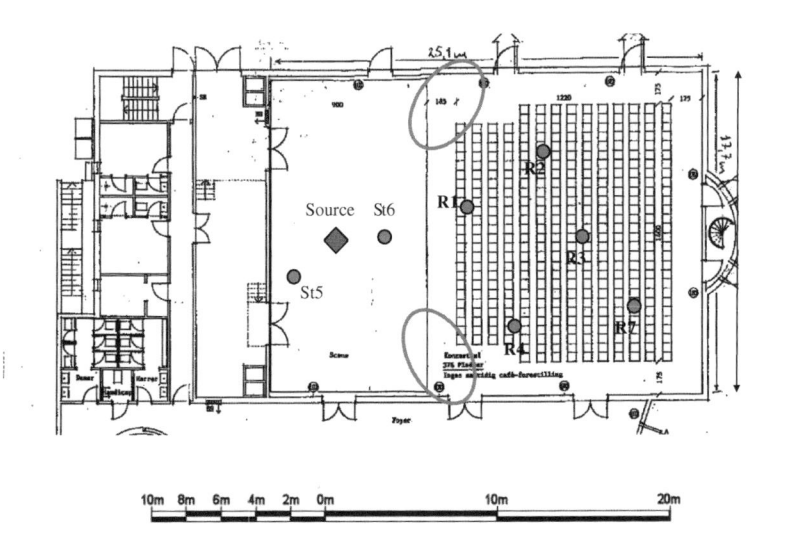

Geometrical data	
Volume (m^3)	3,950
Height (m)	8.9
Surface area stage (m^2)	<125
Surface area audience (m^2)	330–400
Total surface area of hall (m^2)	1,650
Capacity of standing persons	700

Acoustical data—omnisource	
Audience area	
T_{tom}—(500–1k) (s)	1.1
T_{fuld}—(500–1k) (s)	0.8
D50—(250–2k)	0.43
EDT—(250–2k) (s)	1.2
Stage area	
T_{tom}—(500–1k) (s)	1.0
D50—(250–2k)	0.83
EDT—(250–2k) (s)	0.8
$BR_{rock,s}$	2.0

Subjective data	
Audience area	
Rating out of 20	13
Stage area	
Rating out of 20	18

This hall was a good example of a hall with no LF absorption and quite an extensive amount of MF/HF absorption in terms of banners. Some LF absorption has now been added.

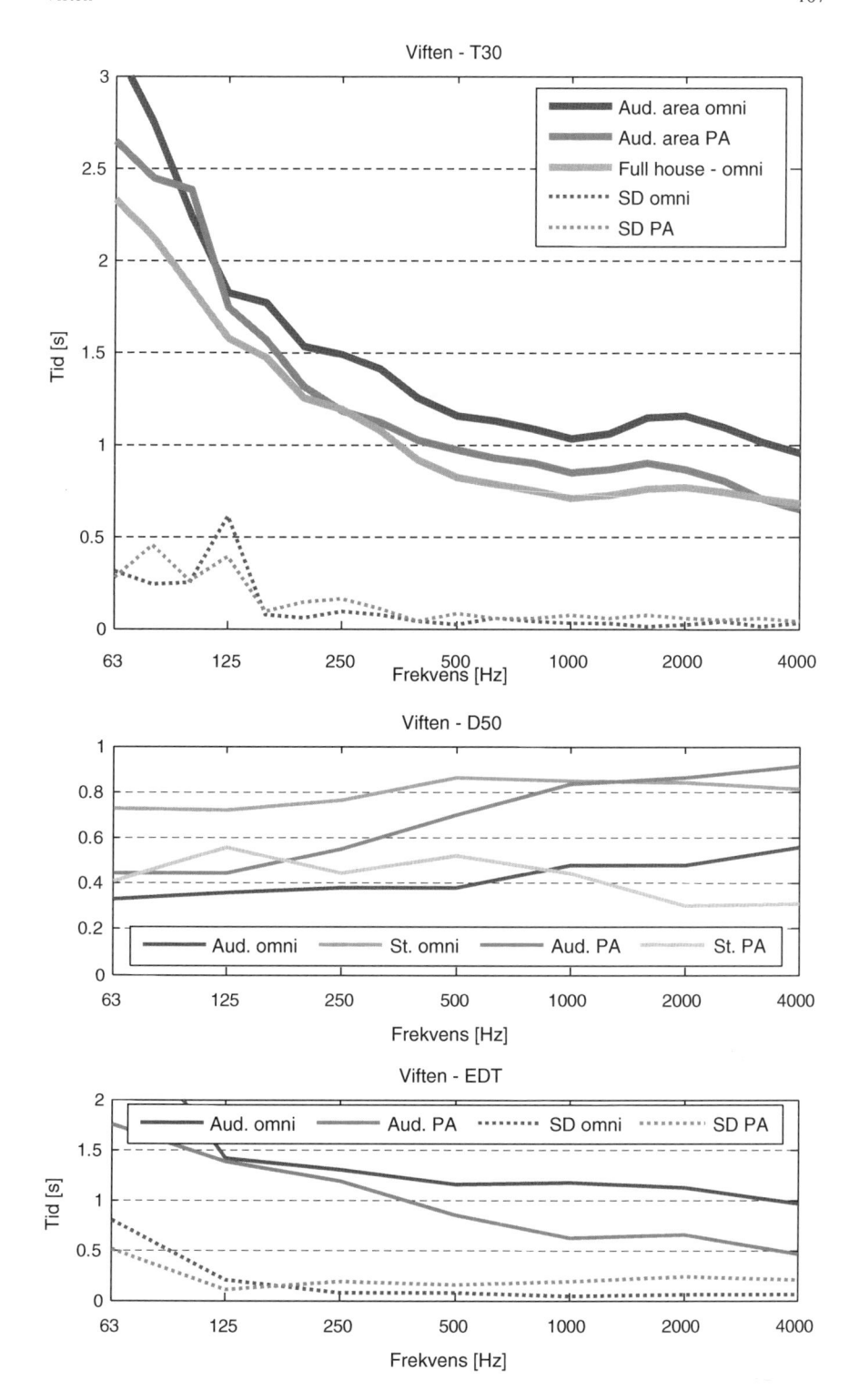

Viften

Viften - T30

Viften - D50

Viften - EDT

Voxhall

BUILT: in Århus, 1999 as a music venue.
  CAPACITY: 500
  NUMBER OF CONCERTS PER YEAR: 150

BUILDING MATERIALS USED: Ceiling: Suspended mineral wool slabs. Gypsum boards on cavity in the center of the hall. Floor: Wood on cavity. Walls: Wooden panels on cavity, concrete.

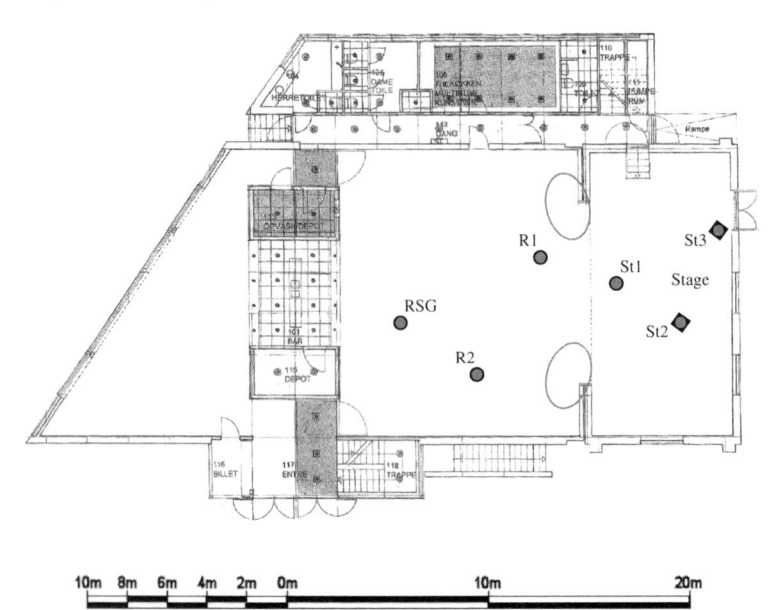

Geometrical data	
Volume (m^3)	1,600
Height (m)	5.6
Surface area stage (m^2)	70
Surface area audience (m^2)	190
Total surface area of hall (m^2)	710
Capacity of standing persons	500

Acoustical data—omnisource	
Audience area	
T_{tom}—(500–1k) (s)	0.6
T_{fuld}—(500–1k) (s)	0.5
D50—(250–2k)	0.76
EDT—(250–2k) (s)	0.8
Stage area	
T_{tom}—(500–1k) (s)	0.6
D50—(250–2k)	0.86
EDT—(250–2k) (s)	0.5
$BR_{rock,s}$	0.9

Subjective data	
Audience area	
Rating out of 20	5
Stage area	
Rating out of 20	8

4 Assessments of 20 Halls

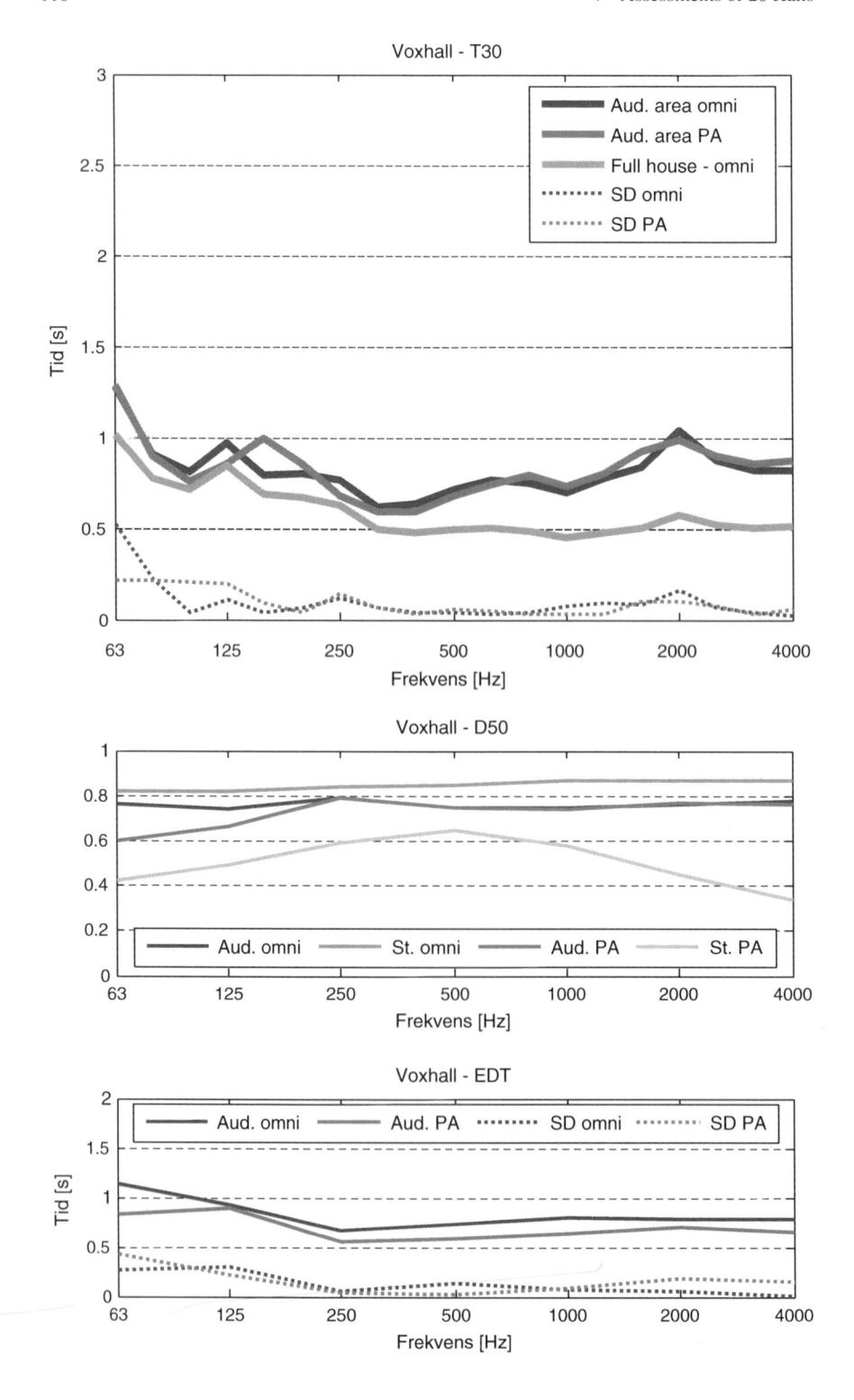

Voxhall - T30

Voxhall - D50

Voxhall - EDT

Chapter 5
Recommended Acoustics for Pop and Rock Music

As pop and rock evolved during the 1950s, '60s, and '70s dedicated venues for this music were built. Buildings formerly used for other purposes such as cinema or factories found new life accommodating live, amplified dance music, from bars and clubs for a few hundred people to actual music halls for more than a thousand people at the other end of the scale. And since the 1990's there has been an improved focus on the acoustics of sports arenas, that are used to house some of the most popular pop stars, with regards to amplified music.

Knowledge gathered from a large number of halls indicates that a fair share of acoustic consultants have been aware of what kind of acoustics is needed for amplified music. But other halls have not had the same luck and there have been a few typical misconceptions and pitfalls when designing for this purpose. Despite good efforts, the recommendations have not been complete regarding acoustics for amplified music. Some authors, however, briefly include the topic. For instance, the late, great architect Russell Johnson is referred to by Ahnert and Steffen in their book, *Sound Reinforcement Engineering* 1999, to recommend a reverberation time $RT_{(500-1k)}$ of 0.8–1.2 s in halls for dance bands. But nothing is mentioned as to which hall volumes these numbers correspond or to recommended RT at other than mid-frequencies. Barron mentions in *Auditorium Acoustics and Architectural Design* 1993, that a T_{30} below 1 s is recommended.

Some textbooks on room acoustics recommend that "halls for music" have an increase of reverberation time at frequencies below 250 Hz (despite the fact that many of the very best rated halls for symphonic music don't show this trait without an audience). Surely, this brings "warmth" to the sound. But this is only true for (unamplified) classical music. For amplified pop/rock music, as shown in this chapter, it is enemy number one! At unamplified music events the acoustics of a hall, together with the sound level produced by the ensemble, are solely responsible for the total sound level in the hall. And with some help from a longer reverberation at low frequencies, the bass sound is acoustically amplified and the overall sound thereby perceived "warmer". At amplified music concerts, however, producing enough level for the audience is obtained just by turning knobs on the instrument, the amplifier on stage, monitors, and of course importantly, by pushing faders and turning gain knobs on the FOH mixing console, adjusting the sound level provided by the PA system. In fact possible level differences because of uneven frequency response are equalized in numerous places as mentioned earlier: many musicians will automatically try to adjust their levels, and in smaller rooms some will equalize their instrument or amplifier to fit the response of the hall; both the PA system and often also the monitor system will comprise graphic equalizers to even out level differences resulting from the sum of the possible acoustic amplification of the hall and electroacoustic amplification of the PA system. And finally the sound engineer will decide both on faders and equalizers for each channel as well as on possible outboard devices such as compressors, the decided level of an instrument, and how it blends in the complete mix.

In this way, in halls for amplified music, the effect of reverberation time on sound level is not in focus. Moreover, delay speakers can be applied to enhance levels farther back in the hall, should it be inadequate. But as shown earlier, a low reverberation time gives a long critical distance and thus few members of the audience experiencing an overall reverberant sound. So is it not as simple as suggesting that for the greatest possible share of the audience to get "good," defined, direct sound, then as short a reverberation time as possible should be chosen. Are the "outdoor conditions" with no reflections what we need to bring inside the hall? The answer is a definite NO. Close to anechoic (little reflections) conditions would be chosen from such a strictly logical reasoning and chapters on critical distance found in, for instance, the *Sound Reinforcement Handbook* by Davis and Jones 1990, can surely leave the reader believing this. Some consultants have chosen almost anechoic acoustics for amplified music halls assuming that the only focus point was freeing the audience from undefined reverberant sound. And that hypothesis has in some cases been taken to the extreme sometimes even without enough focus on absorbing low frequencies. But as shown later demonstrate, this is not a correct path to pursue. In this chapter, we will see what values of reverberation time are recommended for a given hall volume; it must be relatively short, but not too short, and within limits it can vary with frequency.

The Basis of the Recommendations

In 2005 a study regarding recommended acoustics for pop and rock music was conducted in Denmark. To this day it seems to be the only proper research ever made for this purpose. The results from the survey were unambiguous, therefore recommendations have been made on this basis and they form the platform of this book. The author of those research papers and of this very book served 15 years in the music industry as a jazz and rock drummer, and played more than 1,200 concerts. A large share of those concerts was performances with the same band, playing the same music in the same venues with the same sound system and sound engineer over several years. That experience made the author certain that halls actually leave an acoustic imprint in the memory of at least some musicians and sound engineers. The author had a good network among Danish musicians, sound engineers, and venues. It was therefore a manageable task to conduct an investigation where a number of musicians and sound engineers were asked their opinion about the acoustics in the 20 most commonly used venues in the country. By looking in the venues' calendars from previous years it was determined which bands and musicians had played most often in a large number of the halls. A questionnaire was sent to 50 musicians and 18 sound engineers of whom 25 musicians and 8 sound engineers responded.

In a letter to the musicians and engineers accompanying the questionnaire the test persons were instructed only to fill out the sheets if they felt sure about their responses and to omit the halls they were not very familiar with or for other reasons felt uncertain about judging. The letter to the musicians said:

> As a musician, one evaluates venues—consciously or subconsciously—based on factors, such as: how good is the visual contact with the audience, is the temperature appropriate, is the service good etc. In this anonymous survey, the focus is on the acoustics of the venue for the performers. This means: how does the hall respond to the music that is played—judged independently (as far as possible) of the PA-system, the monitor technicians etc.

Then the first page of the survey included questions about what kind of monitors the band uses (in-ear, headphone, stage monitors, other), whether the respondent used to discuss the acoustics of halls with their colleagues (yes/no), how important acoustics are for the respondent (very, somewhat, a little, not important), whether the respondent had chosen not to play in certain halls on the account of the acoustics (yes/no), and whether the respondent found that possible negative effects of the acoustics could be mitigated through the use of in-ear monitors (very, somewhat, a little, no).

Then the respondent was asked to complete a questionnaire for each hall, asking for ratings of the halls on several acoustic aspects. This part of the questionnaire, that had to do with each hall, was based on the questionnaire used by Barron in his 1988 paper, "Subjective Study of British Symphony Concert Halls." Some of the parameters used in Barron's questionnaire were changed to better fit a rock setting. It was expected that the subjective ratings of Clarity, Reverberance, and Bass Balance would be correlated to the objective measures D_{50}, T_{30}, or EDT and BR. Figure 5.1 shows the questionnaires that were sent to musicians and sound engineers, respectively.

The respondents were free to set a mark anywhere on the continuous line. There was an "optimal" mark at the center point of the line for all but the Clarity rating. The positions of the respondents' marks on the line were measured assuming a linear scale and the data were gathered for statistical and correlational analysis in order to investigate how they corresponded with objective measurement data of the 20 halls.

The 20 halls were acoustically measured according to standards (ISO 3382:1997) with an omnidirectional source (dodecahedron). Obtaining the relevant data also in the 63 Hz band was a focus point wherefore an omnidirectional subwoofer was used together with the dodecahedron. For another round of

Musicians:

Clarity:	Muddy		Clear	☐ No Response
Reverberance:	Too Dead	Optimal	Too Live	☐ No Response
Audience Contact:	Too Little	Optimal	Too Much	☐ No Response
Bass Balance:	Boomy	Optimal	Weak bass	☐ No Response

General Rating:

☐ Very Poor ☐ Poor ☐ Mediocre ☐ Reasonable

☐ Good ☐ Very Good ☐ Excellent

Sound Engineers:

Clarity Bass:	Muddy		Clear	☐ No Response
Clarity Mid/Treble:	Muddy		Clear	☐ No Response
Reverberance:	Too Dead	Optimal	Too Live	☐ No Response
Bass Balance:	Boomy	Optimal	Weak bass	☐ No Response

General Rating:

☐ Very Poor ☐ Poor ☐ Mediocre ☐ Reasonable

☐ Good ☐ Very Good ☐ Excellent

Fig. 5.1 The questionnaires sent to musicians and sound engineers differed slightly

measurements, the PA system of the hall was used as the sound source in conjunction with the exact same microphone positions used for omnisource measurements.

Results of the Interviews

The First Page of the Questionnaire

The results of the study, and a precise description of it, were published in the *Journal of the Acoustical Society of America* in 2010[1] with the help of Dr. Eric R.

[1] "Suitable reverberation times for halls for rock and pop music." *JASA*, 127(1), Jan. 2010, Adelman-Larsen et al.

Thompson and Dr. Anders Gade. As mentioned, about half the people to whom the questionnaire was sent actually answered it. Those who did not return it may have felt unable to answer and may not be as conscious of, or as affected by, the acoustics as those who did. Or there may have been other reasons. Of course, not all halls obtained an equal amount of questionnaire returns, so the statistical certainty for correct ratings is not the same for all halls. Among the 25 musicians who responded there were eight drummers, seven bass players, five guitar players, three keyboard players, and two singers. It is very possible that different instrumental groups prefer somewhat different acoustics. More test people than 25 are needed in order to achieve significant knowledge about this. In any case, the average obtained in this study is relevant because all these instruments are regularly represented on any stage for pop and rock music.

On the question, "How important are the acoustics of a venue to you?" Seven out of eight sound engineers and 17 out of 25 musicians answered "very important," the remaining sound engineer and 7 musicians said that acoustics are "important," and the remaining 1 musician said that acoustics are only "slightly important." Two of the eight sound engineers had considered not playing, and 8 of the 25 musicians said that they had chosen not to play in certain venues because of inadequate acoustics. All sound engineers and all musicians said that they discuss the acoustics of specific halls with colleagues.

Five sound engineers responded that their bands used in-ear monitoring, seven reported using onstage monitors, and one reported using headphones and monitors (note that the respondents could choose more than one monitor type). Fourteen musicians reported using in-ear monitors, 19 used on-stage monitors, and three musicians (all drummers) reported using headphones. On the question of whether in-ear monitors can help mitigate the possible bad effects of a hall's acoustics, four sound engineers and nine musicians responded, "very much," three sound engineers and eight musicians responded, "somewhat," and one sound engineer and three musicians responded, "a little." The remaining five musicians either responded, "don't know" or did not respond. These responses are of importance; the direct sound experienced when using in-ear monitoring certainly to a large degree masks possible unwanted reverberant sound, but only at frequencies above some 250 Hz. The in-ear/closed headphones do not block lower frequency sound that the musicians seem to be able then to hear, both with their ears and from vibrations leading to sound perception by the inner ear through bone and body conduction. Some musicians are capable, in a positive way, to focus on the higher frequency direct sound rather than reverberant, undefined lower frequency sound. In order to try to mask the reverberant low-frequency sound with some direct low frequencies, some musicians, especially bass players, get a vibration-plate to stand on and drummers sometimes invest in a so-called "butt kicker," a vibration transducer that can be mounted on their drum seat. None of the musicians in this survey used those tools.

To the question whether musicians choose not to play in certain halls, there were cases where one member of a band said no, and another said yes. Maybe the one answering yes is involved in the booking process and the other one is not.

Or the first one plays in other bands also with other preferences on this subject. Overall, these results showed that acoustics are certainly important for rock musicians and sound engineers although in-ear monitoring lowers the importance to some extent for musicians.

PA System Versus Omnidirectional Source Measurements

In the Fig. 5.2 definition, D_{50} is shown as a function of frequency as an average across all halls, for both omnisource and PA measurements and both with measurements in the audience area and on stage. It is seen that the highest definition is achieved with the omnisource on stage somewhat more defined than the PA sound in the audience area. Of course the microphone positions farthest away from the PA speakers often account for a lower D_{50} than those close to the speakers where the direct sound is usually louder relative to the reverberant, not so defined, sound. In most of these halls those more distant measurement positions pull the curve downwards. Some speaker configurations seek, as earlier mentioned, to compensate for this effect. On stage the distance cannot get as long as in the audience area but on the other hand the PA speakers are more directive than the omnispeaker at mid–high frequencies. The effect of this is seen on the curve of the omnisource measured in the audience area.

Not surprisingly the least defined of these groups of sound is encountered on stage as a result of the reflected higher frequency sound emitted by the PA speakers. If that PA sound becomes too loud on stage the musicians have no choice but to turn up their monitoring. And if the monitors are open monitors on stage (and not in-ear monitoring) they may get so loud that the sound engineer operating the PA system feels obligated to turn up the PA level because the loud monitor level masks the correct mix in the PA system, even at the sound engineer's position among the audience. This is a well-known phenomenon, an evil spiral, leaving both musicians and audience with too-loud sound levels and worse sound quality because of monitor sound leakage into open microphones on stage, as well as possible inappropriate monitor sound in the audience. Furthermore, because the low-frequency sound emitted by the PA speakers is omnidirectional the graph shows higher values of D_{50} on stage. Probably the 250 Hz band is just omnidirectional enough to get a high-definition value whereas on stage the 125- and 63 Hz values decrease due to a higher reverberation time at these frequencies in the average hall. The later, low-frequency reflections are, as we show, actually the primary cause for poor acoustics as perceived both by sound engineers and musicians.

Referring to Fig. 3.7, it is a fact that the more sound the PA system shoots onto the walls and ceiling the more the reverberation of the hall is evoked. The recommendations in this book cannot take the effect of the different PA configurations in different halls into account. The ratings in the following section are a grand mean of many responses to many halls, therefore it is believed that the effect of different

Fig. 5.2 D_{50} as a function of frequency, averaged over 20 halls in the audience area and on stage for both omnidirectional source and PA system

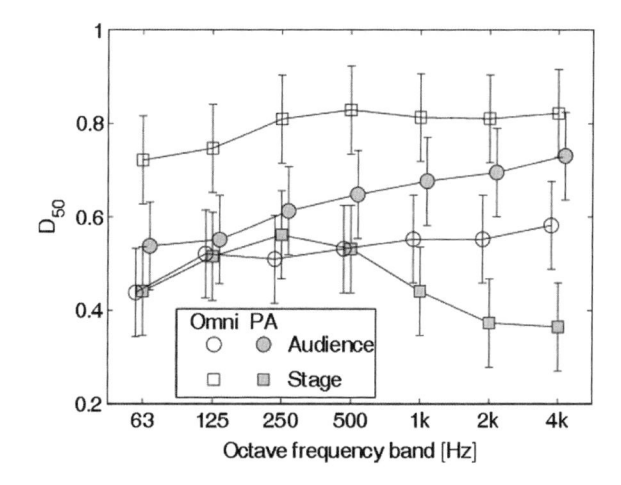

PA systems is somewhat evened out and that the ratings are indeed applicable and shall be employed in conjunction with an appropriately designed PA system.

General Ratings of the Halls

Each musician and sound engineer assigned a general rating to each hall constituted by a number from 1 to 7, where a 1 corresponded to "Excellent" and 7 corresponded to "Very Poor." The mean general rating for each hall was then calculated for the group of musicians and for the sound engineers, and the combined rating was calculated as the mean of the two groups. The ordinal rank of the halls' ratings from 1 (best hall) to 20 (worst hall) for each group and the ordinal rank for each hall are shown in Table 5.1. The halls are sorted by volume in order from smallest to largest, and it is interesting to note that there is no correlation between the size and the overall rating.

Interestingly, the driest hall, Stars, is in the tenth place in the musicians' ratings category but is the favorite of the sound engineers, which moves it to the fourth best rating overall. Stars was also rated the driest on the "Reverberance" scale (the only hall rated by the musicians as "too dry"). So even though the group of sound engineers in this survey liked the recording studio quality of the hall, it is a good example that a hall can be too dry for musicians. Later interviews with other sound engineers have revealed that another group of sound engineers actually prefers acoustics much like those favored by the musicians. The four lowest-rated halls have a relatively high T_{30} and typically a longer reverberation time at lower frequencies. Viften has an extraordinarily long reverberation time in the 63 Hz octave band (over 3 s) and in the 125 Hz band and much shorter reverberation above 500 Hz (around 1 s) due to banners on the walls. This is also the hall that the sound engineers rated the lowest on "Clarity Bass."

Table 5.1 Overview of the 20 investigated halls that form the basis of the recommendations in this book

Name	Volume (m³)	Audience capacity	$T_{30,B}$ (s)	$T_{30,M/T}$ (s)	EDT (s)	D_{50} (s)	Bass ratio	Musicians	Sound engineers	Combined
Rytmeposten	655	300	0.8	0.8	0.3	0.6	1	14	10	11
Lille Vega	785	500	0.5	0.7	0.4	0.7	0.7	1	6	3
Loppen	890	350	0.9	0.8	0.5	0.7	1.2	5	13	9
Skråen	1100	375	1.5	0.8	0.9	0.4	1.8	13	12	13
Paletten	1420	375	1	0.9	0.7	0.7	1	8	8	8
Stars	1440	400	0.6	0.6	0.3	0.8	0.9	10	1	4
Voxhall	1600	500	0.9	0.6	0.5	0.7	1.3	7	5	6
Sønderborghus	1600	420	1.2	1	0.8	0.6	1.3	20	18	19
Musikhuzet	2080	700	1.1	0.9	1.1	0.5	1.2	9	17	12
Godset	2150	700	0.7	0.8	0.5	0.6	0.8	6	4	5
Magasinet	2540	525	1.9	1.3	1.3	0.3	1.4	12	19	18
Pumpehuset	3000	600	1.2	1.1	1	0.6	0.9	16	15	15
Forbrændingen	3050	450	1.1	0.9	0.5	0.8	1.2	19	11	14
Train	3300	900	0.8	1	0.4	0.7	0.9	3	2	2
Slagelse	3800	700	1.8	1.6	1	0.5	1.1	17	20	20
Viften	3950	700	2.6	1.2	1.1	0.6	2	13	14	16
Amager Bio	4500	1000	1.2	1	0.8	0.6	1.1	4	9	7
Torvehallerne	5400	700	1.2	1.5	0.9	0.5	0.8	15	16	17
Store Vega	5800	1430	1.4	1.2	0.7	0.7	1.1	2	3	1
Tobakken	6500	1200	1.5	1	0.8	0.6	1.4	11	7	10

The values of EDT and D_{50} are averages of the octave bands 63–2 kHz. $T_{30,B}$ is averaged from the 63- and 125 Hz bands and $T_{M/T}$ is averaged from 250–2 kHz. The BR is the ratio of the average reverberation time in the 63- and 125 Hz bands to the average reverberation time in the 0.5–2 kHz octave bands

Musicians' Preferences

It is first of all important to note that all musicians' taste regarding acoustics is not the same. There may be, as mentioned, some instrumental groups that want a more reflective hall than other groups, and there certainly is a degree of personal taste involved. The recommendations in this book are a grand mean of instrumental groups and individual preferences. Therefore it is safe to construct venues from these, but it is also almost certain that someone will not fully agree. Also there is some influence stemming from what exact genre within amplified music the hall is to be used for; a Brit-pop band has a different frequency content than an electronic music act.

The survey of 2005 proved among other things that musicians need halls not to be too acoustically dead and not too lively either. Probably the most frustrating for musicians is hearing the music reflected from the audience area loud compared to the earlier reflections from the stage surroundings including their own direct sound and that of the monitors. It gives a distancing sensation; the musician feels detached from his or her own playing and thereby disengaged from the situation. It is often thought that this can be eliminated with the use of monitors, but neither open monitors nor in-ear/closed headphones can sufficiently mask the sound of the hall if it is dominant. There is a need for the early sound being enveloping for the musicians who often move around the stage and this calls for some early reflections from the stage surroundings. This is indicated by halls with an overall quite long reverberation time and their quite bad rating such as Torvehallerne and Sønderborghus. Moreover this was confirmed by some musicians who in the survey, as a comment at the end of the questionnaire, specifically stressed that "the worst thing is a small enclosed stage detached from a large hall."

Some musicians are better to cope with this situation than others. For instance, one of the world's greatest jazz pianists of all time, Keith Jarrett, stopped his concert twice during the first set of his 2011 appearance in Copenhagen and announced that he was unable to play certain tempos as he "did not receive any sound back from the hall." In the intermission a reflective curtain was therefore lowered covering the huge hole in the proscenium of the hall behind the musicians. Not only did Jarrett and his trio play without further disturbances through the second set, just as important, the sound engineer was now able to turn up the PA level considerably for the benefit of the audience, because he did not have to worry any longer that the PA level would mask the direct monitor sound and early reflections on stage. So Jarrett got more of a feeling of the music he and his trio were producing. He was unspecific about what sound he was missing in terms of where the reflections should come from, but two important lessons can be learned from that concert: the monitors on stage consisting of both one monitor on the floor for each of the three musicians and a side-fill system of two loudspeakers at a greater distance from the musicians were not able to deliver enough sound. Early stage-based reflections were what the musicians first and foremost needed to feel good about their playing. Second, they needed a certain idea of what imprint their music had in the audience area, that is, enough late, hall-based reflections from that area to be audible on stage. They like to be able to "hear how the music lands" at the audience because the audience is their primary concern. The louder PA level provided that in the second set; but very, very important were the late reflections not louder than what the earlier reflections from stage surroundings could partially mask. Musicians live for giving audiences a great experience.

Without reflections from the stage area, even with a complete monitor set-up, the musician will not experience a sensation of being enveloped in his and his colleagues' sound. With too much sound coming back from the hall he certainly will feel enveloped in sound but, too-strong late reflections will make the musician feel disengaged from his playing and will tend to affect his timing. Of course musicians experience these defects frequently and they cope with them by being somewhat conscious about the sound and navigate accordingly to get timing correct. But that does not make defects acceptable or recommended. On the contrary, if both stage and hall are too dead, small and natural timing differences between the musicians become very clear which can lead to uncertainty and for them to lose confidence. And ironically, when the confidence is there, there will probably be no timing issues.

So it is seen that, according to musicians, the acoustics on stage mustn't be dead compared to those in the hall, and the acoustics in the hall must be neither too dead nor too lively.

Sound Engineers' Preference

Sound engineers are responsible for the sound during concerts given the equipment and band at hand. The sound engineer is placed in the audience area and therefore has perfect possibilities for knowing what sound impression the audience perceives. It's safe to say that sound engineers are trying to give their audience as good an overall experience as possible. Unless really well prepared for, they have little or no possibility to enhance the acoustics of a hall before a show because this implies quite dramatic changes in large areas of the hall. The sound engineer sees it as her job to create as defined and transparent sound as possible and to add suitable effects such as artificial reverberation into the mix. If the hall does not add much reverberation itself, or rather if the combination of hall and PA system does not add much reverberation, the engineer has quite a lot of freedom in playing with artificial effects.

When the question about acoustics in halls is debated, it seems that sound engineers can roughly be divided into two categories: those who want the hall to give some envelopment, as is the preference of the musicians, and those who like more control over their outboard effects to be added to the mix. The largest share of the group of top sound engineers, who were asked in the above-mentioned survey, had the driest of all halls, Stars, as their favorite. They do admit an element of "selfishness" to this preference; after many difficult concerts in inadequate halls they see a concert in Stars as their "shining hour;" a possibility of having complete control and freedom because no reflections interfere. Later interviews with other, equally acclaimed sound engineers have shown that they don't want the hall to be completely unresponsive. They too, enjoy being surrounded by sound, just like the musicians and probably also the audiences as long as they can provide a nice transparent mix too. Sound engineers are trained in the fine art of making a great mix. That is a completely different métier than room acoustics.

Many sound engineers like the stage to be quite dead: this prevents sound from instruments and monitor speakers from being reflected to the audience area or to leak into open microphones on stage. In this way they can keep as much unprocessed sound as possible out of the total mix that meets the audience. The stage reflections entering the open microphones on stage are delayed and possibly out-of-phase with the direct signal. These reflections of course harm the total mix. Much in the same way regarding the audience area, a relatively dead hall will make the sound engineer capable of forming a sound experience to his taste on the PA system.

Debate

So here is in fact often a dilemma between what most musicians want and what many sound engineers find recommendable. A dead stage leaves the musicians unable to hear themselves, each other, and the audience sufficiently loudly and with enough envelopment. Moreover it requires from the point of view of the musicians a dead hall, for the stage not to be more dead than the hall, and that they don't want either. This of course opens the debate about who should decide on what acoustics is appropriate for a venue for pop and rock music. For classical music there is no sound engineer but there is a conductor who often brings valuable insights into play when discussing recommendations for classical music halls.

Evidently the audience plays the key role. They actually buy the tickets that pay for the band and sound engineer. So what does the audience want? The most encountered opinion on this is that the audience wants to experience the fantastic ambience and incredible moods that are often connected with pop and rock concerts. They want to be drawn into a special atmosphere that is made during the concert. The better the musicians feel on stage, and the less they worry, the better their chances of creating a great performance, possibly even unforgettable for themselves and the audience. And remembering that musicians like to hear their music "land" at their audience, it is regarded safe to say that when the musicians are pleased, the audience is pleased too. A musician is not content if her audience is not. And as we saw, musicians too need quite a clear sound, although not overly defined, in order to enter a state of togetherness with their colleagues and the audience who then in return share a common bond with the band and each other. And everybody will praise the sound engineer as well for having participated in creating such an event.

Therefore it is believed to be correct to follow the taste of the musicians which is identical to that of some sound engineers. Other sound engineers may find these conditions on stage as well as in the hall a little too reverberant in order to create the perfect sound they had in mind. These slightly reverberant conditions make their job a little harder if the band sounds harsh or unprofessional, but it must be remembered that there are also other interests in play that benefit the whole event. Many performers have reported their most memorable concerts to have taken place in halls that were not dedicated halls for music. An acoustics signature of such a space leaves them with a good impression, as long as they were able to adapt to the conditions at hand.

Spectral Analysis of Surveyed Data

From all the concerts the author had experienced as a musician he knew that a long reverberation time at low frequencies was particularly disturbing. This is not so peculiar and has been known by some acoustical engineers and other professionals for decades. This is due to different factors: the bass sound is amplified by thousands of watts at pop and rock concerts. By far the biggest share of electric amplification energy is used below 200 Hz and reaches considerable levels as seen in Fig. 3.2. The audience does not absorb much low-frequency sound (Fig. 1.16). Because of a low Q value of bass loudspeakers emitting bass sound the critical distance becomes very short leading to much reverberant bass sound almost everywhere in the venue (Chap. 1, Eq. 1.8). Therefore an overall undefined sound will most often stem from reverberant bass sound that, because of the loud level, will partially mask even the direct higher frequency sound. Only a controlled reverberation time at low frequencies can make up for this.

From this knowledge the average reverberation times of the 10 highest and the 10 lowest rated halls were calculated and presented in the same diagram as a function of the octave band. To eliminate the factor that bigger halls can admit a longer reverberation time, the reverberation time of each hall was divided by its volume. With that normalization the effect of volume was eliminated. The ratings for sound engineers and musicians were for this purpose averaged into one combined rating hoping to find a factor that would be important for both groups of professionals.

The result is seen in Fig. 5.3. In this figure the upper line shows an expression of the average RT of the lowest rated halls for each octave band and the lower line is the average of the best halls. The vertical lines around each point show the statistical confidence levels. The results of the two groups of halls cannot really be differentiated from one another above the 250 Hz octave band. But it also, more notably, means that what actually distinguishes the best from the not-so-well-liked halls is a shorter reverberation time at low frequencies. This is believed to be the most important finding in the survey. Furthermore, inasmuch as the variances do not overlap in the two (or three) lowest octave bands, the diagram shows statistical significance, whereby it constitutes a scientific proof that must be accepted by any

Fig. 5.5 Measurements before (*black*) and after (*grey*) installing tuned membrane absorbers in a hall

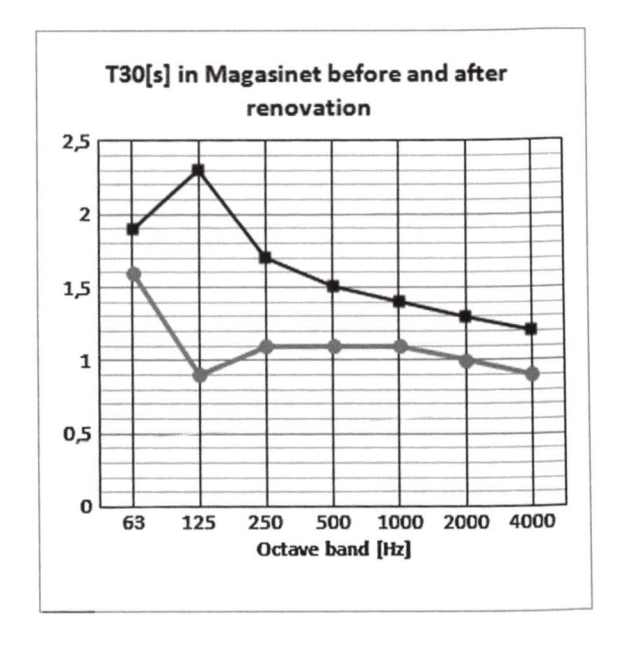

Fig. 5.6 Approximate factors of T_{30} in the octave bands 63–4 kHz. Factor 1 refers to the relevant value in Fig. 5.4

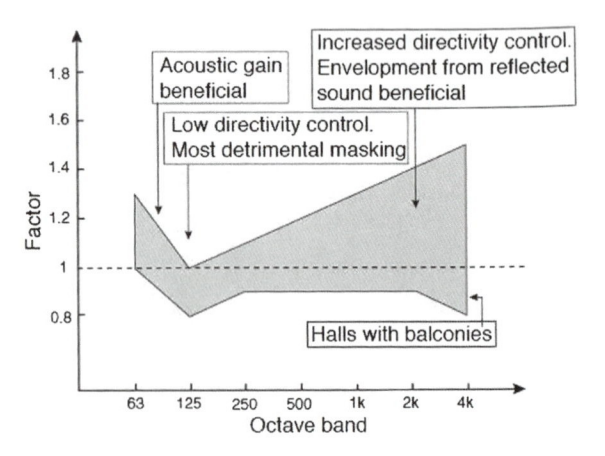

These investigations lead to an understanding that RT in the 63 Hz band can be a factor higher than that of the 125 Hz band. It can even be of advantage that the hall helps bring forward these power-demanding very low frequencies. Possible third octave upper tolerance factors of T_{30} relative to the T_{30} at 125 Hz are: T_{30} at 50 Hz :1.8; 63 Hz :1.4; 80 Hz:1.2. If the hall has very low RT at higher frequencies the rise at lower frequencies will be audible more easily and therefore not recommended. A rise at these very low frequencies means that the hall helps that sound to be acoustically amplified; a doubling of RT gives an extra 3-dB sound pressure level. It is important to note that the factor of 1.3 shown in Fig. 5.6 only corresponds to the situation mentioned above where the factor increases with the

Results of the Interviews 123

Fig. 5.3 Averaged RT divided by hall volume for the 10 best and the 10 worst rated halls as a function of the octave band

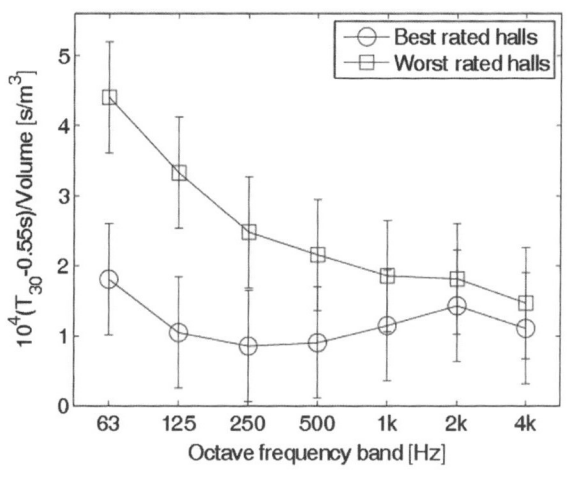

Fig. 5.4 Solid line shows recommended RT for an empty hall at various hall volumes. The dotted line is for the average including the 63 Hz band. The line is linear in the small interval from 1,000–7,000 m^3, but certainly cannot be extrapolated linearly to larger volumes. Applying a logarithmic scale on the x-axis over large volumes, recommended RT would approach a straight line

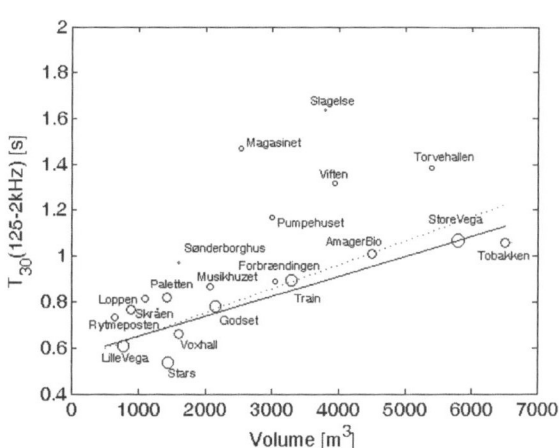

scientist or acoustical consultant. This is one factor that has to be fulfilled in making a recommended hall for pop and rock music. This is the single most important message of this book.

Furthermore, for the best halls the diagram shows a small increase in reverberation time in the 63 Hz band. Of course this cannot per se be taken as a recommendation because it is just a result of the average of the halls at hand and this rise is difficult to avoid. But the increase is an indication that this is acceptable in the 63 Hz octave band. As mentioned, the best halls have a significantly lower RT in both the 63 Hz and the 125 Hz octave bands compared to the worst halls. It therefore is at least hypothetically possible that an increase in just one of them is acceptable. Because of this ambiguity, some other venues with an increase of RT mainly in the 63 Hz band were studied (Figs. 5.4 and 5.5).[2]

---

[2] "On a new variable absorption product and acceptable tolerances of T_{30} in halls for amplified music;" convention paper, ASA, San Diego, 2011, Adelman-Larsen et al.

lower one third octave band. A factor of 1.3, 80 Hz is not appropriate. The higher value of acceptable RT in the 63 Hz octave band compared to the 125 Hz band is believed to have something to do with the human ear's relative insensitivity to sound at these low frequencies (Fig. 1.13).

Recommended Reverberation Time for a Given Hall Volume

A safe choice when designing a venue is to choose a reverberation time that is constant over a frequency according to the values represented by the solid line in Fig. 5.4. In this figure the combined ratings of sound engineers and musicians are the basis of the size of the circles for each venue; larger dots mean a better combined rating. The line is a best fit within the largest circles; the five best halls were given double weight compared to the halls rated numbers 6–10. The 10 worst halls were not included in the equation of the line.

This solid line is an average of the frequencies 125 Hz–2 kHz. Often in acoustics such diagrams only include the mid-frequencies, but because, as pointed out above, control of the low frequencies is important in pop and rock music halls these must be incorporated in the recommendation. The dotted line shows the recommended RT with the 63 Hz band incorporated in the average. This was originally done in the *JASA* paper (see footnote 1) but was later left out (see footnote 2) because as mentioned above, the 63 Hz band can be admitted to have higher values of RT.

The acoustics on stage must not differ too much compared to the acoustics of the hall. In small clubs the sound level is a major concern. If the stage area is equipped with a lot of sound-absorbing material then the rest of the hall must be given a similar design. And then RT is apt to drop below the recommendations in Fig. 5.4.

Acceptable Tolerances of T_{30} in Pop Rock Venues

In the town of Odense in Denmark, the author was asked to design the acoustics of two different venues. Both were to accommodate pop and rock concerts. Where one, Posten, would exclusively be used for this purpose, the other one, Magasinet, was also planned to host more acoustic acts, such as a solo folk guitarist or stand-up comedy, theatre, and so on. Posten was a completely new building whereas Magasinet was already a music venue and actually rated number 18 out of 20 in the survey. Both venues are approximately the same size. With two similar venues close to each other in a medium-sized town, instead of making identical acoustics, it was suggested to give each hall its own sound. Posten was therefore built with quite tamed acoustics in the entire frequency span (63 Hz–4 kHz).

In Magasinet the only acoustic change made was bringing down the overly long RT in mainly the 125 Hz band (from 2.3 s to 0.9 s) while leaving the hall with a relatively long RT at higher frequencies (1.4 s with some upholstered chairs dispersed around the room). In that hall there is no mid–high frequency absorption material other than that provided by the upholstered chairs (!). The chairs are removable, and when very popular bands are playing the hall holds about 700 audience members on two levels. The ceiling height in that hall is about 6 m whereas the balcony and other construction details of the old factory make it quite diffusive at all frequencies. No large single portion of the back wall is apparent because the audience area is somewhat sloped and partitioned by the balcony. This eliminated the possibilities of an echo effect.

The halls are very well liked according to musicians and owners. Some sound engineers say that Magasinet, which is not dampened at higher frequencies, has a too long RT at higher frequencies but most musicians love it because they enjoy a phenomenal acoustic contact with the audience as well as with their own sound, both through strong early and later reflections. The stage room is only dampened at mid–high frequencies by a backdrop woolen curtain. That venue takes a good sound engineer and a professional band, but with that at hand magic can happen. The town is pleased with having such acoustically different but very functional venues.

Completely omitting mid–high frequency damping material in Magasinet was not planned. It was the idea to install a woolen curtain to be drawn in the opening of the balcony that would make up for the presence of an audience there when the balcony was not in use. That was never installed due to lack of financing after the complete restoration of the venue in 2007. Furthermore, it had been planned to install just a little porous absorption in the perimeter of the ceiling in both the audience and stage areas but this has largely proven to be unnecessary in as much as the hall owners are overly happy about the result due to the positive feedback they get from most musicians and audiences. The RT was brought down primarily in the 125 Hz band but also somewhat in the 63- and 250 Hz bands by installing tuned membrane absorbers in the entire ceiling, also in the stage area, as well as on the large stage area. The before and after curves can be seen in Fig. 5.5. The change of RT at higher frequencies (light grey ellipse) is due to a higher number of upholstered chairs during the "after" measurement.

The author is convinced, that envelopment and "togetherness" in general shall be obtained from a higher value of RT at higher frequencies, not necessarily just to create a frequency-independent reverberation when the hall includes the audience. There can possibly even be a rise with the audience in place. This is also where a unique sound for venues can be obtained without jeopardizing the overall acoustic impression. Magic will happen in such halls. Also there will be songs that work less well, but never to a degree of the unacceptable. It's like red wine: a $12 Australian Shiraz will do the job. That resembles a flat frequency response according to Fig. 5.4. But with a $50 red wine, chances are you will get an unforgettable experience, although it may not be appreciated to its full potential with certain dishes.

Derived from that experience it seemed appropriate to suggest a set of acceptable tolerances around the recommended T_{30} shown in Fig. 5.4 also at higher frequencies in halls above some 1,000–2,000 m^3. These tolerances are shown in

Fig. 5.6. The recommended T_{30} values in Fig. 5.4 correspond to a factor of 1 in Fig. 5.6. This yields the following recommendations for empty halls of volumes between 1,000 and 7,000 m³:

1. T_{30} in the 125 Hz octave band should be in accordance with Fig. 5.4. This octave band is extremely dominant; ask any experienced sound engineer. It is by far the band most often encountered as problematic.
2. At higher frequencies, T_{30} can be higher according to Fig. 5.6. This is due to the high degree of absorption provided by the audience and the air, and due to higher directivity of loudspeakers at higher frequencies. It is also a fact that at amplified concerts usually artificial reverberation is added to these frequencies by the sound engineer partly to compensate for little natural hall reverberation. These exact T_{30} in each band should be chosen by the acoustical engineer according to what the hall owner is striving for in terms of genre and taste and to what the general architecture of the building suggests. It is a fact, though, that higher frequencies easily get overdampened making the low end stand out more easily. The hall will appear unbalanced.
3. Acceptable tolerances for the factor of T_{30} in the 63 Hz band are as follows: 50 Hz: 1.4; 80 Hz: 1.2 times the recommended value at 125 Hz. A tolerance of, for instance, 1.4 in the entire 63 Hz octave band is thus not recommended. These tolerances are particularly acceptable if there is a similar increase at higher frequencies that will help balance the 63 Hz band rise. The reason a higher value of T_{30} in this octave band is acceptable is partly that the masking effect here is less broad (Fig. 1.13) and that the A-weighted sound level in pop and rock music is usually somewhat lower compared to the 125 Hz band. Also from Fig. 1.9 it is seen that a sound decay in the 63 Hz band becomes less audible to humans sooner than a decay in the 125 Hz band because the higher threshold in quiet at 63 Hz (see Fig. 1.9).
4. Tolerances lower than a factor of 1 from 125 Hz and up are to be used in halls with large balcony areas. It is acceptable here to place absorption material in the ceiling areas underneath the balcony whereby the RT will drop to lower levels.

It must be remembered that the extra reverberation, above the factor of 1, at higher frequencies than Fig. 5.6 allows for, calls for a higher level of early reflections at these frequencies on stage too. If the stage is very big or if there is a very high ceiling above the stage, for instance, a stage tower, it can be recommended to supply these missing early reflections by adding a set of reflectors that may be mobile and arranged according to the size of the band. Such reflectors can preferably be diffusive. It is also important that the higher T_{30} at mid to high frequencies should not be applied in smaller venues due to the risk of ear fatigue unless the room is very diffusive indeed at these frequencies. On the other hand, if T_{30} is chosen lower than unit 1 because of balconies it is still fine to leave the stage not too acoustically dead.

The fact that higher frequencies can attain higher values of RT can also be seen in the light that the dynamics of music mostly is expressed at these frequencies.

Although examples of this from pop or rock recordings do not exist, it is true for a symphonic orchestra or simply an acoustic guitar: when increasing the level from, for instance, *pp* to *ff* the higher frequencies above some 2 kHz increase much more in level than mid and low frequencies (Pätynen and Lokki 2013). The hall should be able to make these dynamics come forward.

The tolerances given in Fig. 5.6 should be useful for companies manufacturing electronic reverberation systems that emulate real acoustics of halls. Still it must be noted that stage acoustics must be similar to the acoustics in the audience area.

Suitable Reverberation Times in Larger Halls and Arenas

Based on the knowledge that RT in the 125 Hz octave band is the most critical parameter for the acoustic quality of a venue for pop and rock music, as well as on values of RT actually obtained in certain acclaimed venues in Chap. 7 of this book, a graph of suitable RT over a greater span of volume (stretching beyond 7,000 m^3) has been made (Fig. 5.7). This recommendation has no subjective studies associated with it and is only to be regarded as the author's best estimate. It is believed that the cautious acoustic engineer can employ the tolerances in Fig. 5.6 in halls with volumes from approximately 2,000–50,000 m^3, evidently with special attention in large volumes (critical distance) and in very small volumes (need of diffusion). It is also possible that volumes larger than 50,000 m^3 can benefit from higher values of RT at higher frequencies. The RT values given in Fig. 5.7 may seem difficult to obtain especially at 125 Hz. However, compared to values mentioned in a conference paper from 2007 where RT especially,[3] where RT especially for "smaller" volumes of, for instance, 50,000 m^3 are extremely strict, the recommendations in Fig. 5.7 are manageable and shown to be more practically applicable.

Fig. 5.7 Best estimate of recommendable values of RT in the 125 Hz octave band; a function of volume in empty halls and arenas that present pop and rock music

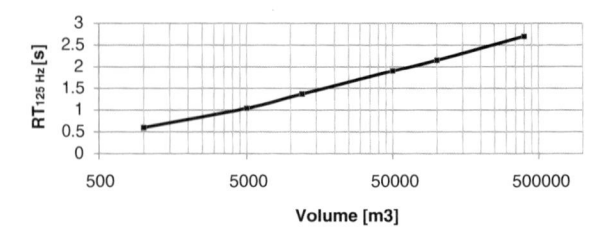

---

[3] "Acoustics for large scale indoor pop events;" ISRA, Seville, 2007. Lautenbach and Vercammen.

Chapter 6
Design Principles

Often pop and rock venues are opened in already existing buildings that were constructed for another purpose such as a factory or a cinema. In later years dedicated buildings for the purpose of musical performance have been constructed. In the first case, the overall building geometry is fixed, leaving limited room for acoustical optimization, whereas the latter leaves many parameters open for the architect, acoustical consultant, and the designer of the loudspeaker system as well as the lighting system. Of course the hall owner, often the community or state, but also in some cases, private investors, have a general idea of the size needed in terms of number of audience members that the hall shall accommodate, as well as the budget for the complete finished venue. If there is already a team of people who have formerly worked for pop–rock venues associated with the new project, these individuals often possess valuable knowhow that should be drawn upon as much as possible in the planning and design process of the logistics of the building.

There are many important non-acoustical factors to be alert about in the early architectural planning of the venue. It is not within the scope of this book to go into detail about these, but nonetheless some important parameters that the author has often encountered are listed below.

Sound insulation must also be sufficient at low frequencies for the area in which the venue is to be placed. Future development of the neighborhood must be considered. Lightweight building structures have the advantage of not fully reflecting low frequencies whereby a low RT is more easily obtained at those frequencies. But the sound simply is not fully "blocked" and passes through directly to the outside. This is very apparent in tents where all low-end frequencies pass through, leaving a high degree of mid-frequency sound to form the sound field inside (the solution regarding acoustics here of course is to treat the inside of the tent with molton, also known as duvetyne, curtains, for example). Someone, for instance, the hall owner, must demand a report from a certified acoustical consultant regarding sound insulation.

There needs to be ease of transportation of musical instruments and so on from the truck and load-in doors all the way to the stage. The heavy gear is carried in flight cases on wheels thus even doorsteps must be avoided.

Correct placement and design of the loudspeaker clusters in the ceiling is another necessity. Also the attachments need to carry an enormous weight and they must be carefully designed in an early stage.

It may be a good idea to let an independent specialist do the sound system design. Some acoustic consultancy firms offer this service too. Some halls have ended up with unnecessarily large sound systems. Too many loudspeakers can lead to unwanted late reflections because they will have to point directly into the upper part of the back wall in order to be fitted into the array or cluster. Line arrays are most suitable in very big venues or outdoors. Also cardioid subconfigurations are not necessarily better than normal subconfigurations especially from the point of view of musicians. Sometimes they are worse.

Some of the lighting and other technicalities are controlled from above the audience. A correct planning of catwalks may be vital.

The cables from the mixing board to the stage are placed in a small "tunnel" underneath the floor. Regarding planning of wiring and the like, a professional installer should be consulted very early in the design phase.

Backstage facilities must be adequate for various artists' different needs.

The acoustic design must also be planned in the earliest phase because it can then be totally integrated in the architectural design of the hall. Insufficient attention to the acoustical design in the early design phase may prove very expensive and difficult, if not impossible, to overcome at later stages of the design and/or construction.

The number, size, and placement of escape doors in case of emergency determine the maximum audience capacity. Terrible fire accidents still take place around the world in discothèques and halls for amplified music. Nothing justifies the loss of life.

Hall Size

The size of the hall is probably the first consideration to make if not already given. How large an audience must it be able to hold? Are they sometimes supposed to be seated? The number of audience members that a venue can accommodate of course depends on the audience floor area but is also a question of how the escape routes from the building can be planned. Audience densities of up to 3.0 standing audience members per square meter of audience floor area have been encountered in the author's survey. The number is much lower if the audience is to be seated.

Considering the ceiling height of the venue, different factors are in play:

- A lower ceiling by itself leads to a lower reverberation time and thereby to smaller areas that have to be covered with absorption. Investigations by the author show that there is a better correlation between T_{30} (even in empty halls) and ceiling height, than between T_{30} and hall volume. Because the floor area is usually covered by the (mid–hi-frequency absorptive) audience, the height of the hall is the main factor for reverberation.

- A low ceiling will make the sound from the PA system too loud for the people in the front and weaker for people in the rear. This may call for more speakers with a time delay. Some basic geometrical considerations will show that if the hall is long, the ceiling should not be too low. Also the direct-to-reverberant sound ratio will differ a lot depending on listener position.
- A low ceiling will bring the PA system closer to microphones whereby acoustic feedback will occur more easily.
- A too-low ceiling (below approximately 4–5 m) opens the possibility of standing wave problems between the two parallel surfaces of ceiling and floor, especially when not filled by the audience. A fully absorptive ceiling, also at very low frequencies, reduces this challenge.

Ceiling heights from 6 to10 m give good opportunities for good speaker coverage in halls with audience sizes of 500–1,800 people (the larger the hall, the higher the ceiling).

In order to avoid a very big difference in the distance to the closest and the farthest away audience members, placing the PA speakers high above the ground is crucial. Almost inevitably the audience in front will be exposed to too-loud levels if the speakers are placed at a lower height because for the sound to be loud enough in the back of the hall the level at the speakers needs to be considerable. Most hearing damage at concerts is caused by sudden extremely loud sounds very close to a speaker. In very long rooms a low ceiling can make the installation of delay speakers necessary but difficulties can arise regarding even coverage for different audience groups.

One way to house more audience members in a given floor area is by adding one or more balcony levels. Of course this demands a certain height of the auditorium and it is seen that halls can obtain 35 % more capacity by adding one balcony. Some halls with a ceiling height of only 6 m have successfully incorporated a balcony although they don't have a large overhang. The direct sound of the PA system must be able to reach even the people underneath balcony levels. Otherwise small delay speakers can be mounted here to compensate for the possible lack of direct sound.

However, acoustic considerations are not the only ones in play when planning the height of a hall. Lighting equipment also calls for a considerable height, and it may well be that the final decision is not taken from an acoustic perspective.

In clubs smaller than some 1,000–2,000 m^3 the recommendations in Fig. 5.6 are not eligible. The risk of a harsh sound, due to dominant room modes even at higher frequencies, can make a higher reverberation time at mid and higher frequencies unwanted, especially if there is a very low ceiling, unless abundances of diffusive objects on wall and ceiling surfaces are present. In smaller clubs, the instruments and amplifiers themselves, and not the PA system, are the important sound sources. A smaller degree of directivity of the sound becomes apparent. The Q factor of instruments is usually lower than that of PA speakers at similar frequencies.

Hall Shape

Much attention must be addressed to the shape of the hall. The acoustical consultant will instruct the architect and hall owner in some ground rules concerning this prior to the first sketches. As mentioned earlier the concept of critical distance means that audiences far away from the sound source will experience a higher degree of reverberant sound. The logical consequence of this is (also for visual reasons) to shape the hall so that as many people as possible are near the stage. This immediately disqualifies long and narrow halls with a stage at one end and opens the possibility of a sloped audience floor.

In cases where old premises with this elongated shape are going to be renovated as halls for amplified music it is a good idea to find out if the stage could be put halfway down one of the side walls. Perhaps part of a side wall can be torn down so that a stage can be fitted partly behind the line of the side wall. If this is the case, care must be taken so not more than some 35 % of the total depth of the stage is behind the wall line, remembering that the stage room must be part of the concert room and not act as acoustically coupled rooms with their separate acoustics. This also calls for a big stage opening, preferably in the full height of the hall, as well as carefully planned loudspeaker coverage on the front corners of the stage. If there is no other option than to place a stage at the end of a long narrow hall, the hall-end opposite the stage must be dampened completely at all frequencies also in order to free musicians from too loud, very late reflections. This dead end will actually be an appropriate location for a bar because the employees will not be exposed to overly loud sound, and customers can take a break here from the turmoil. The narrow parallel walls will lead to a high level of reverberation due to standing waves bouncing back and forth between them. This should probably be somewhat controlled with partial coverage with absorption or diffusive structures not just in the rear.

It is much better if the room shape in front of the stage is more quadratic. Because there are usually PA speakers on each side of the stage front, the audience area could also be somewhat rectangular and stretch out a little longer to the sides. This may lead audiences at each side to feel a little left out visually because the lead singer, who is responsible for direct communication with the audience, is often placed in the center of the stage. Therefore the close to square-shaped audience area is good, preferably with the stage stretching throughout the entire width. Bar areas are typically located as far away from the PA system as possible, if not in adjacent rooms, also in order for the employees to not suffer from too large doses of sound.

Many cinemas were built fan-shaped (distance between side walls increase with increasing distance from the stage) in the mid-twentieth century and such venues often find a new life as rock clubs. The fan-shaped hall has not survived for classical music because lateral reflections are difficult to maintain at a loud enough level and because the nonparallel side walls generate less reverberation. For pop and rock concerts they seem to work very well indeed as long as some envelopment is maintained in the hall despite the angled sidewalls.

Hall Shape

Of course a suitable speaker-to-audience distance is also obtained if the hall perimeter describes a circle with its center at center stage front. Such concave shapes are unsuitable acoustically because of the focusing effect unless treated with absorptive material or, maybe even better, diffusive shapes.

Also important are audience sightlines and the possibility of creating the desired visual effects with spotlights, and the like. This must all be incorporated in the first stage of the design phase. A sloped audience area can be incorporated beneficially as done, for instance, in the O13 in Holland. This will also lead to a closer to constant distance from loudspeaker to audience.

Not as crucial is symmetry. But a somewhat symmetric design around a center line that runs down through the middle of the hall is appropriate: that will make the task of an even loudspeaker coverage easier.

Stage and Its Surroundings

It has become common practice to place the subwoofer speakers underneath the stage front. A first concern is therefore to make sure that the stage platform is placed high enough above the floor to make room for the subs. The low frequencies radiate almost omnidirectionally from the cabinets, therefore the stage easily becomes a victim for bass sound and associated structural vibrations. The hollow room underneath the stage can become a giant, unwanted resonator box. It is of major importance to prevent too much of this sound energy to transmit from below the stage to the upper side of the stage. Therefore it would be a good idea to do the following.

- Make a strong sound barrier between the subwoofers and the hollow space behind them if a cardioid subwoofer configuration is not adapted. For instance, concrete 15–20 cm thick can be used, not light or porous concrete. Mass is a good insulator against even bass sound.
- Make the stage floor very heavy and stiff. In a new building a concrete slab can be made as the stage floor on top of which a wooden floor on joists can be placed. Otherwise, in smaller venues, heavy joists or I-beams must be placed very close (40 cm), and several layers of heavy material such as plywood, gypsum boards, and the like must be applied to reach a thickness of no less than 45 mm. This will also prevent, for instance, parts of the drum kit from moving around the floor due to vibrations caused by the musicians' movements on the stage.
- Fill the hollow volume underneath the stage platform with damping material such as mineral wool. There is no need to unwrap it from its plastic packaging. Approximately two-thirds of the total cavity volume must be filled predominantly at the inner circumference.

Hall owners' tastes differ regarding the preferred height of the stage. As mentioned, there must be room for the subwoofers, and in big halls with a large audience it is recommended to make it difficult for audiences to fall onto the stage because of pressure from the standing audience behind them. Usually there is a fence at a distance from the stage front keeping audiences away from foot pedals and other gear on stage. In small clubs of a couple of hundred people it looks awkward with a very high stage. Usually stages range from 70–120 cm in the type of halls that accommodate audience numbers from about 400–2,000.

As a rule of thumb, the acoustics of the stage should, as mentioned in Chap. 5, be similar to the acoustics in the rest of the hall. The stage area often takes up a considerable part of the hall because when designing the building, the building owner usually needs to take into account that acts with a considerable number of musicians, such as a big band, can also fit onto the stage. As mentioned earlier, the acoustic need, in the event of smaller bands playing, is the opposite: vertical surfaces closer to the players supplying them with early reflections. Mobile reflectors can be a possibility in the case of a big stage and a small band. Parallel surfaces, we know, may create flutter echoes so the sidewalls on the stage should either be slightly angled towards the front of the stage or be made diffusive or even both. Of course the sound engineer's concern is to get as little "leakage" as possible from the stage into the processed PA sound among the audience so the two side walls should not be angled more than 4–5° each. Both walls should be angled equally. The musicians would like to hear as much of themselves and their colleagues as possible from the supportive early reflections from walls. If the side walls are very far apart this is of no importance and in that case early reflections should be provided by other means. In order to create a space with a well, and more evenly distributed sound energy from all instruments, all walls can be made diffusive.

As stated, the acoustics of the stage area must not be deader than the acoustics of the hall area. A diffusive stage surrounding, apart from providing early, good-sounding reflections well distributed over the area for the musicians, also makes the sound that leaks into open microphones more manageable for the FOH engineer compared to more harsh specular reflections. As mentioned before, many sound engineers prefer the stage quite dead with a lot of absorption in order to achieve a high degree of control over leakage and to lower the risk of feedback among microphones and PA system and open monitors. Unfortunately this does not correspond to the musicians' wishes.

If, for instance, in the audience area there is a thick, porous absorptive false ceiling with a large cavity behind, it can also be so on stage. The stage area must certainly not be a small box detached from the audience area. That is the worst nightmare for musicians because it will allow for much deader acoustics on stage than in the actual hall. For musicians, it is a strange sensation to be playing their instrument on stage but to hear the music 20 m in front of them. They are simply detached from their own music, and won't have a chance of sensing the connection between what they're playing on their instruments and how their music "lands" in the audience. These two areas should not be separated in the ceiling either. If some structural support here is unavoidable, it should not take up more than 5–6 % of the height, for example, a 50–60-cm slab for a 10-m-high ceiling.

Usually there is an absorptive backdrop on the back wall. If there is no or only little mid–high-frequency absorption applied in the hall maybe this backdrop can be chosen to reflect sound partly. Some velour qualities include a reflective coating. Of course, only frequencies above some 500 Hz and below 2–4 kHz are efficiently reflected where the lower pass through and the higher are absorbed. But some sound is left for the benefit of the musicians. The potential wall behind the backdrop probably reflects lower frequencies, therefore the velour has finally been somewhat transparent for the sound below 2–4 kHz altogether. If possible the back wall can be made decoratively diffusive with no backdrop or a backdrop made of stage gauze which is almost nonabsorptive. Unfortunately it usually is difficult to get the hall owner convinced about this because of his or her expectations as to what a stage for pop and rock should look like. It is understandable because that wall is after all what the audiences will be looking towards during shows. A normal wool- or molton-type backdrop mimics quite well the absorption properties of an audience. In this way the stage surrounding with relatively few people and an absorptive backdrop in front of the back wall ends up having equal acoustical properties to those of the rest of the hall including an audience, given that ceiling and side walls are treated similarly in the two spaces.

There should not be a carpet on the floor. A carpet absorbs extremely nonlinearly, and only the high frequencies, on which musicians rely for much of their timing and dynamic expression. Such a dull sound is uninspiring and without envelopment. The floor should be left without a carpet also for cleaning purposes.

Surface Materials

The surface materials of any room and how they are mounted makes them responsible for a large portion of the acoustic response of the room. Any material will reflect, scatter, or absorb sound, or a mixture of them, at different frequencies. The absolute focus point, when building or restoring a hall for pop and rock music, is that a lot of low-frequency sound energy must be dissipated. In the design phase not a floor, not a wall, or the ceiling must be left without considering low-frequency absorption. And because the mid–high-frequency RT should preferably not be much lower than low-frequency RT porous absorption material such as mineral wool products, wooden fiber materials and the like must be employed in a way so that there is a very high absorption coefficient in the 125-Hz octave band and to some extent also the 63-Hz band. This of course calls for a thick layer of porous material at a distance from the reflective surface (such as a concrete ceiling or wall). Perforated gypsum board and other resonating absorbers are known to retain some high-frequency sound.

The best and most uniform sound is found in halls with good diffusivity. This means that absorption must be equally distributed around the hall surfaces although this often may not be architecturally possible. At concerts the primary sound source, at least if the concert takes place in an actual hall and not a smaller club, is the PA system. As mentioned, it is very important that not more loudspeakers are used than necessary, and that they are directed properly and, for instance, not pointing towards the rear wall but aimed at the audience.

The rear wall, opposite the stage, is crucial. It needs to be only 9 m away or more from, for instance, the snare drum on stage in order to be able to create a distinct echo. If the rear wall is very close to the stage, diffused reflections from this may be acceptable. But if the distance is much longer, even diffused reflections are largely regarded as unwanted. They arrive late seen from the musicians' perspective and can challenge their performance if too loud compared to earlier reflections and direct sound. On the other hand, a sense of envelopment needs to be present for audience, musicians, and at least some sound engineers. And envelopment stems from reflections.

The author has in some halls designed a "communication bridge" in the middle of the ceiling down through the entire length of the audience area in order to achieve "a push-effect," meaning an impression for the musicians of their own music's impact at different times shortly after it is played. But likewise, this hard reflective surface of a couple of meters of width also makes it easier for the musicians to hear what the public is communicating. The shape of the bridge can also be somewhat triangular with the tip at the rear of the hall.

The only way to turn down the level of the reverberation is by placing absorption (or diffusion) on surfaces. But, as we know, this also shortens RT. So for the musicians to experience a little bit of late response, then some of the rear wall, and other surface areas in the hall can be left without absorption material and made scattering or simply reflective instead. This is a way of picking the reflections one wants without creating echoes. This creates a room with overall good diffusivity.

The normal way about it, of course, is to find the appropriate reverberation time from Figs. 5.4 and 5.6 and calculate the amount of absorption needed in all octave bands from acoustical specifications about different materials' absorption coefficients at various frequencies. The amount of needed absorption that is determined in this way must be evenly distributed around the hall according, for example, to Sabine's equation, because the equation is only valid for perfectly diffuse rooms. Here computer-based acoustical prediction software, such as Odeon, comes in handy, because architectural reality only rarely allows for perfect distribution of absorptive materials. Usually the entire ceiling area is chosen to be covered by the same material. Furthermore, from an acoustical point of view, there are some considerations that lead to a different priority than perfect diffusivity, regarding where to place absorptive materials:

- The rear wall opposite the stage must not reflect too much sound. The part of that surface which extends above the audience should be more or less absorptive at all frequencies (unless the hall is very short in which case it can be made highly diffusive at all frequencies).
- If the wall areas close to the PA speakers are closer than a few meters away from the speakers then some absorption here is good inasmuch as sound propagates around the speaker cabinet and will create unwanted interference with

the reflected sound from the walls (see Fig. 3.7). The absorption placed here should function also at lower frequencies. Approximately 3–6 m² of broadband absorption is sufficient in each of these two areas. A more precise impression of the needs can be obtained by using computer software such as Ease for a given speaker system and placement in a given hall.
- Placement can be in the ceiling, evenly over both audience area and stage.
- Placement can also be in the upper part of the side walls. If the room is very high and a low RT is wanted, then more surface area than just the ceiling and rear wall is needed for absorption. Again it is advised to always give priority to LF absorption rather than MF–HF absorptive materials. LF absorption and MF–HF envelopment can be created on wall surfaces. Parallel side walls are not challenging in themselves if not very narrow. If the hall is quite wide no absorption or diffusion is necessary on side walls other than in the stage area. It's worth noticing that the reverberation will get stronger if there is no diffusion on these parallel side walls. But we should also remember that reverberation to some degree is wanted. On the other hand, a diffusive room always sounds better than a "box" even when the decay is relatively short such as in halls for pop and rock.

Balconies and Overhangs

Balconies can provide room for a significant number of audience members. If enough people appear at enough events the income from tickets and so on can pay the cost and more of building a balcony.

A balcony represents some additional surface area too. The RT in halls with balconies can be lower than that normally recommended for a given volume (see Fig. 5.6). If the balcony front is facing the PA speakers, it must be made either diffusive (e.g., convex) or absorptive because it can be difficult to avoid direct sound, primarily from the PA system, to be directed towards that surface. If there are no upholstered seats permanently present on the balcony level, it can be an advantage to place a retractable curtain above the balcony front, in the attempt to cut off the balcony volume acoustically, when not in use. The absorption coefficient of a curtain in front of a wall is very close to that of an audience. However, a freely hung curtain absorbs even less at mid frequencies. Sound-insulating curtains can then be applied, some of which have up to seven layers and an attenuation of 18 dB.

Floor

When designing the acoustics of a hall for pop and rock according to the guidelines of suitable reverberation times as a function of volume given in Chap. 5 it immediately becomes apparent that large surfaces must come into play and be treated appropriately in order to succeed in the 63- and 125-Hz octave bands. Depending on the height of the room and on what absorption coefficients are chosen and reached in the ceiling area, it is a safe choice to incorporate the floor to obtain some additional low-frequency control. It is a common misconception among non acousticians that "wood sounds good". Well, it depends on what is needed. Wood placed directly on concrete provides almost no absorption and doesn't offer much diffusion either. In this regard wood will maintain a long reverberation in the room. But if wood or any plate such as gypsum boards or plywood is mounted to joists with a cavity behind, partially filled with a damping material such as mineral wool, then the construction will dissipate low-frequency sound energy at a frequency interval that depends on the depth of the cavity, the weight of the plate, the distance between the joists, and so on. Not very high absorption coefficients are attained but because of the large surface area of the floor it becomes a significant factor. Some sought-after control of the 63-Hz band can be provided in this way especially.

Stage

The stage floor must be rigid and completely stable. Because the subwoofers are usually placed under the front edge of the stage, a great deal of low-frequency sound will propagate backwards and can cause unwanted low-frequency emissions through the stage bothering the musicians; in addition, the microphones on stage will pick up this sound and cause trouble for the sound engineer regarding both PA sound and monitors. Therefore stages should be designed to prevent this as much as possible. For instance, will a concrete or brick wall be good for shielding the cavity from the subs as mentioned earlier?

Seating

If anything, the absence of seats is what best describes a hall uniquely used for rock concerts. In multipurpose halls often mobile seating arrangements are incorporated. It is always a question as to whether the chairs should be upholstered. The acoustic advantage with upholstered seats is of course that the reverberation time of the hall can be quite similar regardless of how many audience members are present. During sound checks both the band and the sound engineer will be pleased with somewhat realistic acoustics compared to what they will experience at the show. It makes good sense to use upholstered seating with absorption properties that resemble those of a seated audience. But often halls are being dampened too much at higher frequencies, especially for classical music genres, and great care must be taken when choosing the chairs so that the amount of absorption of the chair does not exceed that of a person, also if a person is seated in it. If the floor is sloped, the seats are more exposed to the sound from the stage and PA system. This can lead to surprisingly high absorption values. The same goes for an audience standing on sloped steps.

Platforms

Podiums and other floor-level differences in the hall have been successfully implemented in many of the visited halls. It is nice for the audience to have choices, and fine for the musicians as well to have different "landscapes" of audience members to address. Small "islands" can also be constructed. One of the 75 visited halls was of special interest regarding interior geometry and layout. The O13 in Holland holds almost 2,000 audience members and still the length and width of the room is only around 20 m in front of the stage. Both the balcony and the floor level are terraced for excellent visibility for every audience member anywhere in the hall. Likewise, no audience member casts a shadow for the (direct high-pitched) sound for any other guest. If the slope is steep there can hardly be room for an actual back wall which otherwise calls for special attention when designing rooms for amplified music.

Sound Insulation

The very loud sound levels at pop and rock concerts mean that special attention is demanded when designing the sound insulation of a hall for that purpose. A certified acoustic consultant should take the responsibility for this challenge. It is important that the building owner take the necessary steps for this to happen. Insulating a hall after its completion is an extremely costly task if at all possible. The future development of the neighborhood may cause unexpected criteria a few years after the opening of a hall; therefore it is safer to insulate to high standards when first designing the hall. Discussing sound insulation is beyond the scope of this book. For the consultant it is practical to know that the sound pressure level at rock concerts sometimes reaches 105–110 dB (not A-weighted) at the FOH many 15 or 20 m away from the speakers in the 63- and 125-Hz octave bands (Fig. 3.2).

Interior Noise Sources

Even pop and rock music is sometimes not overly loud, for instance, during ballads or solo performances. The noise sources within the room such as ventilation, refrigerators, moving spotlights, and the like, sometimes, although seldom, cause disturbances. Jazz concerts are much more vulnerable to this. For rock halls, unlike in classical concert halls, such interior noise issues are not to be given a high degree of attention. But if practical and not overly expensive, the acoustic designer may to some extent take these issues into consideration. When designing multipurpose halls or performing arts centers, among others, surely interior noise sources are a relevant factor.

Multipurpose Halls

All over the world pop and rock concerts are being played in multipurpose halls. One night such a hall presents a rock concert, the next night a theatrical play, and the third night chamber music or a choir can be on the poster. As mentioned in Chap. 2, different musical styles call for different acoustics. Ideal reverberation times, (125–1,000 or 2,000 Hz) in empty halls range for a given volume: for instance, a 5,000-m^3 hall, from 1.0 s for a rock concert to over 1.4 s for chamber music to 1.8 s for a symphony orchestra (although remember the tolerances given in Fig. 5.6 for amplified music). Even longer reverberation can be aimed at for a choir.

For classical music, a somewhat longer reverberation time at low frequencies compared to mid frequencies is sometimes sought after, in order to provide warmth to the sound and enough strength for the instrumentalists and vocalists in the bass register for them to play or sing effortlessly. It is very questionable indeed whether this belief should apply to multipurpose halls because it heavily jeopardizes the acoustics for amplified music. It is a fact that the best symphonic music halls in the world, Grosser Musikvereinsaal, Concertgebouw, and Boston Symphony Hall do not show this trait of a longer RT at lower frequencies. So nothing indicates that, for instance, more envelopment in the bass range is needed. If the will is there to employ a large enough bass section, or even to amplify a smaller one very delicately electroacoustically, the need for room gain seems to vanish. Hence a frequency-independent RT should be employed also for the classical genres.

Mainly three different approaches are used in order to try to adapt the halls' acoustics for different musical settings:

- Physical variable acoustics. Variable acoustic elements can be activated alongside the walls or in the ceiling in the form of acoustic banners, retractable curtains, or movable plates with absorptive material behind. Also movable diffusers have successfully been implemented. A new technology uses the membrane absorption principle for variable absorption.
- Artificial electronic reverberation systems. These will add artificial reverberation through a very large number of loudspeakers to an acoustically somewhat dead hall so that it functions better for classical music.
- Enlarging the volume of the hall to generate longer reverberation. This is rarely encountered because it is an expensive solution. In some halls walls can open to give access for the sound to big reverberation chambers. Also in some halls the ceiling can be raised or lowered.

Usually electronic acoustics are not favored by (classical) musicians who find that the way their acoustic instrument resonates should not be treated electronically. Needless to say, the thousands of reflections from every little surface that acoustics in more lively halls are a result of cannot be substituted by any large number of loudspeakers emitting artificially processed sound. Therefore this may be an idea for halls that only very rarely have classical concerts and it will surely function better than a completely dead room.

Physical variable acoustics seem to be a more logical and straightforward solution because this method copes with the challenge where it starts: on the reflecting surfaces.

Physical variable absorption often suffers from an important flaw: the variability mainly lies at frequencies above some 300 Hz which are proven to be the not-so-important frequencies in regard to pop and rock music. As a matter of fact, the absorption curve for traditional acoustic banners is often almost identical to the absorption curve of an audience. In some cases a relatively high absorption at low frequencies can be achieved with curtains and banners, but this seems to some extent to depend on the diffusivity of the hall among other things. Still, porous absorption will usually absorb more high-frequency sound that then will be "absorbed twice" due to audience absorption. Hence the bass reflections will still mask higher frequency sound. Furthermore, it is very likely that an over dampening of high frequencies makes it difficult for the musicians to express, and for the audience to experience, the dynamics of amplified music. This is true, for instance, for acoustic guitar and a complete symphonic orchestra. High-frequency variable absorption can be used to make up for audience absorption (Figs. 1.16 and 5.6) in cases where seats with absorption properties similar to those of an audience are not installed.

Another flaw is that some devices have a quite high absorption coefficient in one or more bands even when in the off position. It is a major concern that the entire frequency span of the fundamental frequencies of musical instruments (63–1,000 or 2,000 Hz) is altered more or less linearly because it makes no sense that certain registers on some instruments receive a very different acoustic response from the hall than others. Another approach for the acoustical engineer in the design of a multipurpose hall would be to apply the tolerances for amplified music given in Fig. 5.6. This calls for variable absorption that works best in the 125-Hz octave band.

A new technology by the author, the AqFlex system, has solved both of these challenges. It absorbs almost linearly in the frequency span from 63–1,000 Hz and can be implemented in the ceiling with an absorption coefficient of up to approximately 0.5 which permits a lowering of reverberation time of some 40 %, depending on hall height, at the push of a button. It does not absorb mid frequencies more than low frequencies, therefore the technology eliminates the issue of low reverberant sound energy masking higher frequencies and also avoids the problem of inadequate bass response from the room at classical music concerts. Apart from ensuring low-end absorption the invention also helps the high frequencies not to be over dampened by the audience and air because its absorption coefficient rolls off above 1 kHz. A sense of sought-after HF envelopment will be apparent in the hall at any event. Absorption coefficient is close to zero in all bands in the off position.

Music Schools

Most music schools have a larger room of approximately 1,000 m^3 that is used both for ensemble rehearsals and for concerts. Usually all kinds of music are played in this type of room. In order to alter the acoustics to fit a certain genre it is important to lower all the frequencies of the fundamental frequencies of the instruments in play equally. All musical instruments have fundamentals somewhere in the 125–500 octave bands and these are the absolute most important bands to vary. What distinguishes the best from the worst halls for amplified music is a low RT most predominantly in the 125-Hz octave band (see Fig. 5.6). But this is also the loudest register in, for instance, the male voice, and with a choir or other classical genres the hall must carry this frequency band out as much as higher frequencies. So this calls for a long RT in that band. It is actually crucial that the absorption coefficient be as close to constant as possible in frequency bands from 63 to at least 1 kHz in order to alter acoustics depending on genre efficiently. The above-mentioned AqFlex system complies with these demands.

Also of importance is to be able to adjust the hall to whether there is an audience. As we know this is easily done with retractable curtains in front of walls. Curtains on all four walls are a good and easy way of making sufficient variability to make up for a packed audience in such halls. They should be drawn at rehearsals and removed for concerts. Because acoustic adaption to genre should be linear and adaption to whether there is an audience should comply with the absorption curve of an audience, preferably two different variable acoustic systems should be used as described.

Chapter 7
Gallery of Halls that Present Pop and Rock Music Concerts

Ancienne Belgique (AB)

Number of concerts per year in this hall: 120. In all halls in the venue: 320
Founded: 1979
Capacity: 2,000
Architect: Werner E. De Bondt
Acoustician: NA

The Ancienne Belgique (AB) is a concert venue, in the very center of Brussels. As a leading musical venue, AB has already been programming international names, national chart toppers and emerging new talent for more than 30 years. An exceptional amount of attention to local talent and the Dutch language prevails in daily operations. More than 500 bands perform on the stages of the AB in the course of 320 concert days per year.

AB is the first pop and rock temple to receive recognition as a "Grote Vlaamse Cultuurinstelling". This means AB is now one of the key cultural organizations in Flanders, a position it shares with deSingel, the Flemish Opera, and the Museum for Contemporary Art Antwerp (MuHKA). AB received this recognition because—according to the Flemish Minister of youth, sports, culture, and Brussels—it carries out a local and international pilot function (AB stimulates and promotes "more Flanders in the world and more of the world in Flanders") and because their program reaches high standards of international quality.

The present-day Ancienne Belgique is located in a historic spot in the heart of Brussels. It used to be the house of the merchants—overseas traders—of whom the first traces go back to the eleventh century. Three centuries later the complex had evolved into a real center with a sociocultural function. The only visible evidence of that time is the inscription on the façade: "Meersliedenambacht 1781". The Belle Epoque brought along new glory: from 1906 to 1913 the "Vieux Dusseldorf" was very popular with its German style interior, 1,500 seats, and many enthusiastic couples on the dance floor. On December 21, 1913 a renovation (the first one in a row!)

was started. This resulted in "Bruxelles-Kermesse", also a kind of brasserie but now with a variety of artists and theater elements. In the 1920s the basement was turned into the "Caveau Flamand" where young literary talent found a way to the public.

In 1931 the entire building was bought by Mathonet, a man from Liège. This was the beginning of the Ancienne Belgique era. Liège, Ghent, and Antwerp have similar venues but the Brussels one, our AB, will survive them all thanks to the life work of Mathonet's son, Georges. He turned the place into one of the leading European music halls. Driven by its success the building was bound to be torn down. Well, the time had come to build a bigger venue, a building that nowadays is known as the Ancienne Belgique.

In the Second World War Mathonet was more vigorous than ever: he honorably participated in the Resistance. The liberation brought a true explosion in the world of amusement and Mathonet's AB was somehow the center of all of this. On the scene we had Annie Cordy, Charles Trenet, Gilbert Bécaud, Aznavour, Brassens, Piaf, and Adamo. Jacques Brel called it an excellent school. In 1955 Bruno Coquatrix turned an old Paris cinema into the most famous music hall of the world: the Olympia. He and Georges Mathonet became partners. Ten grand years followed.

In the 1960s, Johnny Halliday, Jacques Dutronc, France Gall, and Claude Francois appeared there. After the fire at the Innovation (1967) Mathonet was forced to secure the building with concrete. This huge investment turned out to be fatal. The attempt to turn the Ancienne Belgique into a Paris Lido for the Eurocrats failed. A call for subsidies went unheard. In 1971 the Ancienne Belgique filed for bankruptcy. Georges Mathonet died shortly after and the building fell into ruin. In 1977 it was bought by the Ministry of Finance. Finally there was good news: together with the "Botanique" the Ancienne Belgique was presented to the Flemish and Walloon cultural departments.

The Flemish prefer the AB for its central location, its popular background, and its wide range of entertainment possibilities. You could compare it to the Mallemunt spirit: providing the Dutch-speaking community with a friendly meeting place right in the center of the capital, a creative spot, a place to be for the young. The original name was kept: Ancienne Belgique, but the abbreviation AB became more and more common. Secretary of State Rika Steyaert opened the AB in 1979 with the famous words, "This is a house of hope." The qualities of the venue were tested and some defects came up: the building was crumbling and not soundproof at all. All this and too many young people in that quiet and sleeping street made the City Council close down the AB in 1981. A major renovation was inevitable and started in 1982. Architect Werner E. De Bondt renovated the Bar Américain and then started building a new main hall. He opted for a robust and spectacular interior design in a high-tech dimension. Circles on two different levels guaranteed intimacy and the overall atmosphere of the hall was set in a warm, red color. The opening on December 23, 1984 was a memorable event in the history of the AB. However, the noise continued to be a problem and has had a fatal effect on the bill. The youth-oriented approach of the AB was mocked by policy-makers. Dark years followed, until a new dynamism emerged in the house. Partying through the night made way for a strictly respected closing-time.

Ancienne Belgique (AB)

Despite the power of the crew, with Jari Demeulemeester as artistic director (director-general from 1988), fear reigned at the AB: police interventions, fines, and threats from the biggest law firms constantly reminded them of the danger of yet another shutdown. In 1986 Secretary of State Patrick Dewael ordered a thorough investigation concerning the noise. The report was disastrous but hopeful at the same time and led to a new architectural plan and a new architectural team. In 1991 Secretary of State Hugo Weckx agreed to finance a huge contemporary center for popular quality culture. His successor, Luc Martens, safeguarded the demanding but ambitious project, "The music house Ancienne Belgique, a project of the Flemish community in the capital," as it was called at the opening on December 6th, 1996. The premises were enormously vast and technologically world-class. The main entrance was no longer in the rue des Pierres. There you now have the café and the ticket shop. Equipment can be loaded and unloaded at the loading bay or via the Square Lollepot at the back of the building. Upstairs there is a second, smaller hall: the club. Another advantage: the AB has its own recording studio. From this studio we can go around the world, live, via satellite or through the Internet. Over the years, through all the renovations, the AB profile has always been the same with the goal of presenting contemporary music, made by people of today and about the world in which they live. They want positive press advertising the newest repertoire and the newest act to fans and music lovers; in short: presenting interesting artists to a broad public of people living and thinking today. The belief in a passionate relationship between the artist and the public persists. Recent acts including Alice Cooper, Bon Iver, Brian Wilson, Queens of the Stone Age, Public Enemy, Vampire Weekend, MGMT, Joe Jackson, and Randy Newman have visited AB.

Humour, attitude, and personality. Belgian metal sensation Channel Zero enjoys a sold-out AB (picture: Gino Van Lancker).

"AB is the best venue in Europe... no, in the world!" Mike Patton, Faith No More. There is an option of installing an extra 400 chairs for a total of 700 seats. Many diffusive ducts increase the absorption also supplied by the audience.

Ancienne Belgique (AB)

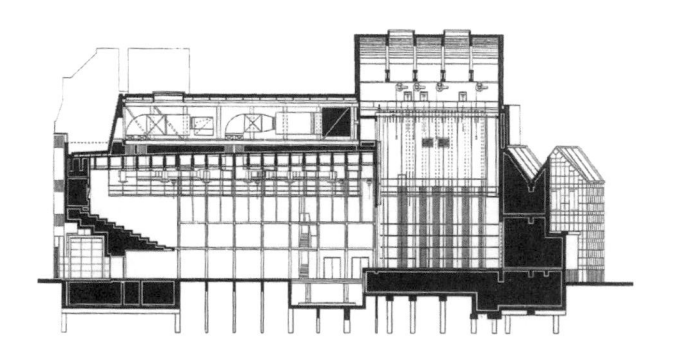

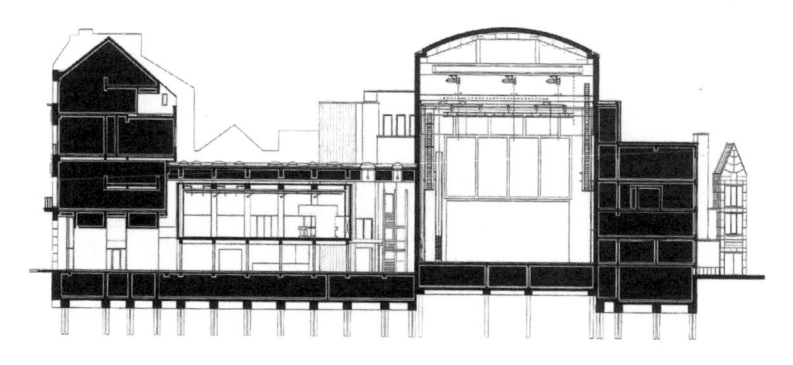

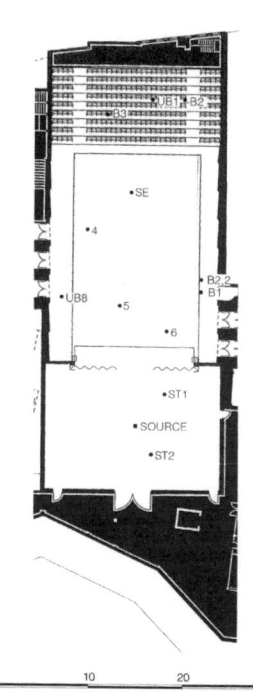

Geometrical data	
Volume	9,500 m^3
Height, audience area	11.7 m
Surface area of stage	270 m^2
$L \times W \times H$	$45.3 \times 18.5 \times 11.7$

Acoustical data	
Audience area	
$T_{30, 125-2k}$	1.41
EDT_{125-2k}	1.41
$C_{80, 125-2k}$	1.49
BR_{63} versus 0.5–1k	1.73
BR_{125} versus 0.5–1k	1.23
Stage area	
EDT_{125-2k}	1.12
$D_{50, 125-2k}$	0.72
BR_{63} versus 0.5–1k	0.87
BR_{125} versus 0.5–1k	1.02

Materials Used

Audience Area

Floor: Concrete.
Ceiling: Concrete/gypsum board on cavity. Many diffusive ventilation ducts, and so on. Porous absorptive plates are placed under balconies.
Walls: Wooden panels on gypsum boards on cavity. Some wooden boards are slit absorbing panels some are plates. 2 cm of cavity behind wooden panels on floor and 1st balcony level. 8 cm of cavity on 2nd balcony.

Stage Area

Floor: Vinyl on wood direct on concrete.
Ceiling: Concrete/gypsum board on cavity.
Walls: Wood wool panels.

State of Hall When Measured

Empty: no additional seats mounted.

T30 in audience area

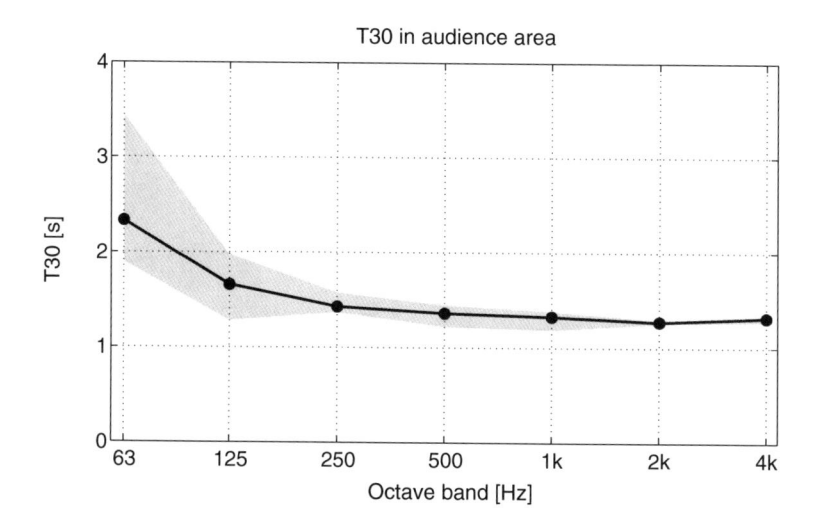

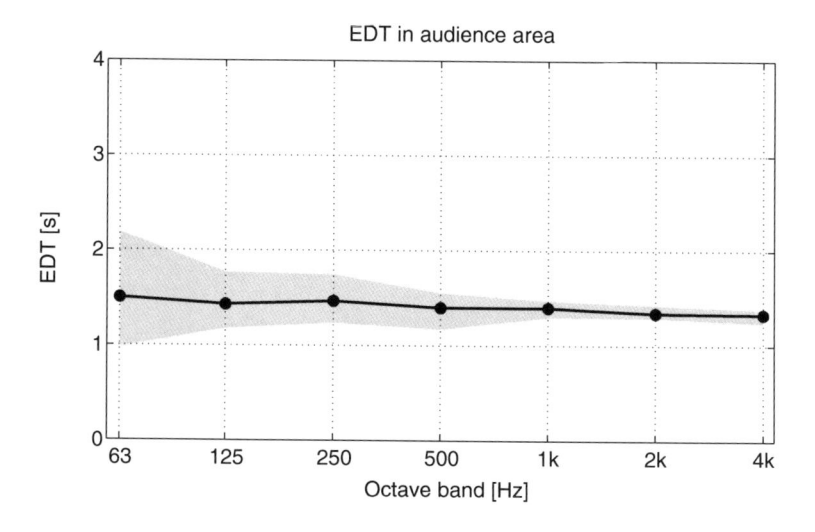

EDT in audience area

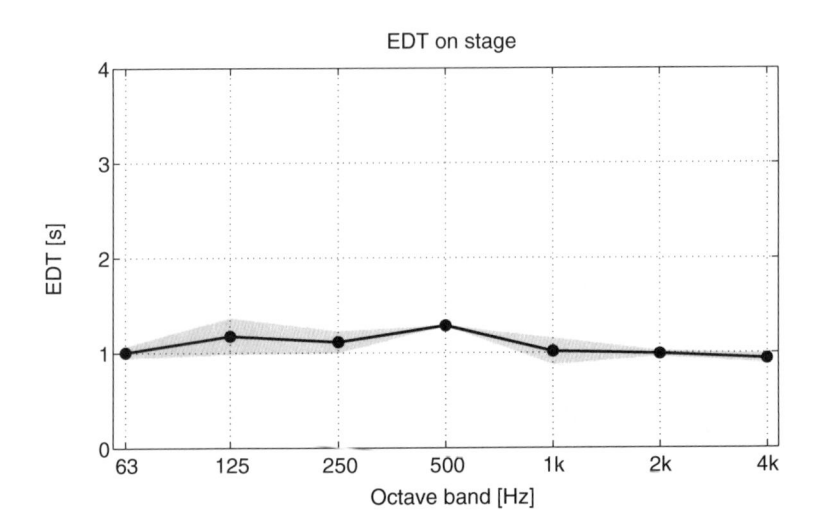

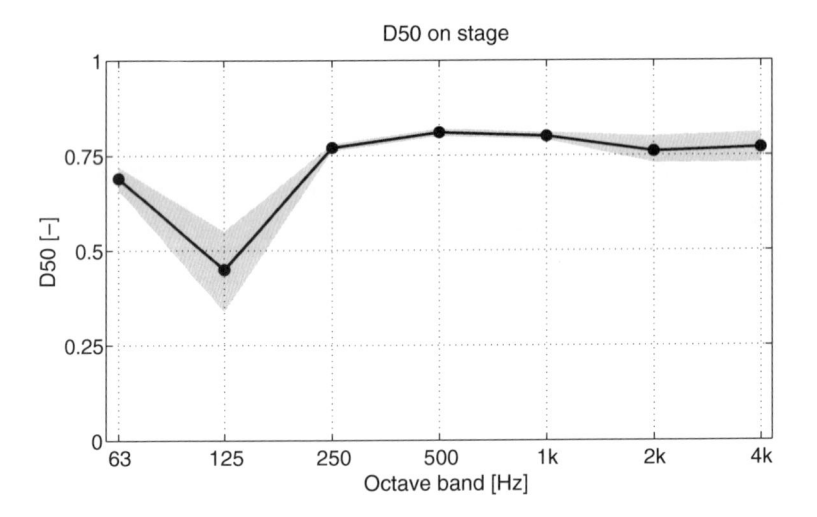

7 Gallery of Halls that Present Pop and Rock Music Concerts

L'Aeronef

Lille
Number of concerts per year in this hall: NA
Founded: 1989
Capacity: 2,000
Architect: NA
Acoustician: NA

L'Aéronef was started in 1989 by Jean-Pascal Reux and Alain Bashung. In 1995 it was moved to new premises in the building d'Euralille which was designed by architect Jean Nouvel. It holds some 2,000 people.

View from stage into the hall mainly made out of concrete covered by curtains.

There is room for an additional 500 people in the balcony areas for a total of approximately 2,000.

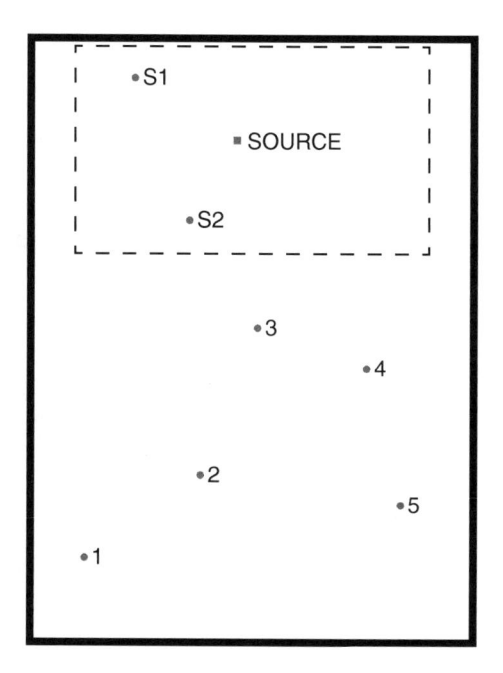

L'Aeronef

Geometrical data	
Volume	8,500 m^3
$L \times W \times H$	32 × 24 × 12

Acoustical data	
Audience area	
$T_{30, 125-2k}$	1.39
EDT_{125-2k}	1.36
$C_{80, 125-2k}$	2.20
BR_{63} versus 0.5–1k	2.54
BR_{125} versus 0.5–1k	1.78
Stage area	
EDT_{125-2k}	0.96
$D_{50, 125-2k}$	0.73
BR_{63} versus 0.5–1k	1.87
BR_{125} versus 0.5–1k	1.65

Materials Used

Audience Area

Floor: Concrete.
Ceiling: Concrete. 430-m^2 convex reflectors of wood fiber plates suspended from ceiling with 5–20 cm of porous absorption on top.
Walls: Concrete with large areas of 5-cm thick porous absorptive panels well distributed in room including back wall. Some of these are behind perforated metal plates. Wool curtain 1 m from back wall.

Stage Area

Floor: Concrete.
Ceiling: Concrete. Lowered somewhat perforated technician's grill of aluminium with 4-cm mineral wool slabs on top.
Walls: 5-cm thick porous absorptive panels hidden behind perforated metal plates. Woolen curtain ½ m from back wall.

State of Hall When Measured

Empty. Two large doors were open. Balcony opening was covered with Molton. Unupholstered seating risers were packed away at the rear of the hall for standing audience performances.

State of Hall When Measured

Stowed away seats against the back wall are an effective broadband absorber if upholstered

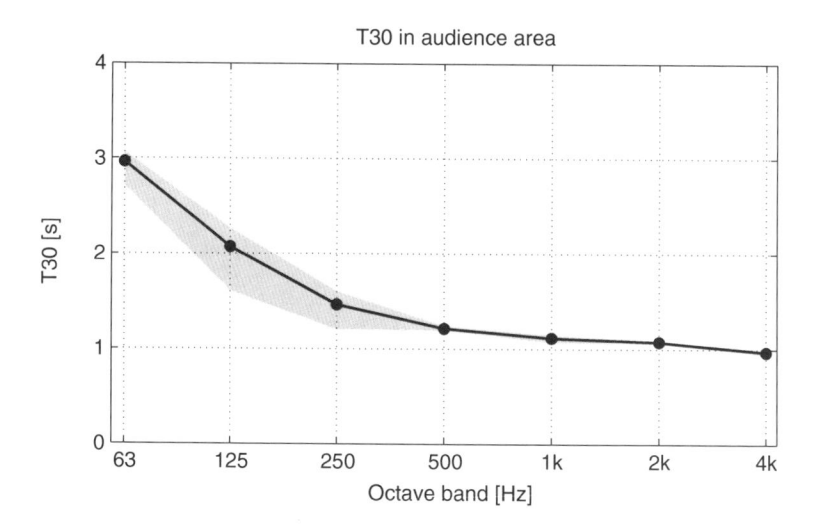

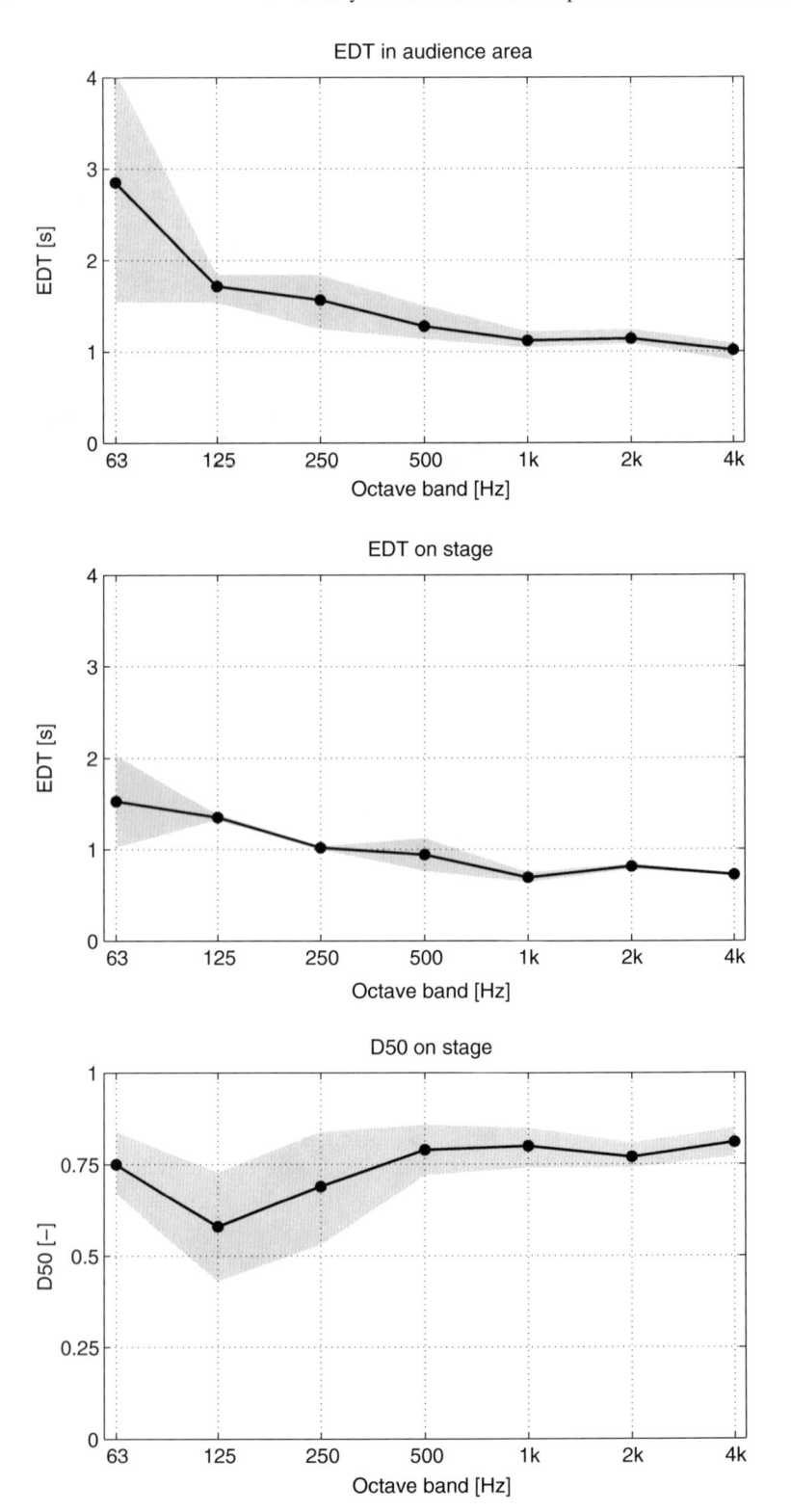

Alcatraz

Milan
Number of concerts per year: 160
Total number of events: 250
Founded: 1997
Capacity: Over 3,000
Architect: Daniele Beretta
Acoustician: N/A
Owner: Roberto Citterio

The building in which it is located, which goes back to 1946, first housed a garage and then a forwarding agency. In 1997 and 1998, the facility was completely renovated, with the precise and ambitious aim of creating an area capable of meeting all the needs linked to events, shows, and musical performances. The project originated with music industry professionals with a decade of experience and unforgettable productions, who felt the need to create a self-sufficient and functional organization. Alcatraz has always distinguished itself by its versatility.

With an overall area of 3,000 m^2, Alcatraz is a multipurpose space that, because of its well-organized set-up and flexible structure, offers endless creative possibilities not only for important events, but also smaller and more intimate events. Within just a short time, it became a location for fashion shows and conferences, private parties, a television program, and a venue hosting performances by the most extravagant artists. Alcatraz is located close to public transport, has private parking, and is located near the Isola district. It is a symbol of the freedom to create when ideas are not subjected to any constraints.

Alcatraz is not just a versatile and accommodating space; it is also made by human capital capable of providing high-level technical, logistical, and organizational support and services at each stage of an event without leaving out a single detail. The facilities are spread out over an area of 3,000 m^2, and are divided into different spaces that can be modified and custom made to the event. The viewing facilities, the technological systems, and the huge air conditioning tubes become an integral part of the fascinating architecture. Because it has no supporting columns, the 1,800 m^2 parterre has no architectural obstructions and lends itself to countless possibilities for creativeness, thanks also to the possibility of dividing spaces through a system of curtains that run on tracks.

Smaller platforms are placed on the sides of the large audience area as well as a balcony level in the back.

Mid- to high-frequency T30 is under control mainly because of a quite thin layer of porous absorption in the ceiling. The porous concrete walls helps dissipate sound energy in the entire frequency span whereas the lower tones probably are never reflected from what seems to be a thin roof construction. This would cause noise problems in inhabited areas.

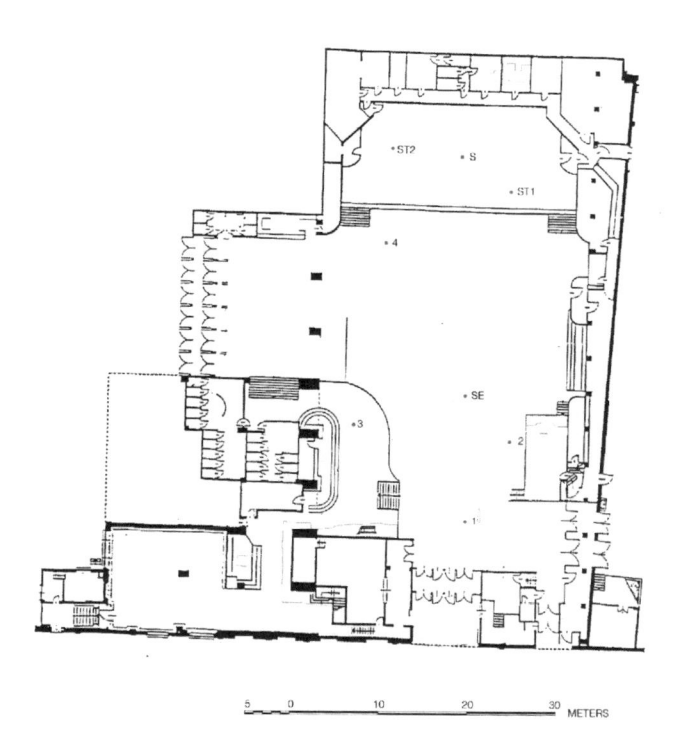

Alcatraz

Geometrical data	
Volume	15,000 m^3
$L \times W \times H$	$52 \times 30 \times 7.8$–13.2 m
Surface area of stage	

Acoustical data	
Audience area	
$T_{30, 125-2k}$	1.47
EDT_{125-2k}	1.35
$C_{80, 125-2k}$	1.30
BR_{63} versus 0.5–1k	0.96
BR_{125} versus 0.5–1k	0.89
Stage area	
EDT_{125-2k}	0.84
$D_{50, 125-2k}$	0.66
BR_{63} versus 0.5–1k	0.82
BR_{125} versus 0.5–1k	0.49

Materials Used

Audience Area

Floor: Concrete.
Ceiling: Panels of porous absorptive material behind perforated metal plates.
Walls: Porous concrete.

Stage Area

Floor: Wood direct on concrete.
Ceiling: Panels of porous absorptive material behind perforated metal plates.
Walls: Porous concrete.

State When Measured

Empty, no curtains.

View from balcony towards the stage.

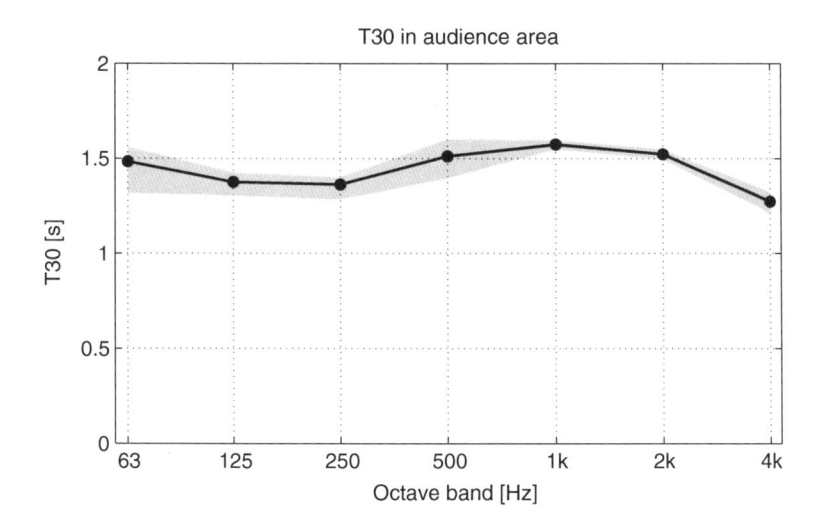

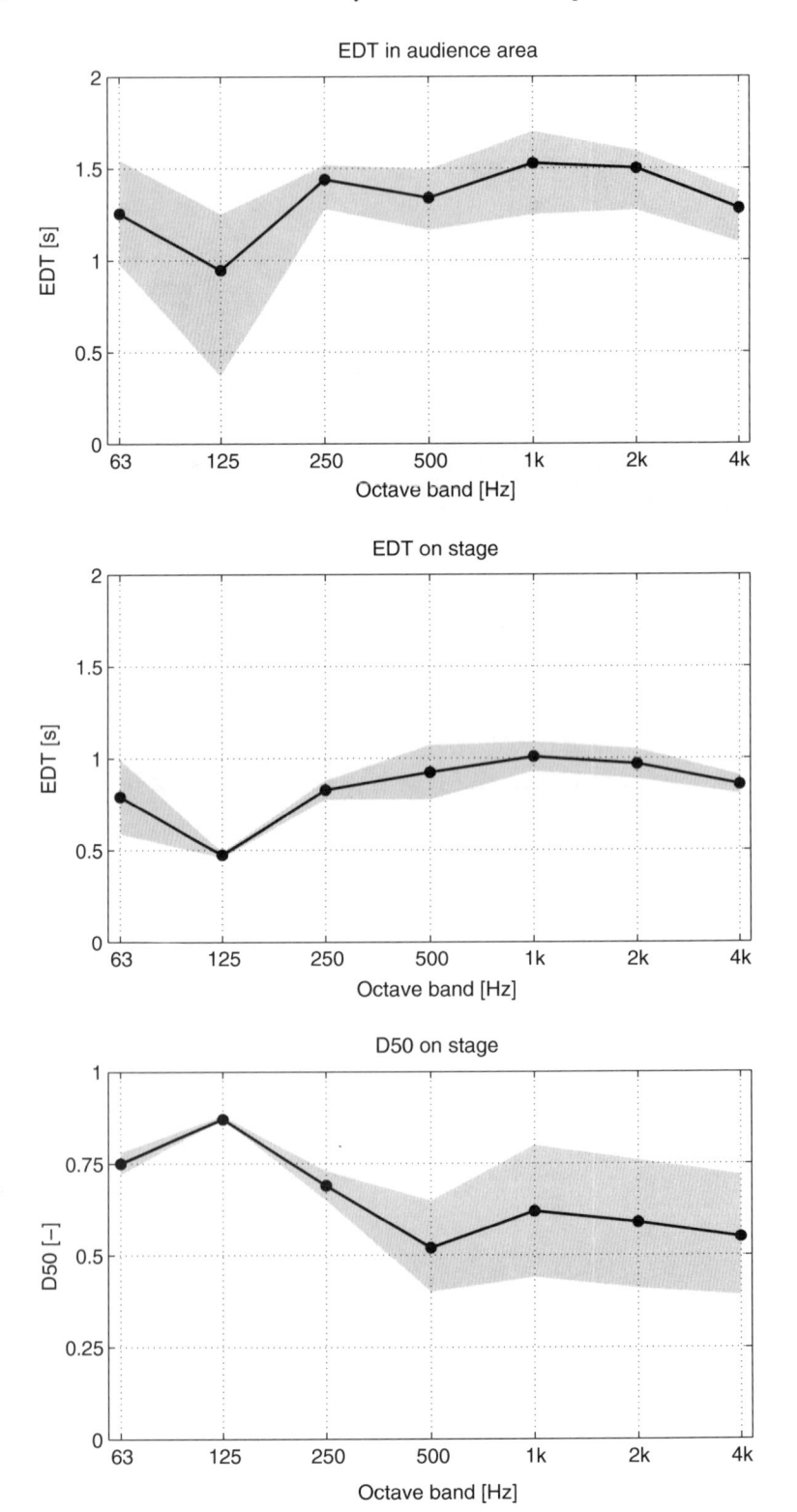

162 7 Gallery of Halls that Present Pop and Rock Music Concerts

Apolo La [2]

Barcelona
Number of concerts per year: 300
Founded: 2006
Capacity: app. 400
Architect: Daniela Hartmann
Acoustician: Jordi Martín

In 2006 a second venue, La [2] de Apolo with a capacity of 400 people was opened. This space is for medium-sized concerts and it is probably the venue that works the most in Spain, with more than 300 concerts per year. Not only has the big venue received great artists, but also Clem Snide, Hayseed Dixie, The Zombies, Wovenhand, Damo Suzuki, Herman Düne, Dick Dale, and Lisa Germano among others.

Both venues have become a referent for the Spanish music community, and the dream is to keep it for years.

The modern look of La [2] is enhanced with the metal lattice panels by the walls. Such structures may, if not carefully designed, like ventilation shafts prove to rattle noisily due to high bass levels during concerts.

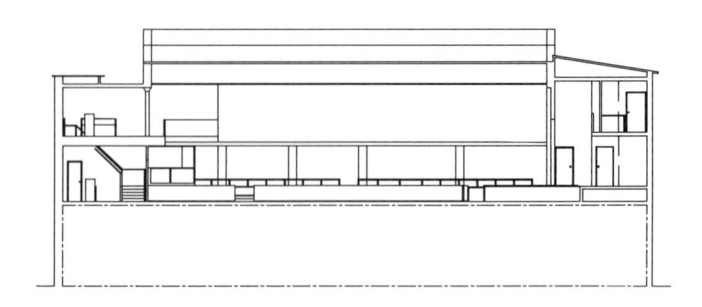

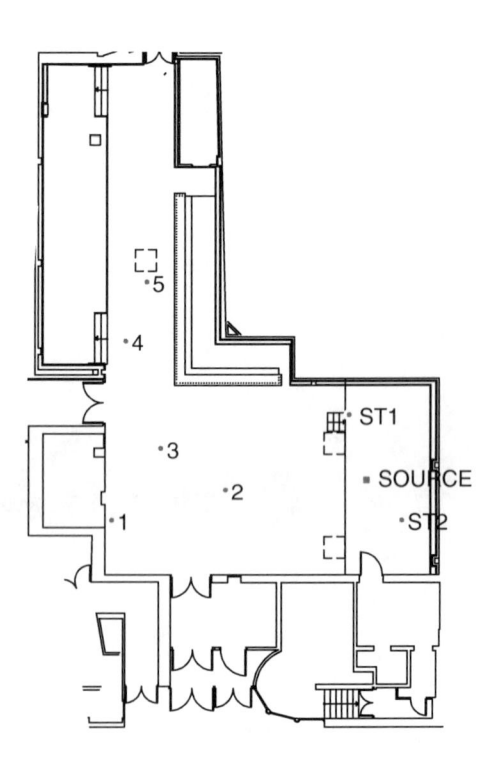

164   7 Gallery of Halls that Present Pop and Rock Music Concerts

METERS

Geometrical data	
Volume	1,000 m^3
$L \times W \times H$	$15 \times 16 \times 4.0$ m

Acoustical data	
Audience area	
$T_{30, 125-2k}$	0.75
EDT_{125-2k}	0.64
$C_{80, 125-2k}$	6.90
BR_{63} versus 0.5–k	1.09
BR_{125} versus 0.5–1k	0.92
Stage area	
EDT_{125-2k}	0.22
$D_{50, 125-2k}$	0.94
BR_{63} versus 0.5–1k	1.75
BR_{125} vs. 0.5–1k	1.02

Materials Used

Audience Area

Floor: Concrete.
Ceiling: Mineral wool product on cavity.
Walls: Several layers of gypsum board on cavity. Decorative metal lattice.

Stage Area

Floor: Wood direct on concrete.
Ceiling: Mineral wool product on cavity.
Walls: Several layers of gypsum board on cavity, curtains.

State of Hall When Measured

Empty.

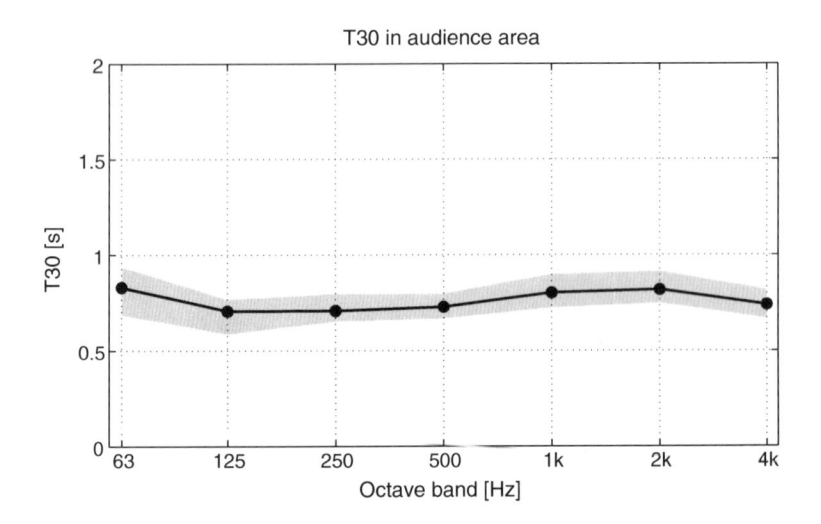

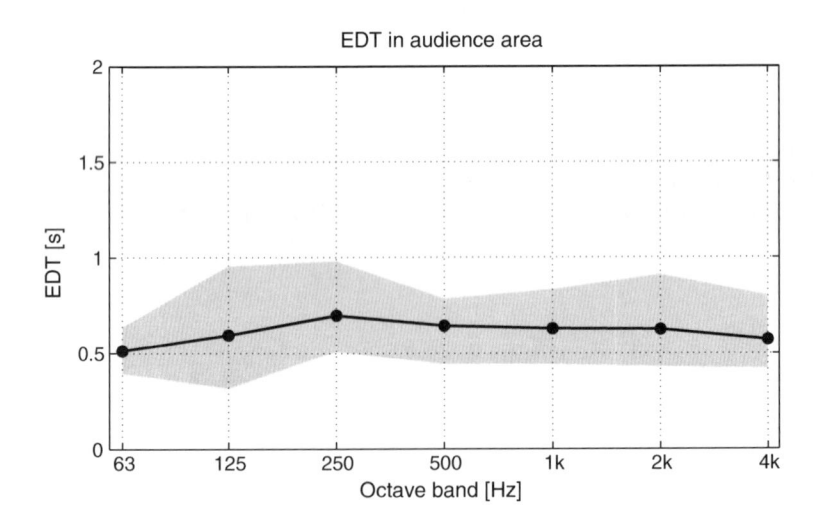

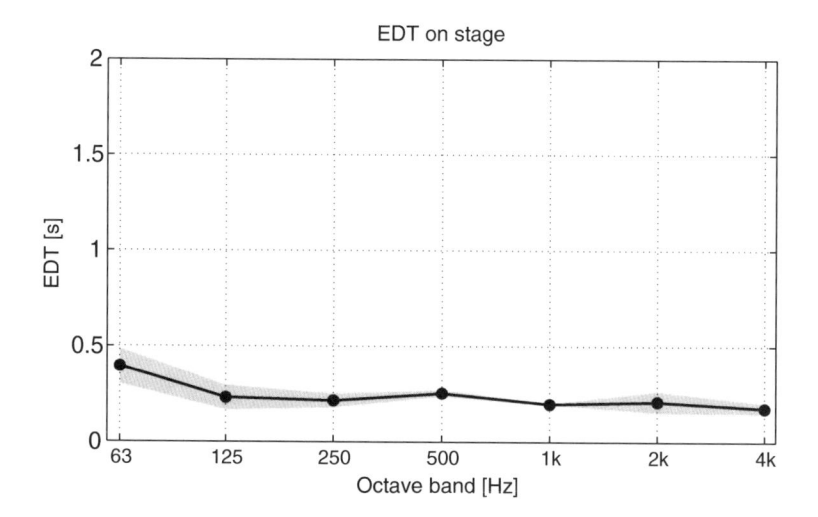

State of Hall When Measured

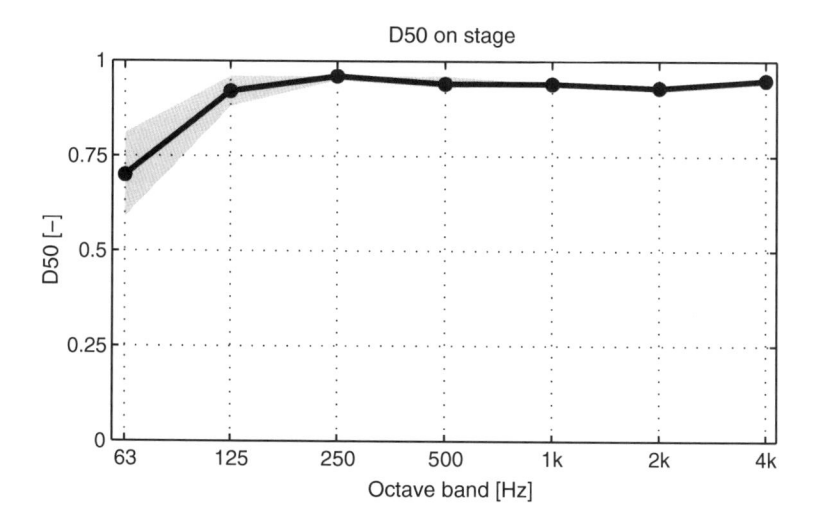

Apolo

Barcelona
Number of concerts per year: 270
Founded: 1936, reopened: 1989
Capacity: 1,200
Architect: Unknown
Acoustician: Jordi Martín

Located in Barcelona's downtown, Apolo opened its doors in 1936 as a ballroom with orchestra. It wasn't until 1989 that the venue became a rock and pop concert hall due to the increasing cultural demand of the city. The inside of the building is still made of wood, giving a beautiful old look to the space according to its age.

The venue has capacity for 1,200 standing people or 700 with seats, and technically, is one of the most well equipped in Spain. This is proved with the approximately 270 concerts that take place here each year. In addition, the infrastructure of the building and the technical materials are constantly being renewed to provide both musicians and promoters with the best conditions for the shows.

Here is a short list with some of the most important artists that have played at Sala Apolo: Smashing Pumpkins, Coldplay, Joe Satriani, Built to Spill, Yo la tengo, Dinosaur Jr, Goldfrapp, Beth Gibbons, Sepultura, Coco Rosie, Edwin Collins, Band of Horses, Tindersticks, Jonathan Richman, Gutter Twins, Solomon Burke, Robben Ford & Bill Evans, Low, and Devendra Banhart.

One of the biggest achievements of this venue is to support national bands that over the years have become important in the scene. Some of them are El Guincho, La Mala Rodríguez, Standstill, Mishima, or Lori Meyers. The aim is not only to produce concerts, but also to create a new and young local scene to improve the cultural life of the city.

Sala Apolo has also collaborated with important festivals such as Primavera Sound since their beginnings.

The main hall at Apolo is kept somewhat in the original style from 1936.

The balcony areas are popular not only at sold-out concerts.

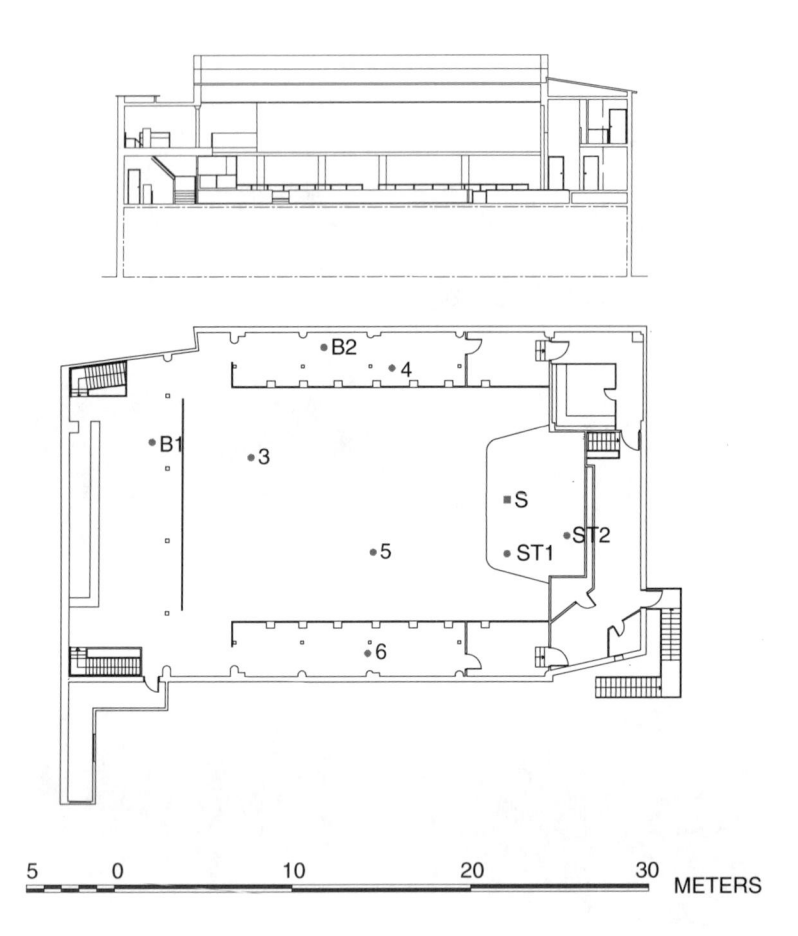

7 Gallery of Halls that Present Pop and Rock Music Concerts

Geometrical data	
Volume	2,800 m³
$L \times W \times H$	24 × 19 × 6.3 m
Surface area of stage	
Height of stage	1.15 m

Acoustical data	
Audience area	
$T_{30, 125-2k}$	0.98
EDT_{125-2k}	0.97
$C_{80, 125-2k}$	3.61
BR_{63} versus 0.5–1k	0.92
BR_{125} versus 0.5–1k	0.95
Stage area	
EDT_{125-2k}	0.43
$D_{50, 125-2k}$	0.86
BR_{63} versus 0.5–1k	1.83
BR_{125} versus 0.5–1k	1.49

Materials Used

Audience Area

Floor: Wood direct on concrete; platforms are wood on cavity.
Ceiling: Lowered (25 cm) porous absorptive ceiling.
Walls: Plates on cavity, one concrete wall.

Stage Area

Floor: Wood on concrete.
Ceiling: Lowered (25 cm) porous absorptive ceiling.
Walls: Plates on cavity, curtains.

State of Hall When Measured

Empty.

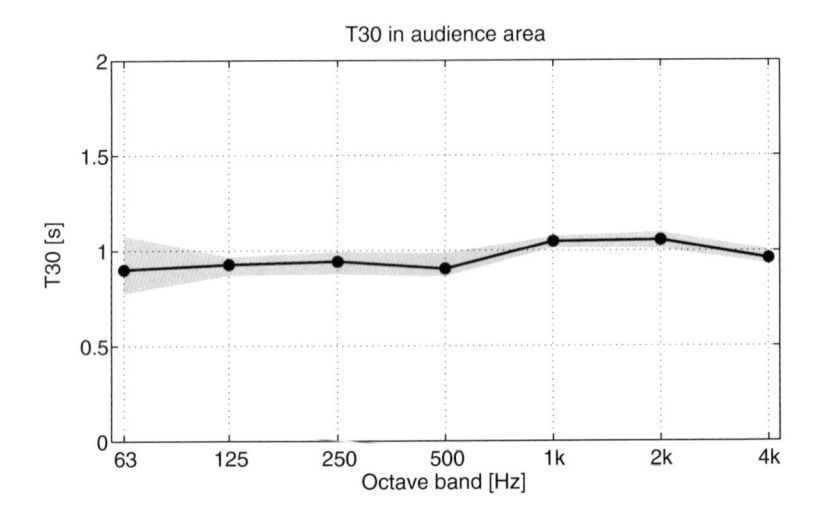

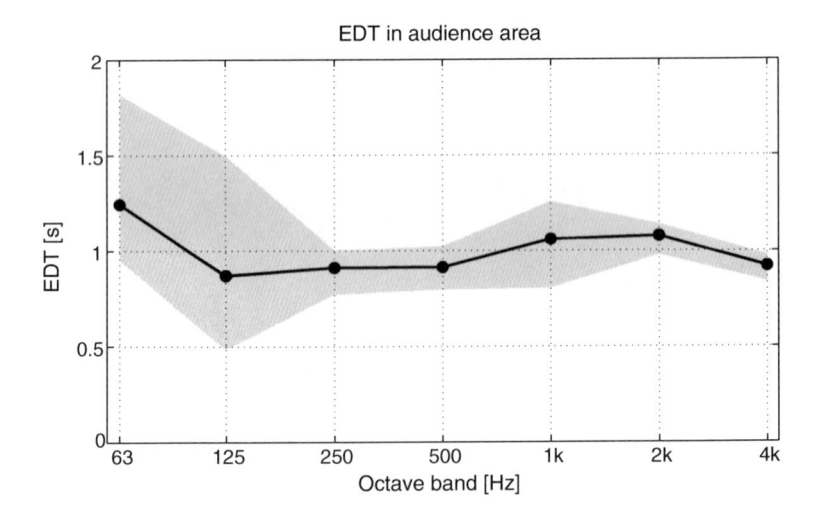

172   7   Gallery of Halls that Present Pop and Rock Music Concerts

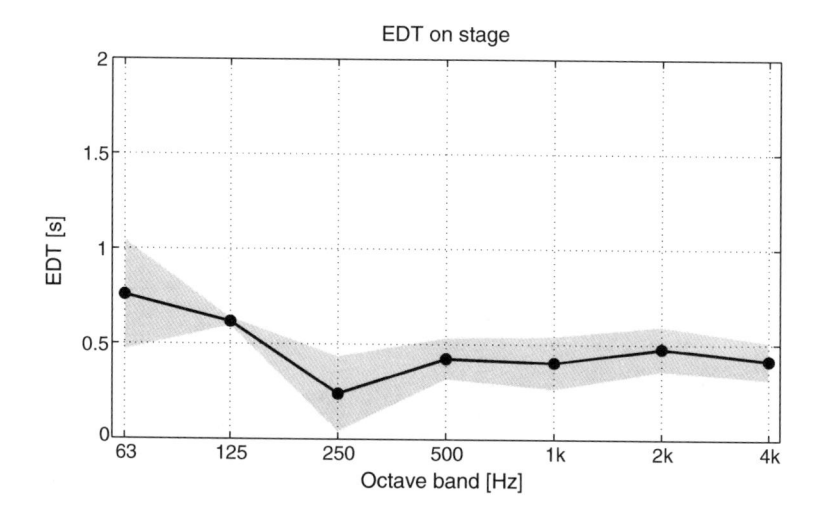

State of Hall When Measured

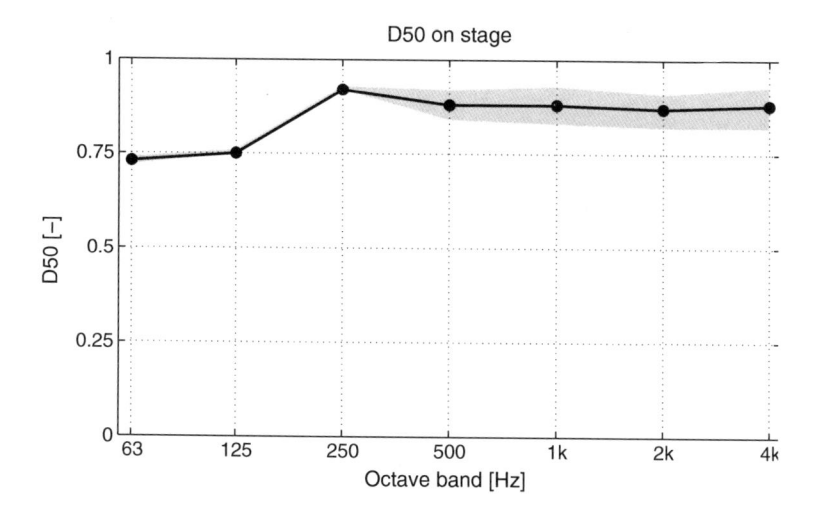

EDT on stage

D50 on stage

Astra

Berlin
Number of concerts per year: 120. In all halls in the venue: 320
Founded: 2009
Capacity: 1,000
Architect: N/A
Acoustician: N/A

The large porous absorption-filled cavity above the suspended lamella ceiling ensures acoustic control even at low frequencies.

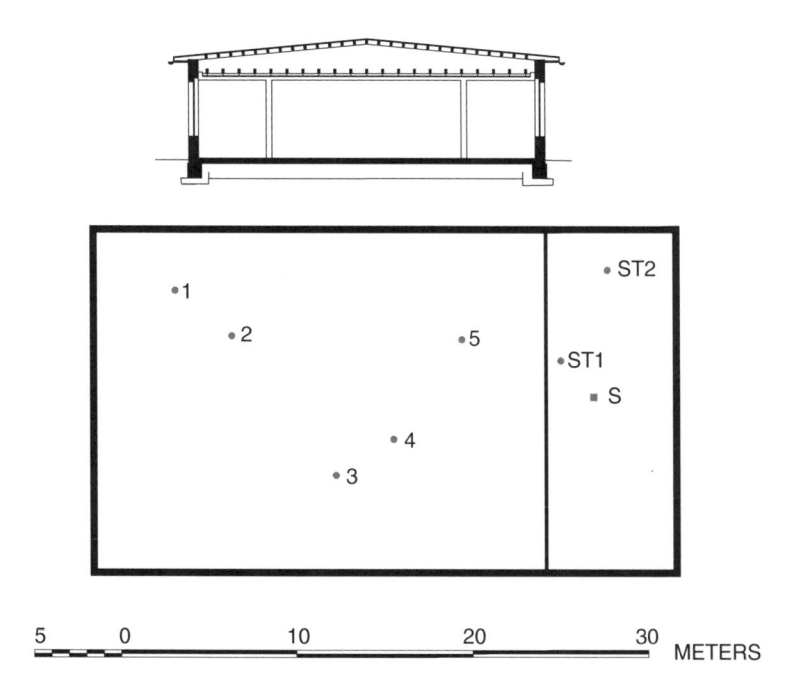

Astra

Geometrical data	
Volume	2,800 m³
$L \times W \times H$	$33 \times 19 \times 4.6$ m

Acoustical data	
Audience area	
$T_{30, 125-2k}$	0.89
EDT_{125-2k}	0.81
$C_{80, 125-2k}$	5.78
BR_{63} versus 0.5–1k	1.03
BR_{125} versus 0.5–1k	0.99
Stage area	
EDT_{125-2k}	0.41
$D_{50, 125-2k}$	0.85
BR_{63} versus 0.5–1k	0.71
BR_{125} versus 0.5–1k	1.62

Materials Used

Audience Area

Floor: Wood direct on concrete.
Ceiling: Slit absorber: thin metal brackets of 10-cm width and 1.5-cm space; 50-cm mineral wool on top.
Walls: Lower region: wooden panels on 1 cm of cavity; painted brick wall.
Upper region: Wooden strips on 2 cm of porous absorption.

Stage Area

Floor: Carpet on wooden plates on cavity.
Ceiling: Same as audience area.
Walls: Curtains in front of brick wall.

State of Hall When Measured

Empty.

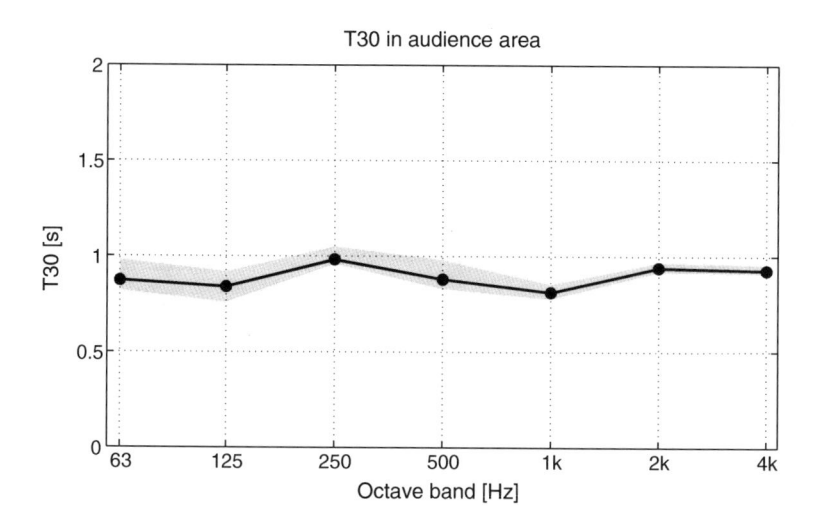

State of Hall When Measured

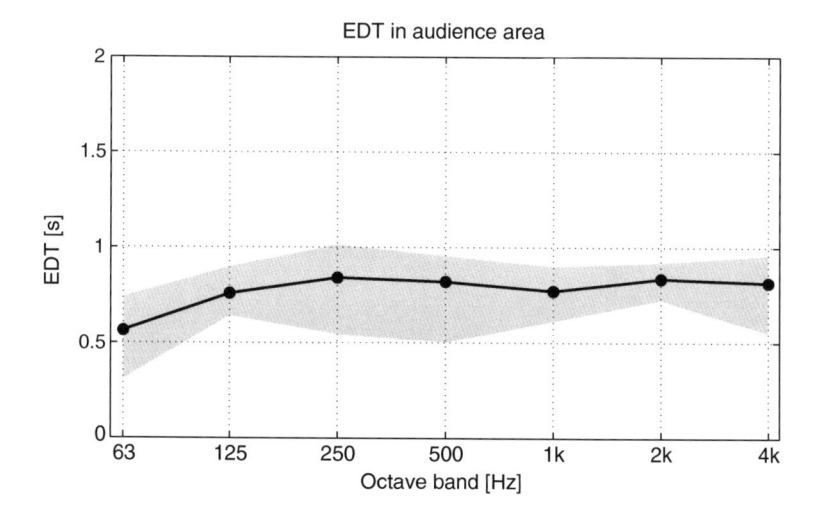

T30 in audience area

EDT in audience area

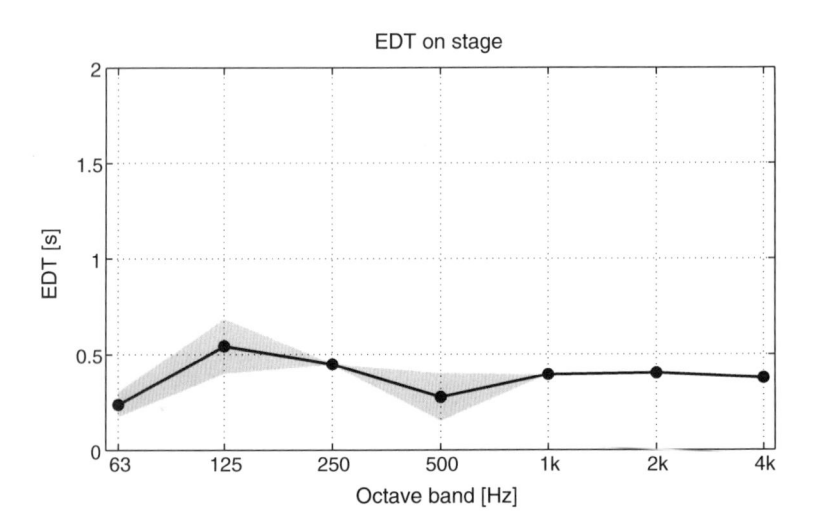

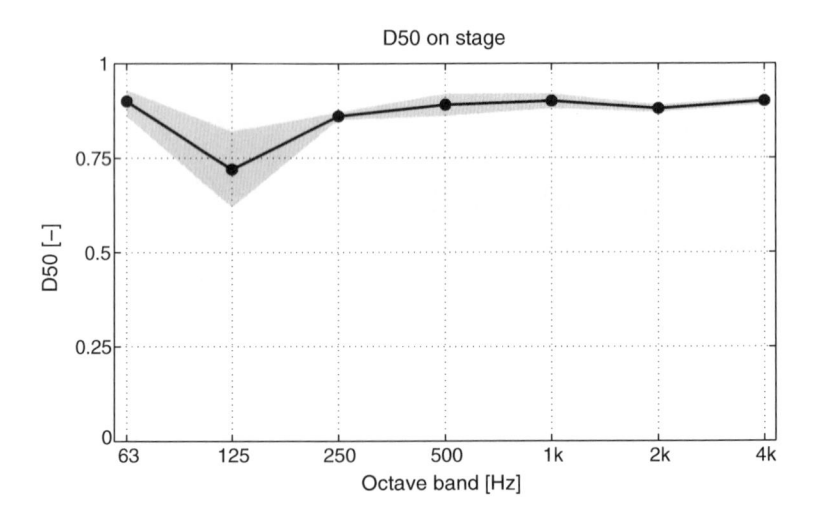

178   7 Gallery of Halls that Present Pop and Rock Music Concerts

Bikini

Toulouse
Number of concerts per year: 150
Founded: 1983/2007
Capacity: 1,500
Architect: Didier Joyes
Acoustician: Christian Malcurt

Le Bikini was originally situated on the border of the river Garonne in the center of Toulouse. It opened on the 25th of June 1983 simply as a nightclub but rapidly transformed into a music venue. Beginning with only a couple of bands a month, it later became very busy and presented a total of 5,000 concerts up until 2001. Le Bikini commenced as the place where local amateurs became hardened in the genres of the time and the taste of the audience.

In 2001 an explosion at a factory next door destroyed the venue. During the following six years Le Bikini arranged about 500 concerts in other venues before rebuilding the completely new spectacular premises used today.

Throughout the following years Le Bikini became one of the best known music clubs in France and an inevitable stage for national as well as international acts. It is also club a where many well-known artists such as Rita Mitsouko, les Stranglers, Pigalle et les Garçons Bouchers, Noir Désir, Mecano, and Lloyd Cole, were born and where stars such as Bérurier Noir, Paul Young, la Mano Negra, Little Bob, Kent, OTH, Jeff Buckley, NoFX, Tool, François Hadji-Lazaro, Muse, Coldplay, Placebo, -M-, Zebda, les Fabulous Trobadors, Arno, Korn, les Pogues, Elvis Costello, LKJ, Louise Attaque, Mickey 3D, Soulfly, Indochine, Jeff Mills, St Germain, and Carl Cox are sure to return.

180   7   Gallery of Halls that Present Pop and Rock Music Concerts

Three layers of balconies ensure that everybody there has a great view of the stage.

Le Bikini is extreme acoustical engineering at work. Reflective surfaces are mainly floors.

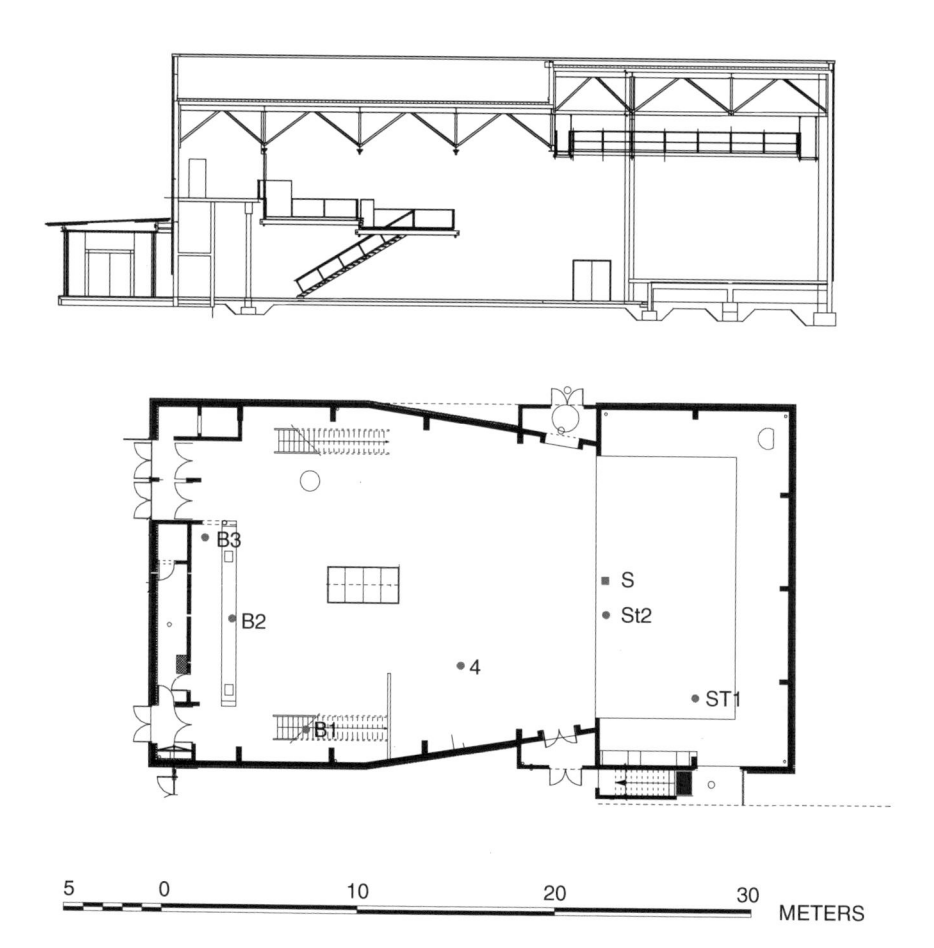

Bikini

- B3
- B2
- B1
- 4
- S
- St2
- ST1

5 0 10 20 30
METERS

Geometrical data	
Volume	7,000 m^3
$L \times W \times H$	33 × 19 × 10.2 m
Surface area of stage	270 m^2

Acoustical data	
Audience area	
$T_{30, 125-2k}$	0.5
EDT_{125-2k}	0.56
$C_{80, 125-2k}$	11.41
BR_{63} versus 0.5–1k	0.72
BR_{125} versus 0.5–1k	0.95
Stage area	
EDT_{125-2k}	0.29
$D_{50, 125-2k}$	0.91
BR_{63} versus 0.5–1k	1.63
BR_{125} versus 0.5–1k	1.59

Materials Used

Audience Area

Floor: Concrete.
Ceiling: Configurations of different layers of mineral wool slabs.
Walls: Extreme configurations of different layers of mineral wool slabs.

Stage Area

Floor: Wood direct on concrete.
Ceiling: Configurations of different layers of mineral wool slabs.
Walls: Configurations of different layers of mineral wool slabs.

State of Hall When Measured

Empty: no additional seats mounted.

T30 in audience area

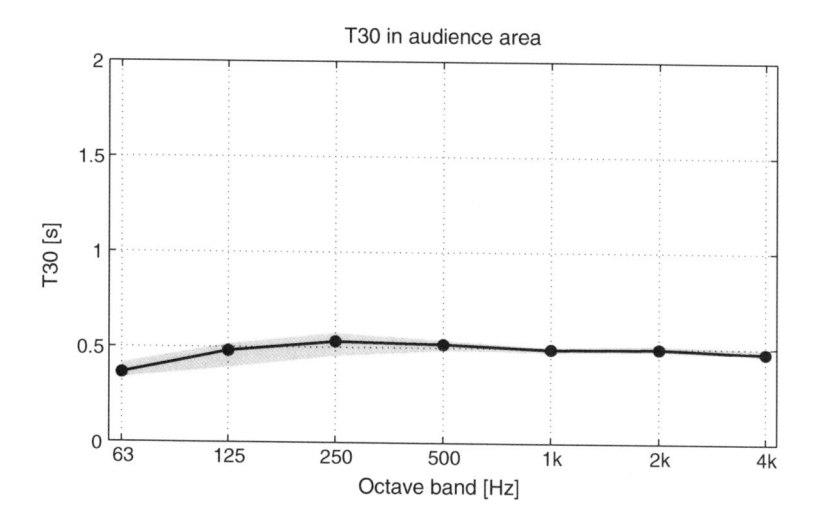

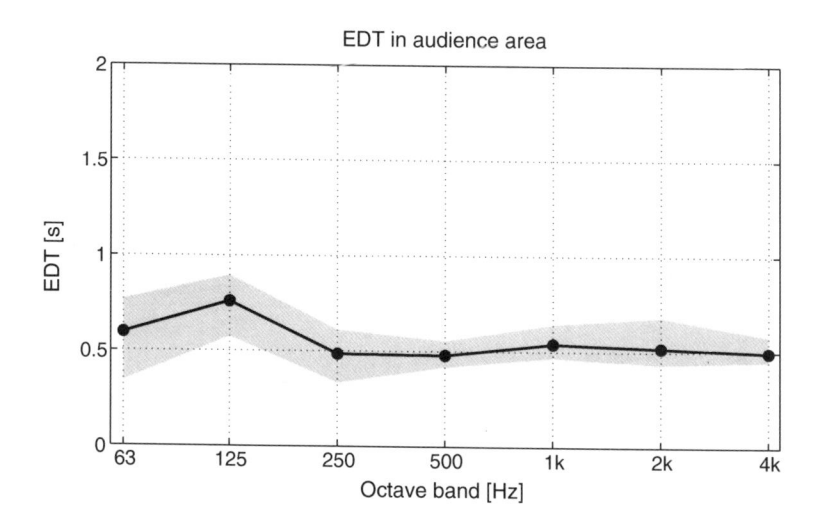

EDT in audience area

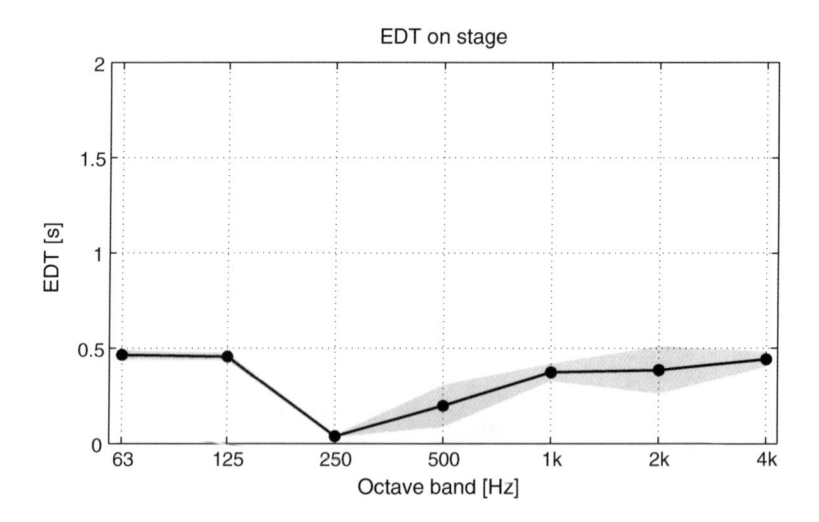

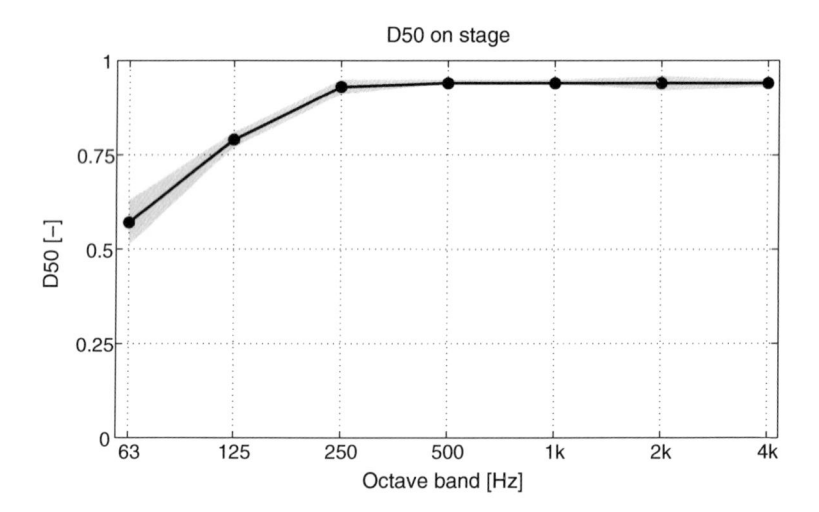

184　　　　　　　　　7　Gallery of Halls that Present Pop and Rock Music Concerts

Cavern

Liverpool
Number of concerts per year on main stage: 500. On both stages in the venue: 800
Founded: 1957/1984
Capacity: 350
Architect: David Backhouse
Acoustician: N/A

    The Cavern Club in Liverpool is the cradle of British pop music. Impressively, so many years after its foundation, it survives and thrives as a contemporary music venue. Through those eventful decades—before, during, and after The Beatles' reign—the legendary cellar at 10 Mathew Street has seen its share of setbacks yet has played a role in each epoch of music. In fact, with the Marquee and CBGB out of business, the Cavern is maybe the best-known rock club today. The front stage located at the end of the central vault with the traditional archways on either side is used every day from the afternoon onwards for soloists and bands playing cover versions of Beatles music and covers of other standard guitar band music. Admission is free Monday to Wednesday and there is a small general admission charge after 8 pm Thursday to Sunday.

    The Cavern was born in a warehouse basement built to service Liverpool's teeming nineteenth century waterfront. Hidden amid a warren of cobbled passages by the city's shopping and business districts, Mathew Street was a dingy crooked canyon unknown to anyone who didn't work in its gaunt storerooms or drink in its only cheerful corner, the tiny Grapes pub. That all changed in 1957 when a local promoter called Alan Sytner dreamed of emulating the Parisian Left Bank jazz clubs, those subterranean dives where femmes fatales and French philosophers met to escape the straight world upstairs. Merely to imagine such a place in mundane Liverpool was a romantic vision indeed, but Sytner's plan was a winner. On the Cavern's opening night, January 16th, 1957, 600 fans crammed inside to see The Merseysippi Jazz Band (like the Cavern, they're still going strong) and about 1,500 were left outside.

    Although jazz was hot in the late 1950s, a new musical mood was gathering force across Britain, especially in Liverpool. Skiffle, the folk style with a rock'n'roll influence, played DIY-style on cheap guitars and domestic utensils, threw up hundreds of teenage acts including John Lennon's Quarrymen, who soon included Paul McCartney. They played the Cavern, as did Ringo Starr in a rival skiffle act. Under the club's new owner, Ray McFall, from 1959 the jazz identity of the Cavern began giving way to the musical revolution now brewing in the city. Beefing up their sound with imported US influences, the groups had evolved a distinctive Liverpool style that would soon be christened Merseybeat.

    The Quarrymen evolved over a four-year period into the Beatles, of course, who became the Cavern's signature act and were talent-spotted here by Brian Epstein, the suave young businessman from a nearby record store. Alongside other

Cavern regulars, the Beatles led a Liverpudlian takeover of British pop in 1963. In turn they inspired the "British Invasion" of America itself, effecting a transformation of global pop culture that shapes the world we live in today.

The Beatles played the Cavern almost 300 times, including lunchtime sessions. Along with Hamburg, it's unquestionably the place where their musical identity was forged and it was the nucleus of an early fan base that was to spread around the globe. The band themselves were always nostalgic about the Cavern. In the fractured final days they attempted, poignantly, to rediscover their lost solidarity as a tough young Liverpool combo. The tune "Get Back" was ostensibly inspired by spirit of the Cavern.

In the wake of the Beatle boom the Cavern became a prestige port of call for everyone from the Rolling Stones to Queen, who each played early gigs here. In truth it was a pretty basic kind of place, a disinfected dungeon. Descending the slimy steps from Mathew Street the visitor was plunged into an underworld whose air was a rancid fug of body odor, cigarette smoke, hamburger smells, and a little something from the toilets. But the venue's triple tunnels gave an almighty acoustic boost to any rock band; in those early days of puny amplification, the Cavern sound was uniquely powerful and its atmosphere electrifying. In the early 1960s it was the most exciting shrine of the youth revolution.

But even at the height of its fame the Cavern was not a secure business. As Merseybeat passed from favor, so the Club's iconic status waned (while the Council grew ever more alarmed at that infamous lack of sanitation). Still, its closure in 1966 came as such a shock that it quickly attracted new investors and was reopened by no less a personage than the prime minister of the day, Harold Wilson. Later years saw the Cavern adapt to modern customs with the introduction of alcohol and the addition of a disco room. Sadly, its historic standing didn't stop the Council closing it in 1973 to allow for work on an underground railway line. The ancient warehouse was demolished and the cellar itself, rubble-filled, lingered like a sealed tomb until 1981.

Yet something in the Cavern's spirit refused to die. A short-lived "New Cavern" opened across the street, was then renamed Eric's, and spawned a whole new wave of Liverpool stars, from Elvis Costello to Frankie Goes To Hollywood. Meanwhile, Lennon's murder in 1980 awoke the city of Liverpool from its apathetic indifference to the Beatle heritage. In 1984 the real Cavern site was reclaimed and an exact replica of the old Club was built in situ, using 15,000 bricks from the original cellar. This is the Cavern of today, proudly back at 10 Mathew Street, an authentic and evocative location that draws visitors and bands from across the world. Obviously it's the ultimate place of pilgrimage for Beatle fans—and much less hazardous than a certain London pedestrian crossing—but it's an important modern venue, too. In recent years the Cavern has hosted memorable shows by Arctic Monkeys, Travis, Embrace, K.T. Tunstall, and Liverpool's own The Coral, to name only a few.

The most memorable show of all, though, was on the night of December 14th, 1999, when the Cavern marked the new millennium with a back-to-basics gig by

Paul McCartney. It goes without saying that the club was packed, but in fact a far wider audience watched as well, thanks to a pioneering webcast that broke new ground in a high-tech medium undreamt of in Paul's early Cavern days. The latterday club has a larger additional stage, as well as its faithful facsimile of the vintage model; it occupies over half of the original space, and the stage McCartney played upon is merely feet away from the site of those first Quarrymen appearances.

Nowadays Mathew Street is a prime tourist destination, lined with bars, shops, plaques, and statues. There is a long-held school of thought that holds it is a place of mystic energy, long-held, admittedly, by people who have spent entire afternoons in the Grapes or the Cavern pub. But close to the club's front door is a bust of Carl Gustav Jung inscribed with his assertion that "Liverpool is the Pool of Life," and many agree with him Across the way, a life-sized bronze John Lennon lounges against the wall, kissed and photographed by a thousand strangers a day. Beneath his hooded gaze the music fans still troop into the Cavern's entrance for an experience they will never forget. Let these songs stand in tribute to a little hole in the ground that really changed our world.

Extract from Paul Du Noyer's book, "Liverpool: Wondrous Place" is published by Virgin Books.

The other stage in the club (Cavern Live Lounge) is primarily used for ticketed shows featuring various tribute bands, established artists, and young contemporary band nights. There is seating for 170 people; alternatively the seating can be taken out for a standing audience of 350 people.

While the Cavern was still a jazzclub. Merseysippi Jazzband late 1950s.

188   7 Gallery of Halls that Present Pop and Rock Music Concerts

The Cavern was reconstructed after it had been demolished and closed down from 1973–1984 due to the construction of a subway. Near field speakers distributed in the ceiling.

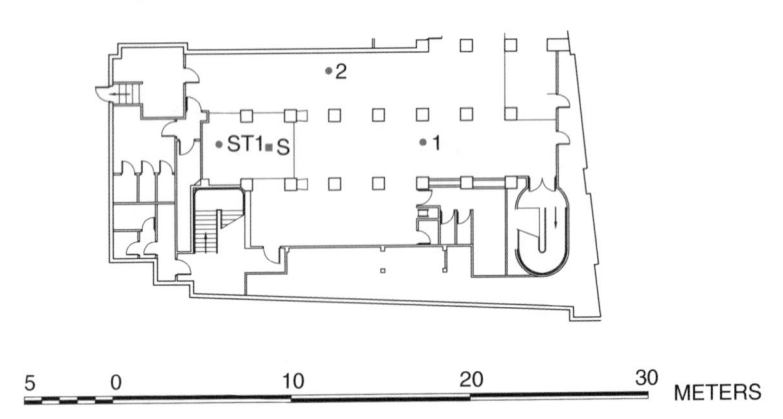

Geometrical data	
Volume	600 m³
$L \times W \times H$	$21 \times 11 \times 2.7$ m

Acoustical data	
Audience area	
$T_{30, 125-2k}$	1.09
EDT_{125-2k}	0.96
$C_{80, 125-2k}$	4.01
BR_{63} versus 0.5–1k	1.06
BR_{125} versus 0.5–1k	1.13
Stage area	
EDT_{125-2k}	0.34
$D_{50, 125-2k}$	0.86
BR_{63} versus 0.5–1k	1.3
BR_{125} versus 0.5–1k	1.23

Materials Used

Audience Area

Floor: Concrete.
Ceiling: Brick.
Walls: Brick.

Stage Area

Same as audience area. Stage is made of wooden plates on hollow cavity.

State of Hall When Measured

Few people at tables; tables and chairs. No curtains at the time of the measurement.

T30 in audience area

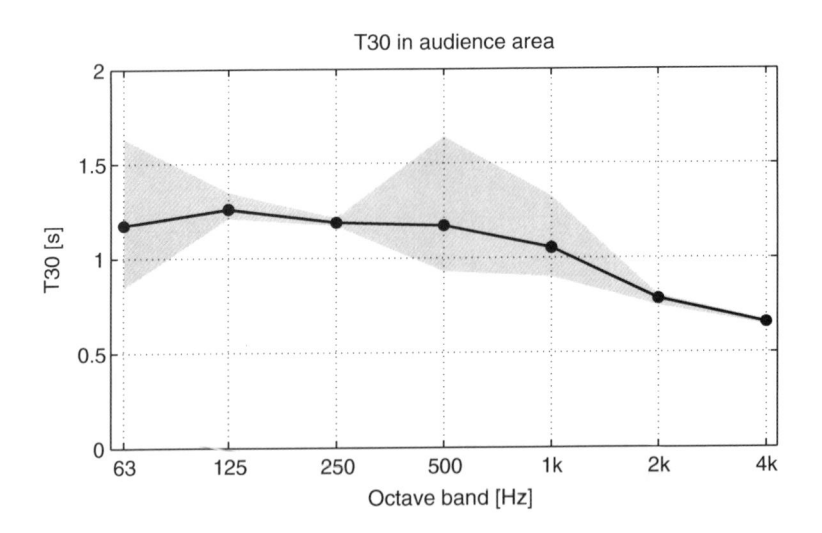

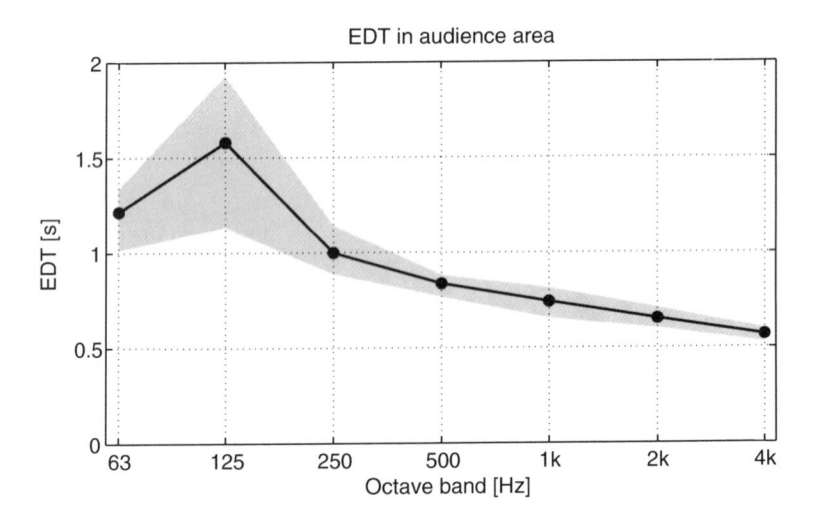

EDT in audience area

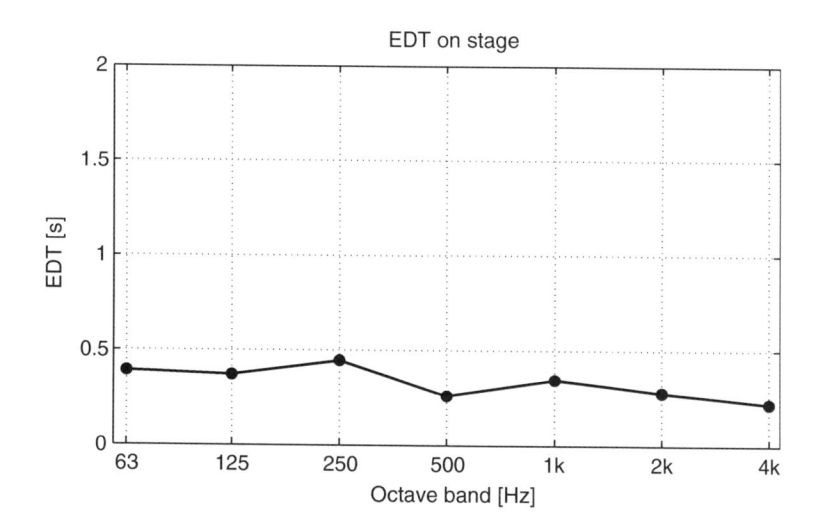

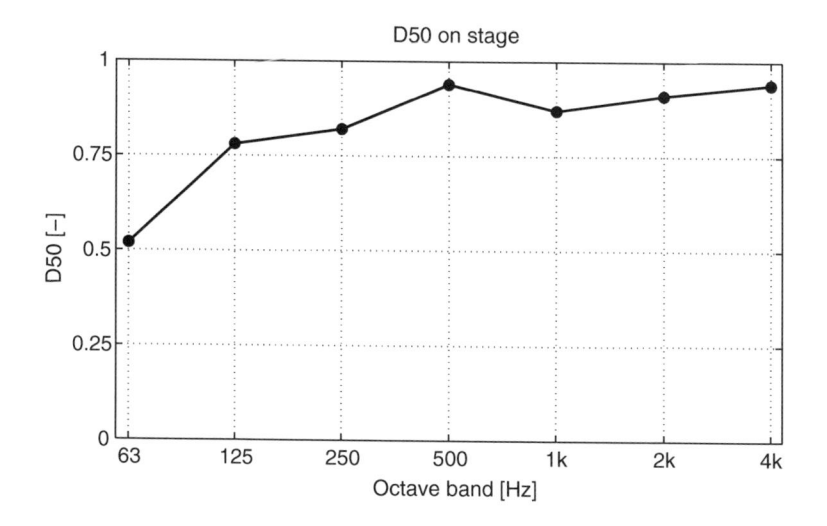

La Coopérative de Mai

Clermont-Ferrand
Number of concerts per year: approximately. 55. In both halls in the venue: 125
Founded: 2000
Capacity: 1,500 standing or 850 seated
Architects: P. Borderie, R. Kander, and J. M. Louviaux.
Acoustician: Emmanuel Giroflet, Thermibel.

La Coopérative de Mai is certainly the main rock venue not only of the city of Clermont-Ferrand but of the whole region of Auvergne. Over 11 years 1,500 concerts have entertained more than 1 million people in two halls. The total surface area of the building that also comprises many offices and so on is 3,000 m^2. Since the opening in 2000 the two stages within the venue had 1,500 concerts with more than 1 million spectators.

"La grande sale" has room for 800 seated or 1,500 standing people, 16 of which are reserved for disabled persons. The total surface area is 740 m^2 and has a modular stage of up to 210 m^2. The staircase construction ensures that everyone can find a spot with an excellent view of the stage. The smaller stage "La petite Coopé" holds 450 people.

The staircase design of the hall ensures that everyone can find a spot with perfect views.

La Coopérative de Mai

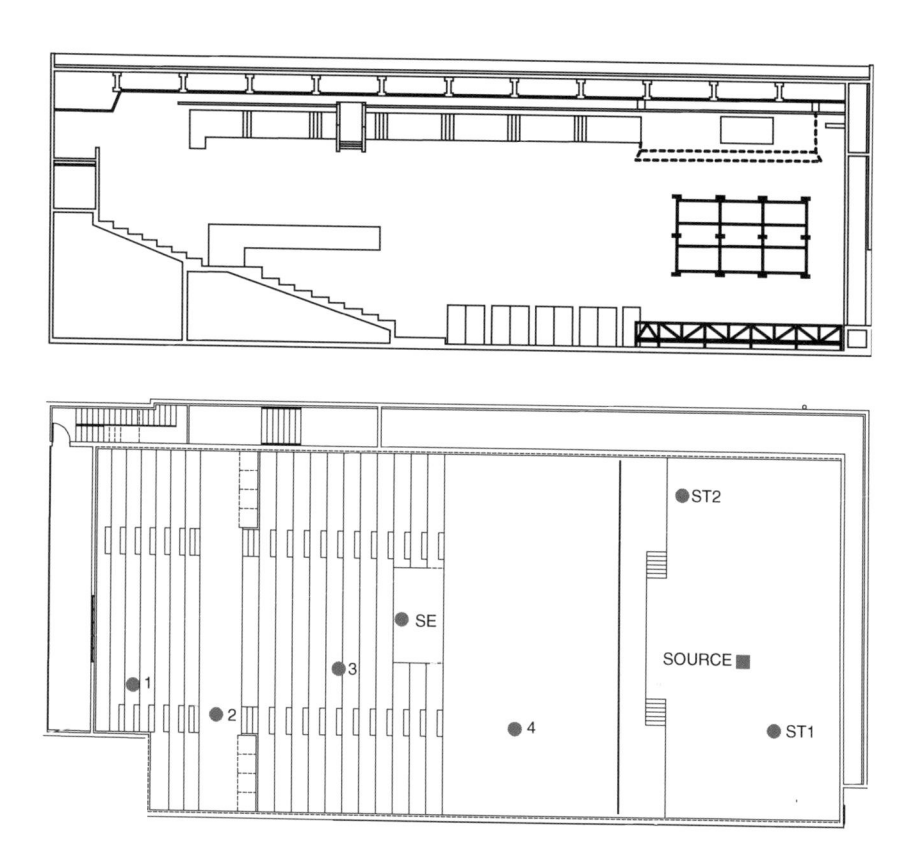

Geometrical data	
Volume	9,000 m^3
$L \times W \times H$	$40 \times 19 \times 13.4$ m
Surface area of stage	210 m^2

Acoustical data	
Audience area	
$T_{30, 125-2k}$	1.09
EDT_{125-2k}	0.93
$C_{80, 125-2k}$	4.43
$BR_{63 \text{ versus } 0.5-1k}$	1.8
$BR_{125 \text{ versus } 0.5-1k}$	1.27
Stage area	
EDT_{125-2k}	0.88
$D_{50, 125-2k}$	0.77
$BR_{63 \text{ versus } 0.5-1k}$	0.92
$BR_{125 \text{ versus } 0.5-1k}$	1.13

Materials Used

Audience Area

Floor: Concrete with antislip treatment.
Ceiling: Suspended mineral wool.
Walls: 2-cm thick perforated panels with cavity behind. Panels in rear of the room not perforated.

Stage Area

Floor: Light stage podiums.
Ceiling: Suspended mineral wool.
Walls: Thin perforated metal plates with mineral wool behind direct on concrete.

State of Hall When Measured

Empty; light rigging equipment on stage floor.

T30 in audience area

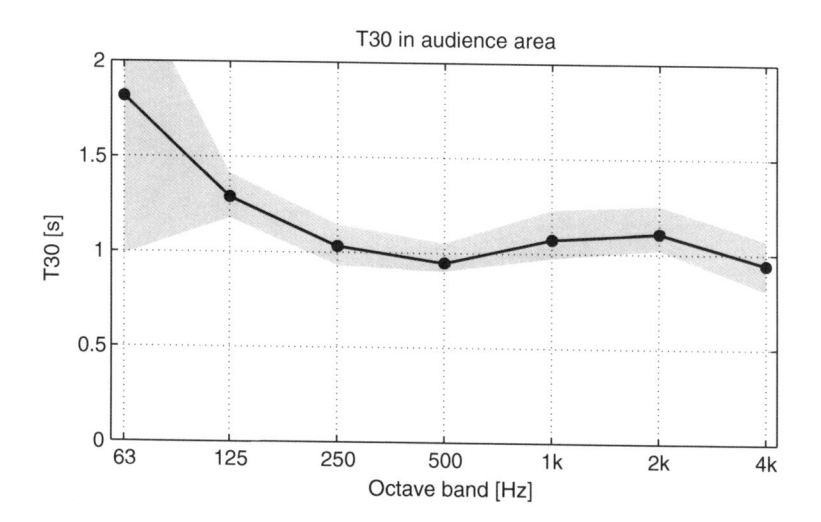

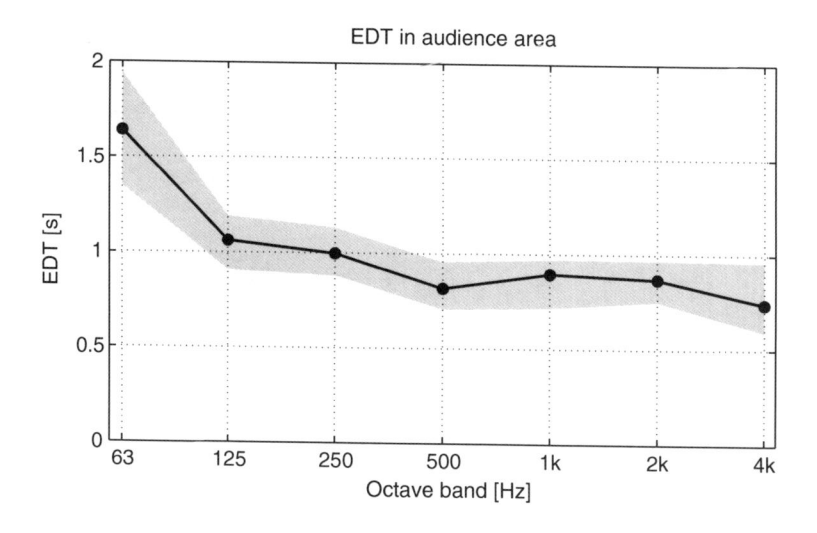

EDT in audience area

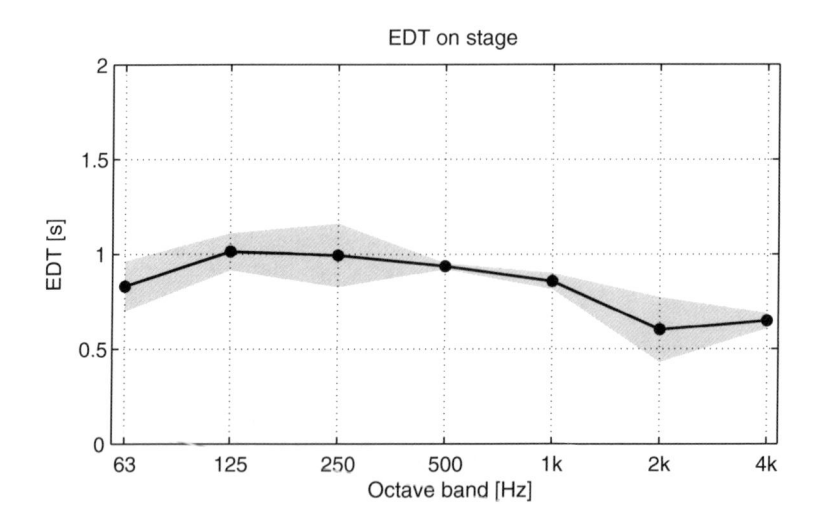

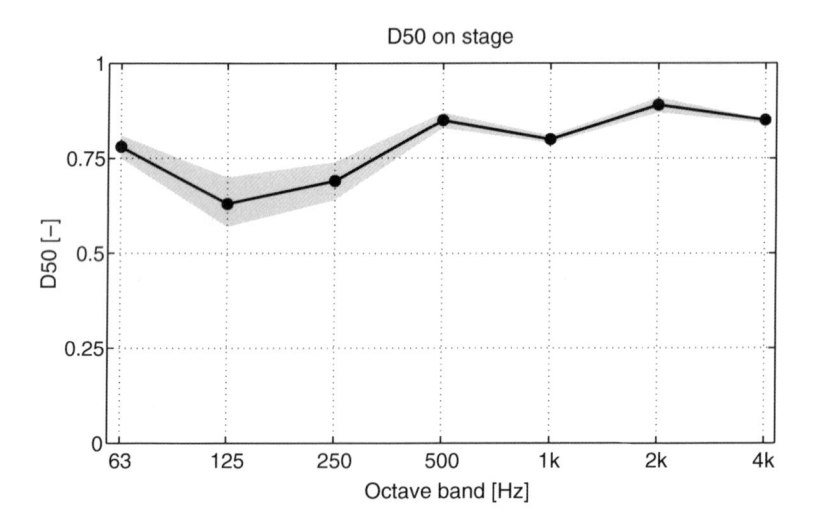

196　　　　　　　　7　Gallery of Halls that Present Pop and Rock Music Concerts

Le Chabada

Angers
Number of concerts per year: 40. In both halls in the venue: 70
Founded: 1994
Capacity: 900
Architect: Architectes Ingenieurs Associes
Acoustician: Acoustique Pierre Poubeau

Le Chabada is a concert hall for popular music in Angers, Pays de la Loire, France. Since its creation in 1994, the venue has settled into an eclectic range of music genres including rap, electronic music, rock, world music, pop, reggae, and so on.

At the end of the 1980s, Angers was home to many musicians, associations, and concert promoters. Back then, the city had the reputation to be the French cradle of hardcore, popcore, grunge, and punk rock. With 63 bands and artists, Angers counted more music creators than any other French city with its most important ambassadors Les Thugs, Dirty Hands, Spécimen, and Lo'Jo Triban. In 1988 the organization, Association for the Development of Rock and Related Genres in Angers (ADRAMA) was created, uniting all "rock-associations" with the intention to start negotiating with the city council to create infrastructure for popular music.

In 1990, ADRAMA obtained the opening of 10 practice rooms for local bands and artists. Petitions and campaigns went on and finally the city council of Angers gave the go-ahead for the creation of a venue dedicated to contemporary music. Like other French cities, Angers had the benefit of the cultural dynamism of these years when Jack Lang was minister for education.

The venue opening was in September, 1994. Housed in the converted former slaughterhouse of Angers, Le Chabada covers over 1,500 m^2 and two floors. The two concert halls occupy the ground floor of the building. The main hall called "Grande Salle" has a capacity of 900 people including 150 fixed seats. The smaller concert hall called "Le club" has a capacity of 300 people and hosts less well-known, upcoming acts.

The city of Angers is owner of Le Chabada and delegates its administration to the nonprofit organization ADRAMA-Chabada.

Each year Le Chabada hosts about 70 concerts of various styles. Noted acts that have played at Le Chabada are: Vanessa Paradis, John Spencer Blues Explosion, Ghinzu, Soulfly, Les Thugs, Fugazi, Buzzcocks, The Ex, The Libertines, L7, Machine Head, dEUS, The Divine Comedy, Frank Black, Roots Manuva, Maceo Parker, Fred Wesley, The Herbaliser....Tindersticks, Noir Desir, The Kills, Femi Kuti, Seun Kuti, Nada Surf, Asian Dub Foundation, Toots and the Maytals, Tarwater, Les Nits, Bauchklang, Sofa Surfers, The Bellrays, Turin Brakes, Giant Train, Gus Gus, The Ex, Rory Gallagher, The Young Gods, Transglobal LKJ, The Descendents, Machine Head, L7, Buck 65, Frank Black, Sharon Jones and

The Dap Kings, Vive la Fête, Roots Manuva, Antibalas Afrobeat Orchestra, Saul Williams, WhoMadeWho, Gojira, John Butler Trio, The Young Good Brass Band, Scott H Biram, Archie Bronson Outfit, and Liars, among others.

Minutes before the French pop act Cocoon enters the stage. Photo © Jordane Chaillou.

The hall holds 900 audiences; here approximately 150 seated. Photo © Jordane Chaillou.

Le Chabada 199

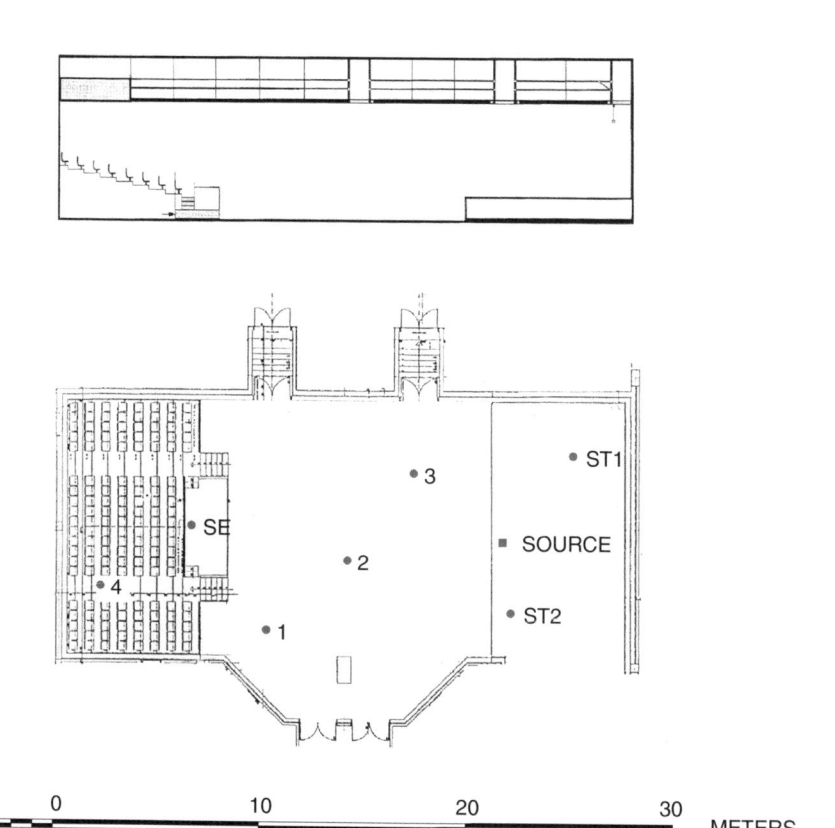

Geometrical data	
Volume	2,800 m^3
$L \times W \times H$	27 × 13.5 × 7.5 m
Surface area of stage	270 m^2

Acoustical data	
Audience area	
$T_{30, 125-2k}$	0.94
EDT_{125-2k}	0.86
$C_{80, 125-2k}$	3.49
BR_{63} versus 0.5–1k	1.54
BR_{125} versus 0.5–1k	1.21
Stage area	
EDT_{125-2k}	0.47
$D_{50, 125-2k}$	0.85
BR_{63} versus 0.5–1k	3.49
BR_{125} versus 0.5–1k	2.04

Materials Used

Audience Area

Floor: Concrete.
Ceiling: Suspended mineral wool.
Walls: Wooden panels mounted in a zig-zag configuration. Every second panel is perforated.

Stage Area

Floor: Wood on cavity on concrete.
Ceiling: Suspended mineral wool.
Walls: Wood wool panels direct on concrete.

State of Hall When Measured

Empty; four smaller curtains on stage.

T30 in audience area

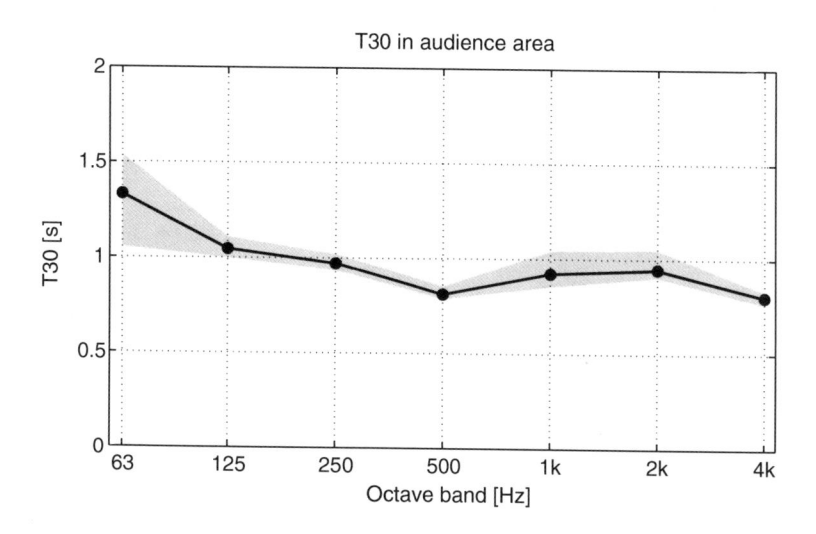

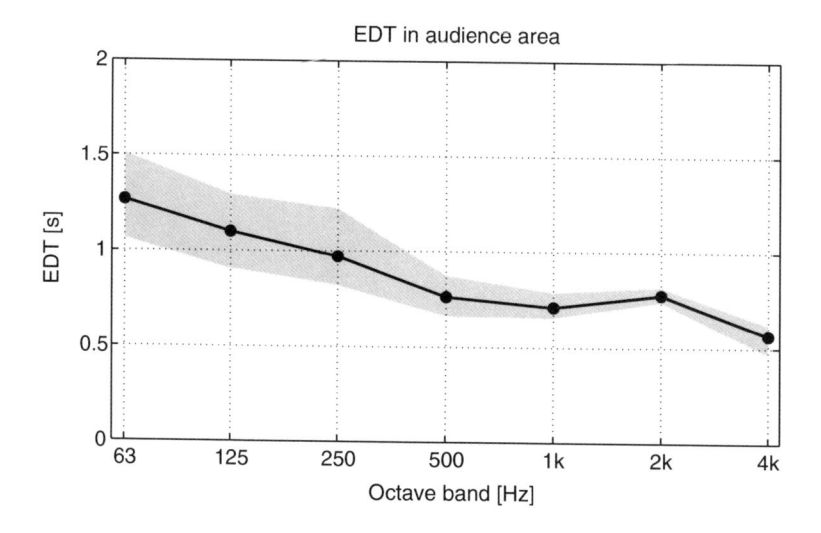

EDT in audience area

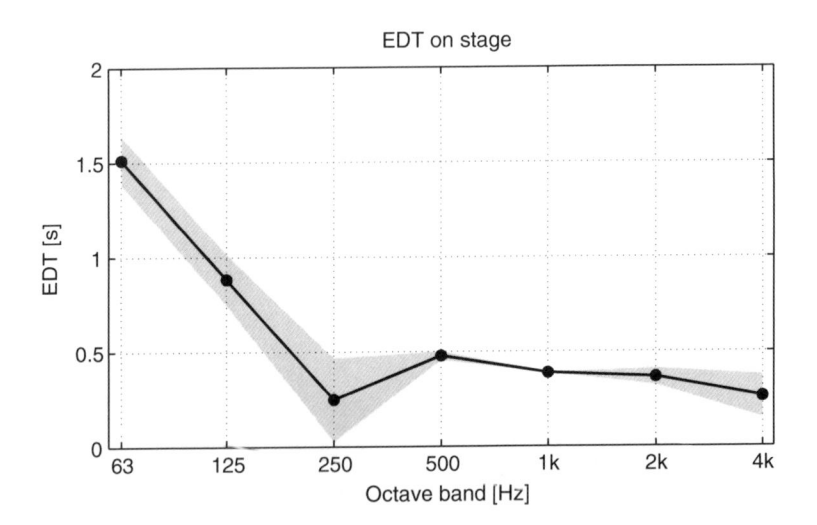

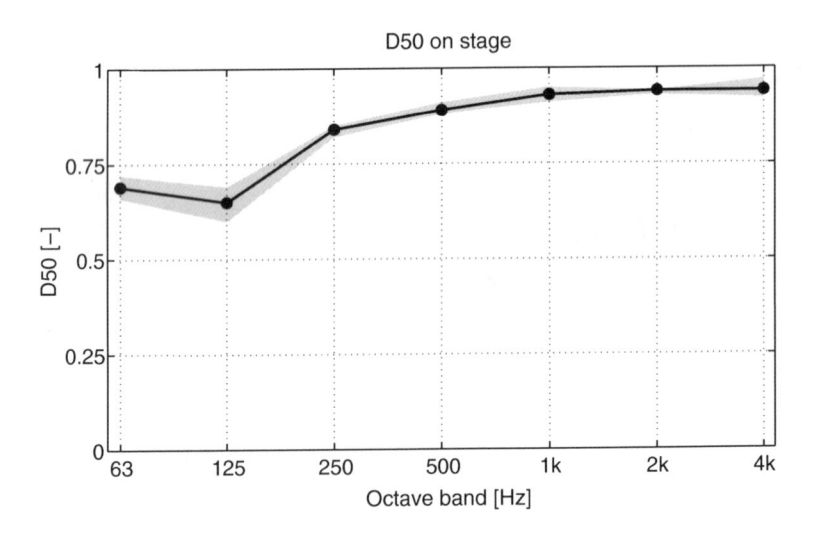

Cirkus

Stockholm
Number of concerts per year: 60–80
Total amount of events including concerts and musicals: 250–275
Built: 1892, refurbished in 1990 and 1997
Capacity: up to 1,800
Architect: Ernst Hägglund

The more than 115-years-old, Cirkus is situated at The Royal Djurgården, close to the very center of Stockholm and yet in a location with a feeling of being in the countryside. The exterior as well as the interior has been fully refurbished in the original design as the building has been declared a historical monument. The building has a very special charm enabling it to embrace very different events although the venue has become most known for housing the great musicals of former ABBA members Benny Andersson och Björn Ulvaeus. The musical Chess entertained over a half million people here in 2002–2003 and Mamma Mia was seen by more than 800,000 people in the years 2005–2007.

There is room in the arena for 1,800 persons, with 1,644 seats giving a feel of intimacy and closeness due to the round shape and the warm colors, brick red and forest green.

The stage is large: 27 m deep and 14 m wide. The arena is very easy to change. The seating stalls are vertically adjustable, the whole floor or just some sections. The floor can be sloping or flat and the chairs can be removed. The floor can also be changed to different platforms with help of hydraulics and show the entertainment on different levels. The Cirkus restaurant has a feeling of a Wienercafé. The kitchen has a big capacity, quantity and quality; a buffet can be arranged for a full arena.

From the stage performers enjoy a sensation of direct contact with even the most remote members of the audience.

The combination of a traditional nineteenth-century seating space with completely up-to-date stage facilities makes Cirkus a sought after venue especially for musicals.

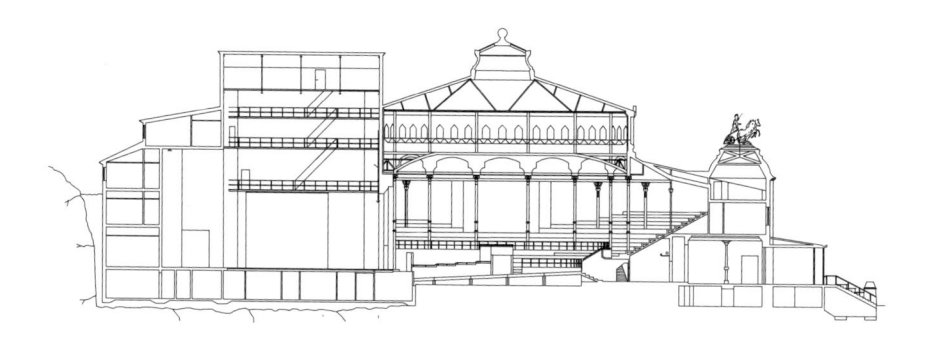

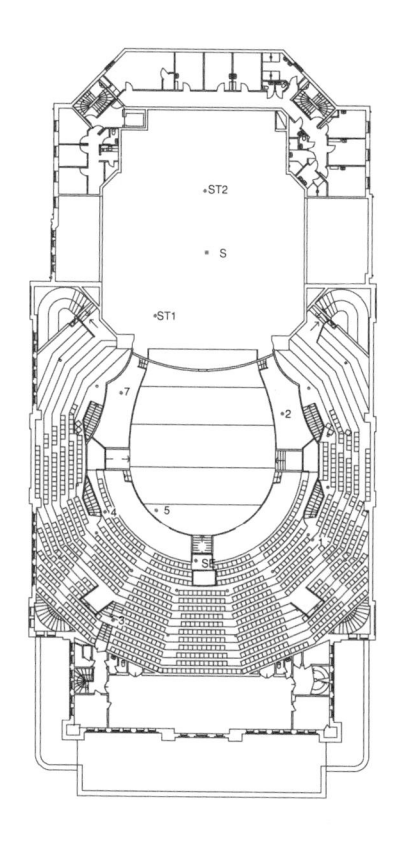

Cirkus

Geometrical data	
Volume	10,000 m^3 + stage volume
$L \times W \times H$	$36 \times 36 \times 7{-}13$
Surface area of stage	375 m^2

Acoustical data	
Audience area	
$T_{30, 125-2k}$	1.27
EDT_{125-2k}	1.18
$C_{80, 125-2k}$	3.32
$BR_{63 \text{ versus } 0.5-1k}$	1.29
$BR_{125 \text{ versus } 0.5-1k}$	1.11
Stage area	
EDT_{125-2k}	0.88
$D_{50, 125-2k}$	0.81
$BR_{63 \text{ versus } 0.5-1k}$	1.23
$BR_{125 \text{ versus } 0.5-1k}$	0.67

Materials Used

Audience Area

Floor: Hydraulic seating platforms. Stationary platforms constructed of plates on cavity. Upholstered chairs.
Ceiling: Probably plaster on lightweight construction on very large cavity.
Walls: Wood on cavity; gypsum board on cavity, gypsum board on brick.

State of Hall When Measured

All additional seats mounted. Shown in photos.

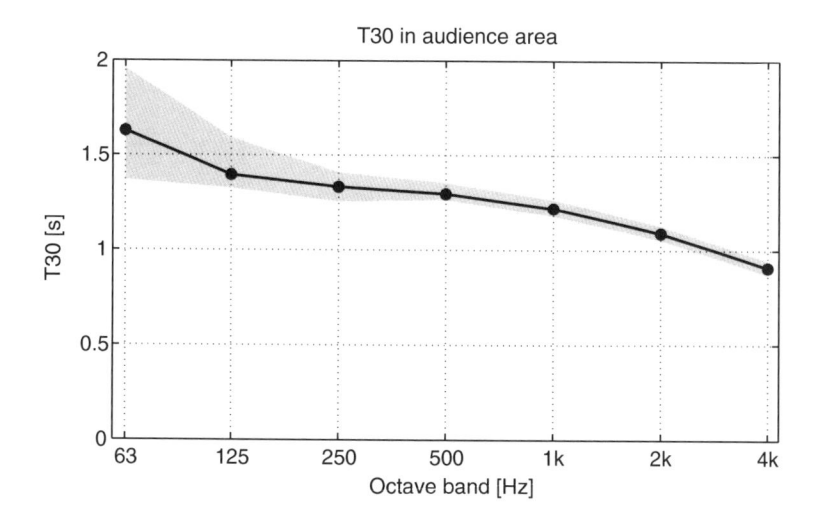

State of Hall When Measured

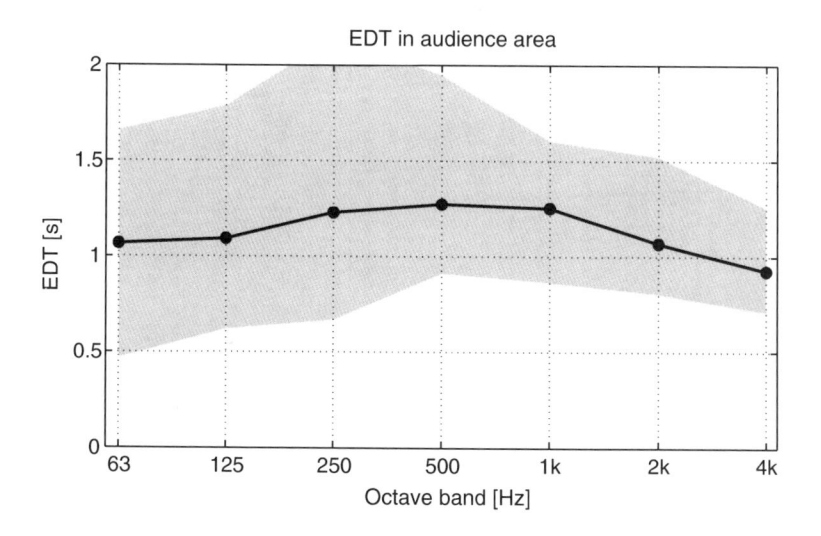

T30 in audience area

EDT in audience area

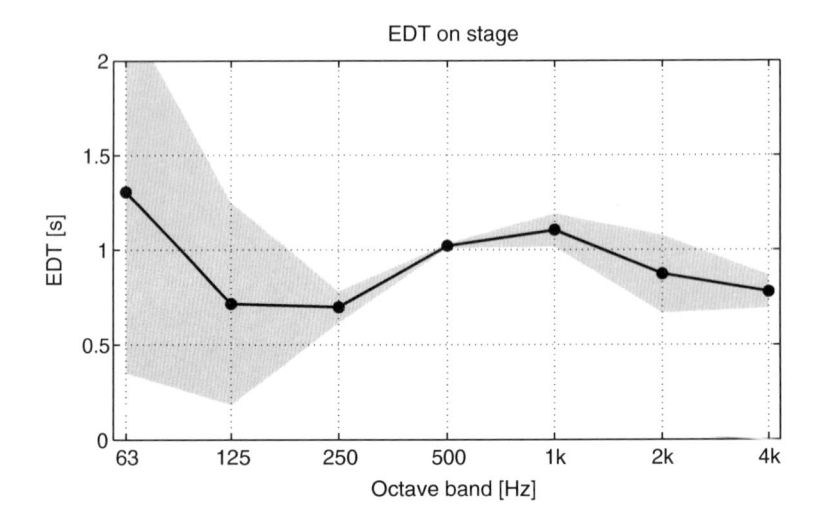

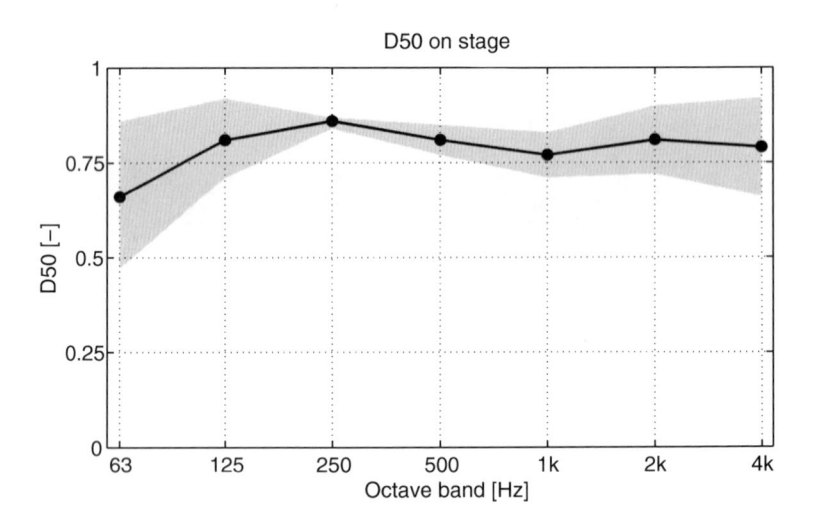

Le Confort Moderne

Poitiers
Number of concerts per year: 60
Founded: 1985
Capacity: 700
Architect: N/A
Acoustician: N/A

In 1979 the Poitiers-based organization L'Œil écoute was renamed l'Oreille est Hardie and arranged 200 concerts up until 1984 all over Poitiers: auditorium Sainte Croix, Amphi Descartes, Théâtre de Poitiers (with, for instance, Glenn Branca, The Residents), Place d'Armes (first-ever concert with Sonic Youth in Europe).

In 1985 a building to house the concerts was found. Francis Falceto, Fazette Bordage, Yorrick Benoist, and Philippe Auvin were responsible for bringing the project forward by renting the old factory building Confort 2000 for creative purposes. In 1988 the city of Poitiers bought the building and made way together with l'Oreille est Hardie for the venue as it stands today.

"High Tone" at Confort Moderne, photo: Yvain Michaud.

210   7 Gallery of Halls that Present Pop and Rock Music Concerts

Confort Moderne in Poitiers has brought many cult punk acts to France.

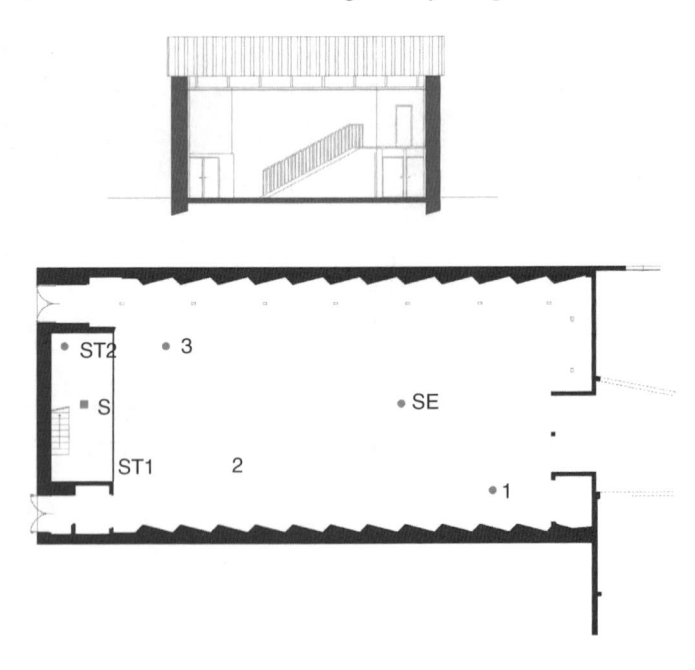

METERS

Geometrical data	
Volume	1,400 m^3
$L \times W \times H$	$25 \times 12 \times 4.8$ m

Acoustical data	
Audience area	
$T_{30, 125-2k}$	0.72
EDT_{125-2k}	0.67
$C_{80, 125-2k}$	6.63
BR_{63} versus 0.5–1k	1.99
BR_{125} versus 0.5–1k	1.8
Stage area	
EDT_{125-2k}	0.29
$D_{50, 125-2k}$	0.91
BR_{63} versus 0.5–1k	1.64
BR_{125} versus 0.5–1k	1.71

Materials Used

Audience Area

Floor: Concrete.
Ceiling: Perforated thin plate on cavity.
Walls: Zig-Zag walls of concrete with unsmooth surface. 10-cm thick porous baffles hanging down alongside the upper part of walls at a distance. Rear wall: plate on cavity.

Stage Area

Floor: Vinyl on wood direct on concrete.
Ceiling: Perforated thin plate on cavity.
Walls: Curtain 1½ m from back and side walls.

State of Hall When Measured

Empty; no additional seats mounted.

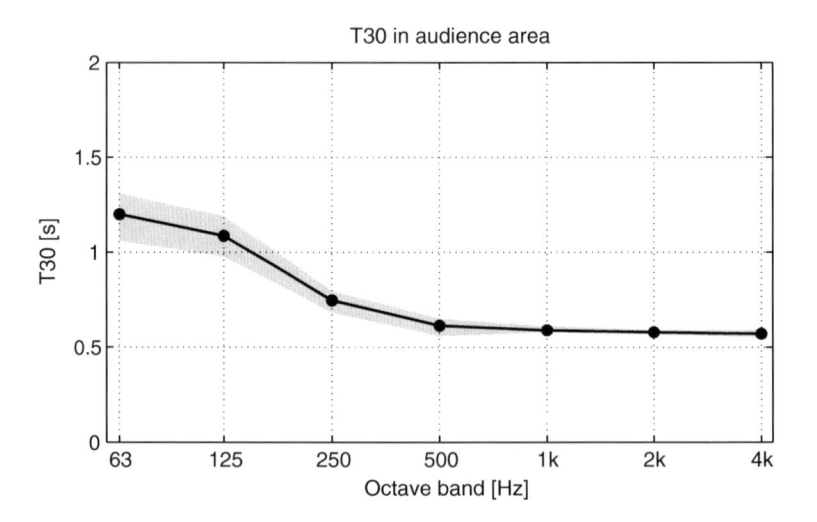

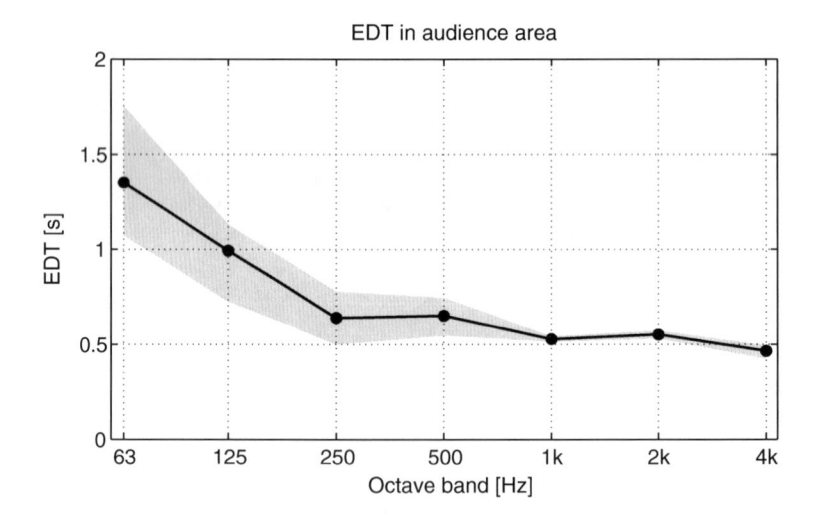

212   7   Gallery of Halls that Present Pop and Rock Music Concerts

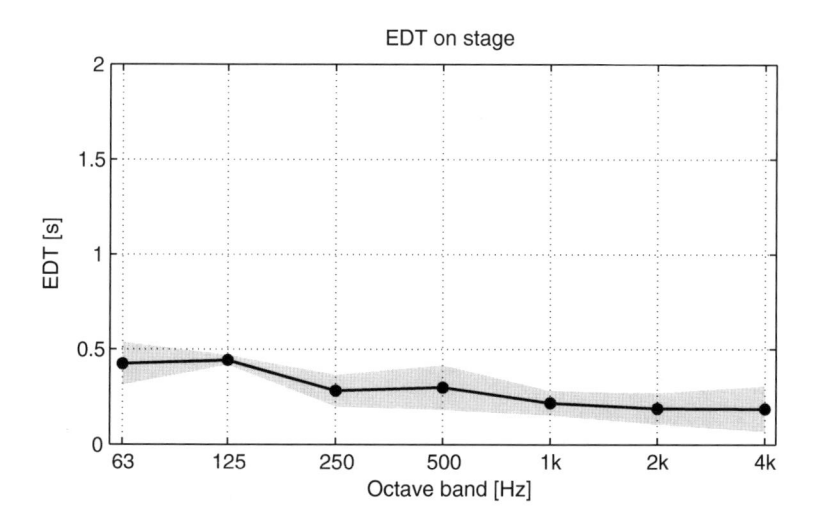

State of Hall When Measured

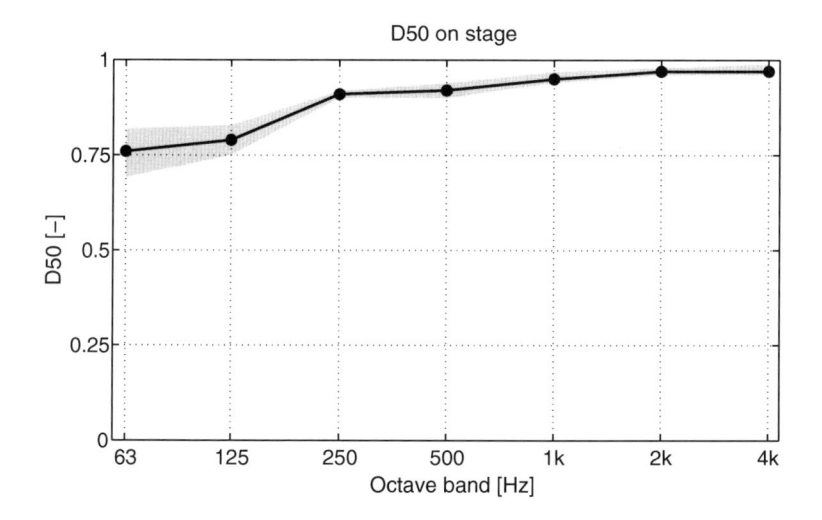

EDT on stage

D50 on stage

Debaser Medis

Stockholm
Number of concerts per year: N/A
Founded: N/A
Capacity: 850 standing + 100 seated
Architect: N/A
Acoustician: N/A

Choosing a diffusive instead of absorptive rear wall also at low frequencies is a possibility if the distance from the stage and PA system is not very big. The diffusers should be constructed to absorb some sound energy in the 125-Hz octave band. Singers and guitarists especially like to get something in return for their efforts.

The cavity under the stage seems to be used for a huge Helmholtz resonator that may absorb some low-frequency sound. A porous absorber may also work and priority must be given to isolating the stage from the subspeakers often placed beneath. The open area underneath the stage is too little to create any impact on the total acoustics of the room.

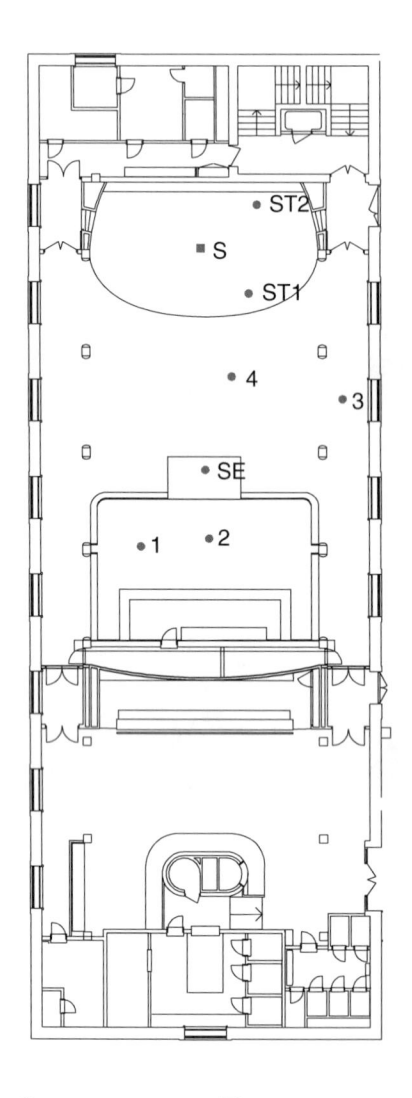

216 7 Gallery of Halls that Present Pop and Rock Music Concerts

5 0 10 20 30
METERS

Materials Used

Geometrical data	
Volume	2,400 m^3
$L \times W \times H$	23 × 16 × 7.3 m
Surface area of stage	50 m^2

Acoustical data	
Audience area	
$T_{30, 125-2k}$	0.73
EDT_{125-2k}	0.75
$C_{80, 125-2k}$	4.76
$BR_{63 \text{ versus } 0.5-1k}$	1.44
$BR_{125 \text{ versus } 0.5-1k}$	1.35
Stage area	
EDT_{125-2k}	0.53
$D_{50, 125-2k}$	0.82
$BR_{63 \text{ versus } 0.5-1k}$	1.25
$BR_{125 \text{ versus } 0.5-1k}$	1.18

Materials Used

Audience Area

Floor: Vinyl on wood on cavity.
Ceiling: Suspended mineral wool plus mineral wool baffles.
Walls: Upper part of walls are concrete with 50-mm mineral wool products on some areas. Lower part is curtains in front of sound-insulating glazing mounted some 30 cm in front of the windows. The rear wall opposite the stage is diffusive with app 50 × 50 cm squares of varying depth.

Stage Area

Floor: Vinyl on wood direct on concrete. Underneath the stage is a custom-built Helmholz horn absorber.
Ceiling: Suspended mineral wool plus mineral wool baffles.
Walls: No walls at the side of the stage.

State of Hall When Measured

Empty; no additional seats mounted.

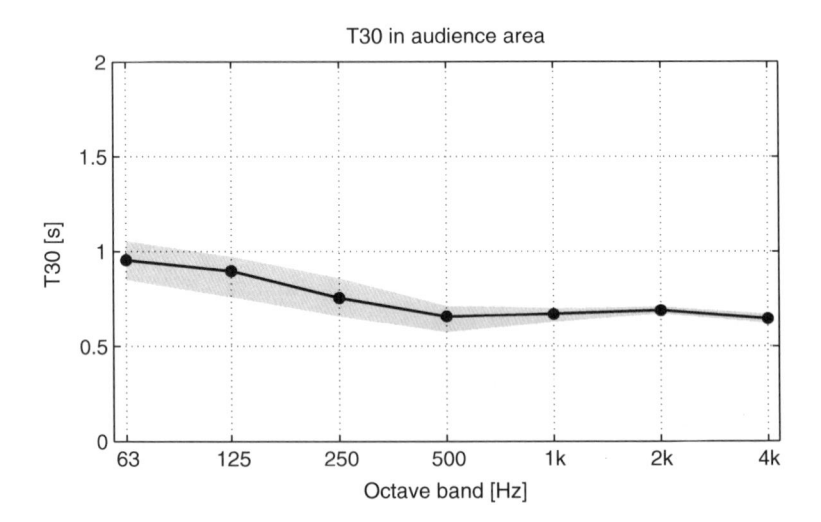

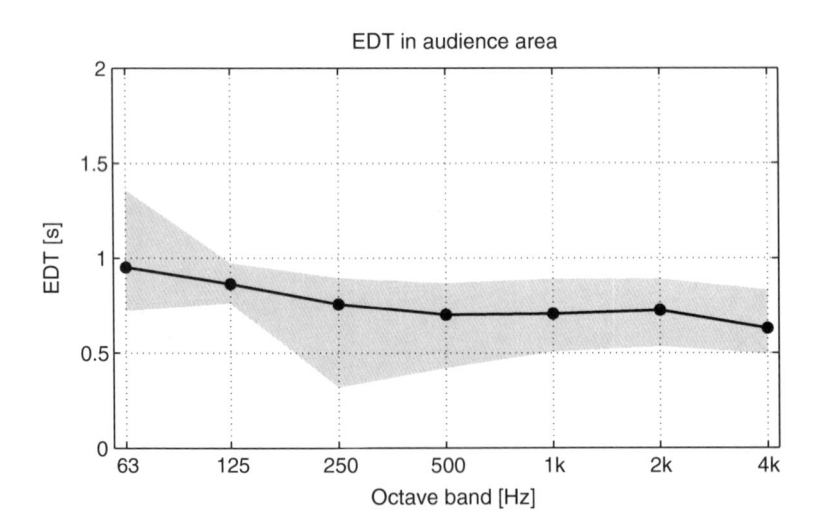

218   7 Gallery of Halls that Present Pop and Rock Music Concerts

EDT on stage

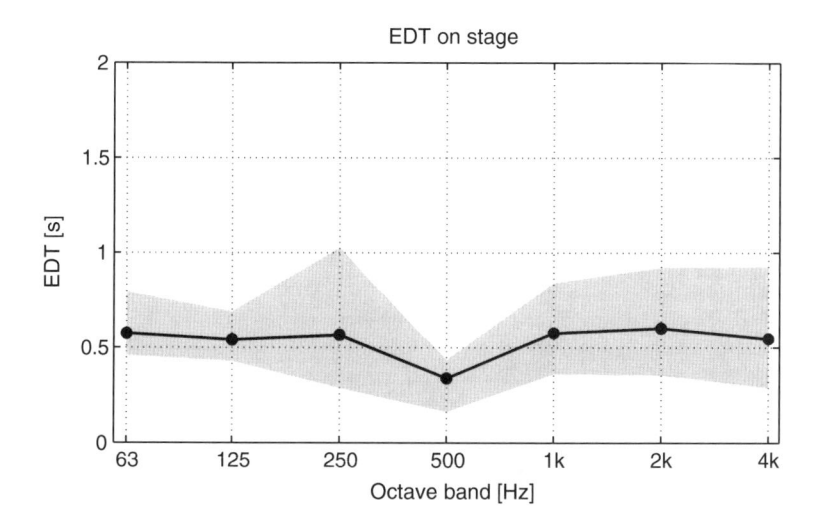

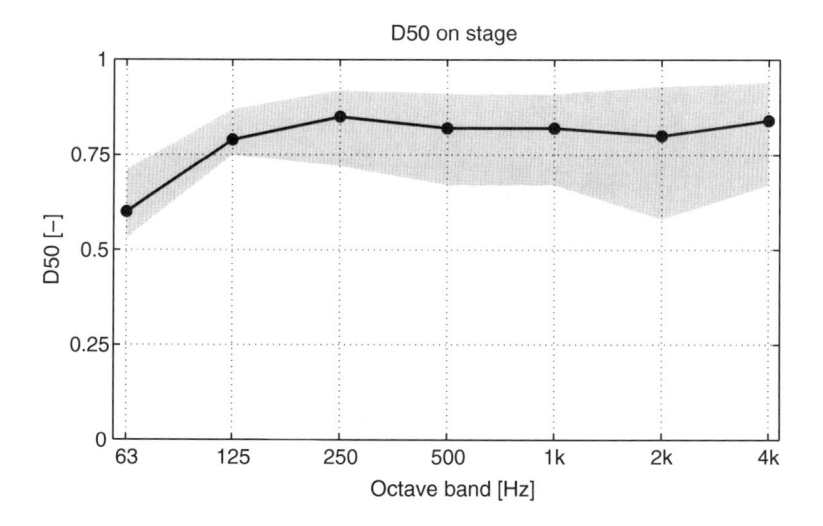

D50 on stage

Elysée Montmartre

Paris
Founded: 1989
Capacity: 1,200
Number of concerts per year: Approximately 120
Architect: N/A
Acoustician: N/A
Owner: Garance Productions

Opened in 1807, the Élysée Montmartre was a dance hall. There they showcased a new dance style: the *quadrille naturaliste* (the naturalist quadrille), or cancan, especially as performed by Valentin le Désossé and Grille d'Égout. The establishment was then composed of three buildings and a large garden. Émile Zola described its façade in his novel (and subsequent drama), *L'Assommoir*. Joseph Oller and Charles Zidler, having heard of the success of the quadrilles at the Élysée Montmartre and who wished to bring in a new audience, high society, for this kind of entertainment, engaged a great number of artists from the Élysée Montmartre for the opening in October 1889 of their new establishment, the Moulin Rouge. Zidler was especially taken with La Goulue, who became one of the most celebrated cabaret dancers. The Élysée Montmartre was also a source of inspiration for painters and artists of the Butte (Toulouse-Lautrec painted numerous posters there). The hall served as the decor for de Maupassant's *Masque* and held the 100th performance of Émile Zola's *L'Assommoir* in 1879. Costume balls such as the Bal des Quat'z'Arts (Four Arts Ball) were also held there. Folowing this, the musical programming of the Élysée Montmartre was diversified and developed.

In the nineteenth century the hall was the site of some of the foremost revolutionary clubs (places where discussions by Utopians and "angry young people" remade the world). In 1894, the garden was torn up to make room for the Trianon-Concert. In 1897, the Élysée-Montmartre was redone by its new owner: a separate café–concert hall. On one side, all the singing, the revues, and other poets and songwriters, and on the other side, dancing and skating. To make this happen, the architect, Édouard Niermans, reused the Pavillon de France structure built by Gustave Eiffel for the *Exposition universelle* (World's Fair) of 1889 [2]. After a fire in 1900, the hall was remodeled with moderistic decorations and a rococo decor.

The Élysée Montmartre held, at the end of 1949, boxing and wrestling matches, and then striptease acts. In 1968, Jean-Louis Barrault mounted *Rabelais* there, a play based on the music of Michel Polnareff. The setting was in a ring. The following year, *Jarry sur la butte* was presented with the music of Michel Legrand. In 1971, Philippe Khorsand put on the play, *O Calcutta*, which was on the bill until 1975.

Elysée Montmartre

Artists such as Jacques Higelin, Patti Smith, Diane Dufresne, and Alain Souchon played concerts there starting in 1976, in addition to many heavy metal groups. In 1983, there was an operetta by Francis Lopez with Georges Guétary.

In 1989, the hall began a new era with the new owners, Garance Productions. They presented rock and reggae concerts, among others. Since 1995, the program has consisted of 15 days of *Le Bal de l'Élysée-Montmartre* enlivened by the GOLEM (Grand Orchestre de L'Élysée Montmartre), thus returning the hall to its first calling.

Beautiful nineteenth-century stucco is apparent in the ceiling.

Numerous delay speakers may lead to a near-field listening experience lacking envelopment and liveliness. But on the other hand, the total sound power in the hall is a sum of the contribution from each loudspeaker.

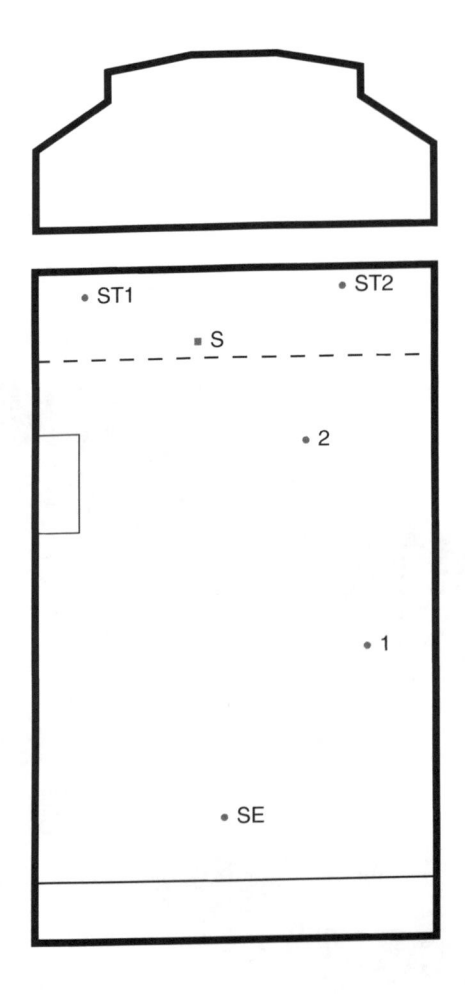

222   7   Gallery of Halls that Present Pop and Rock Music Concerts

- ST1
- ST2
- S
- 2
- 1
- SE

5　0　　10　　20　　30   METERS

Geometrical data	
Volume	6,000 m^3
$L \times W \times H$	$37 \times 23 \times 4.4$–9.6

Acoustical data	
Audience area	
$T_{30, 125-2k}$	0.97
EDT_{125-2k}	0.91
$C_{80, 125-2k}$	4.02
BR_{63} versus 0.5–1k	1.13
BR_{125} versus 0.5–1k	1.34
Stage area	
EDT_{125-2k}	0.37
$D_{50, 125-2k}$	0.89
BR_{63} versus 0.5–1k	2.52
BR_{125} versus 0.5–1k	1.84

Materials Used

Audience Area

Floor: Wood on joists.
Ceiling: Masonry, stucco.
Walls: Masonry, 30 % covered with drapes. Some pillars covered with drapes.

Stage Area

Floor: Wood on joists.
Ceiling: Masonry.
Walls: Backdrop, bare concrete walls on side at the time of the measurement.

State of Hall When Measured

Empty; lots of lighting equipment on stage.

T30 in audience area

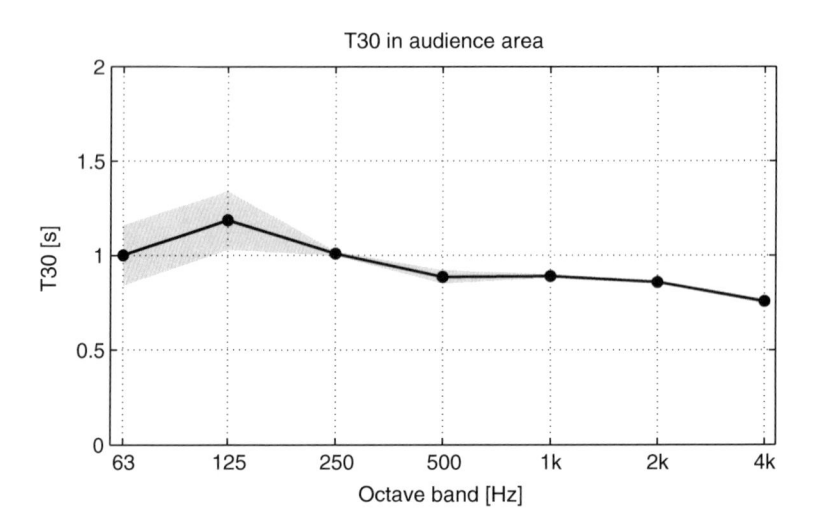

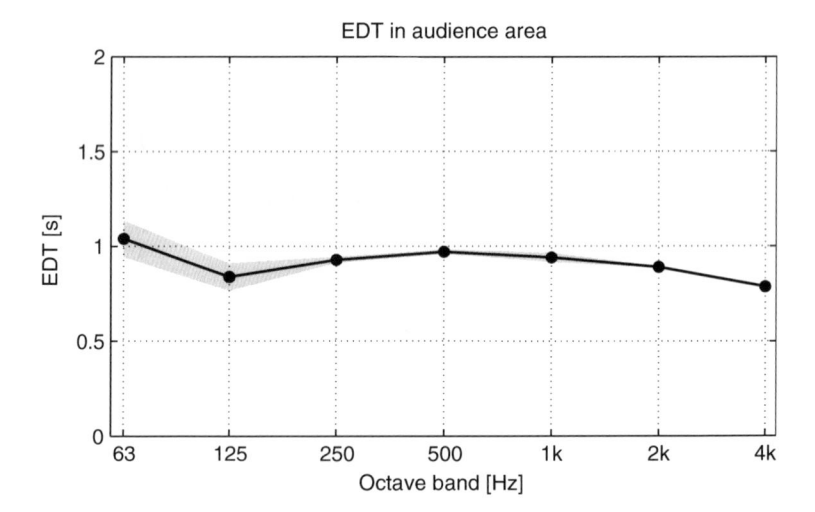

EDT in audience area

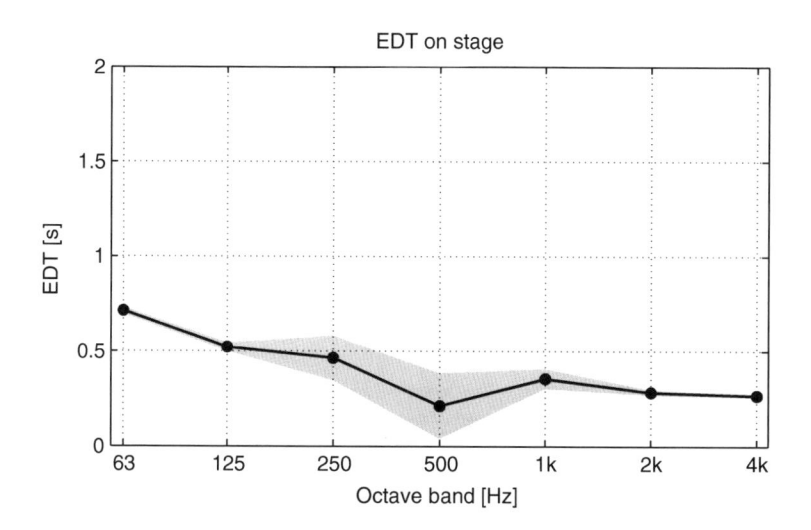

State of Hall When Measured

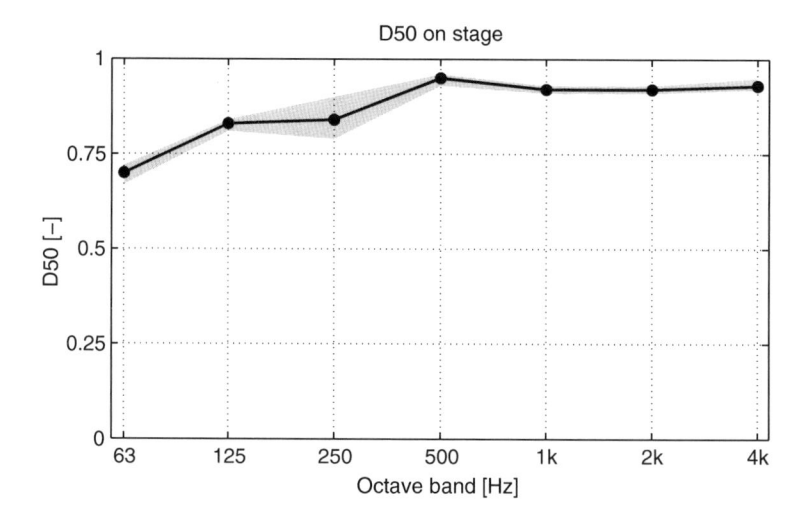

Festhalle

Frankfurt
Number of nonclassical music concerts per year: N/A
Founded: 1909
Capacity: 13,500 without chairs
Architect: Friedrich von Thiersch
Acoustician: N/A
Owner: Stadt Frankfurt (40 %)/Land Hessen (60 %)

On May 19th 2009, it was exactly 100 years since Frankfurt's Festhalle welcomed its very first guests. From rock concerts, operas, sporting events, and circus performances to international conventions, trade fairs, AGMs, and gala balls, the Festhalle is versatility itself, housing all kinds of events with practiced ease.

With the opening of the Festhalle a hundred years ago, the city of Frankfurt finally boasted a suitable venue for hosting large-scale exhibitions and events, signaling an end to expensive makeshift solutions. At the same time, it marked the beginning of a long tradition of outstanding events of all kinds.

When Kaiser Wilhelm II made it known in Summer 1905 that he would be willing to have the traditional singing contest transferred permanently to Frankfurt as well as having the Eleventh German Gymnastics Festival held there three years later, the municipal authorities lost no time in taking action. At a meeting of town councillors, funds were allocated for a general architectural competition. The invitation to tender was issued in April 1906 and was won just one year later by Friedrich von Thiersch, who designed a 6,000 m^2, daylight-flooded hall, suitable for playing host to all manner of exhibitions, musical performances, and other events.

Construction work began on June 11th, 1907 and was supervised directly by the Frankfurt City Council. Responsibility for marketing and operating the hall was given to "Ausstellungs- und Festhallen-Gesellschaft", known today as Messe Frankfurt. Exactly 13 months later, the still-unplastered Festhalle hosted the opening ceremony for the Eleventh International German Gymnastics Festival. When the gymnasts had returned home, the construction was completed and officially opened on schedule.

Even at the ripe old age of 100, the "grand old lady" of event halls is still going strong. After all, nothing is too outlandish or too extravagant for the Festhalle.

From rock concerts, operas, sporting events and circus performances to international conventions, trade fairs, AGMs and gala balls, the Festhalle is versatality itself, housing all kinds of events with practised ease.

Uwe Behm, a member of the Messe Frankfurt Board of Management, sums up the unique role of the Festhalle: "As the historical core of our exhibition grounds, it is an integral part of the company's success story. We are delighted to be able to use the Festhalle for all kinds of events. The building combines event expertise, glamour and historical ambience—which is what gives it its own special atmosphere."

Festhalle 227

Visually, certainly the most spectacular of all visited venues. One's senses are sharpened.

Hard surfaces ensured the unamplified sound of the early twentieth century to be carried over the vast distances inside the hall.

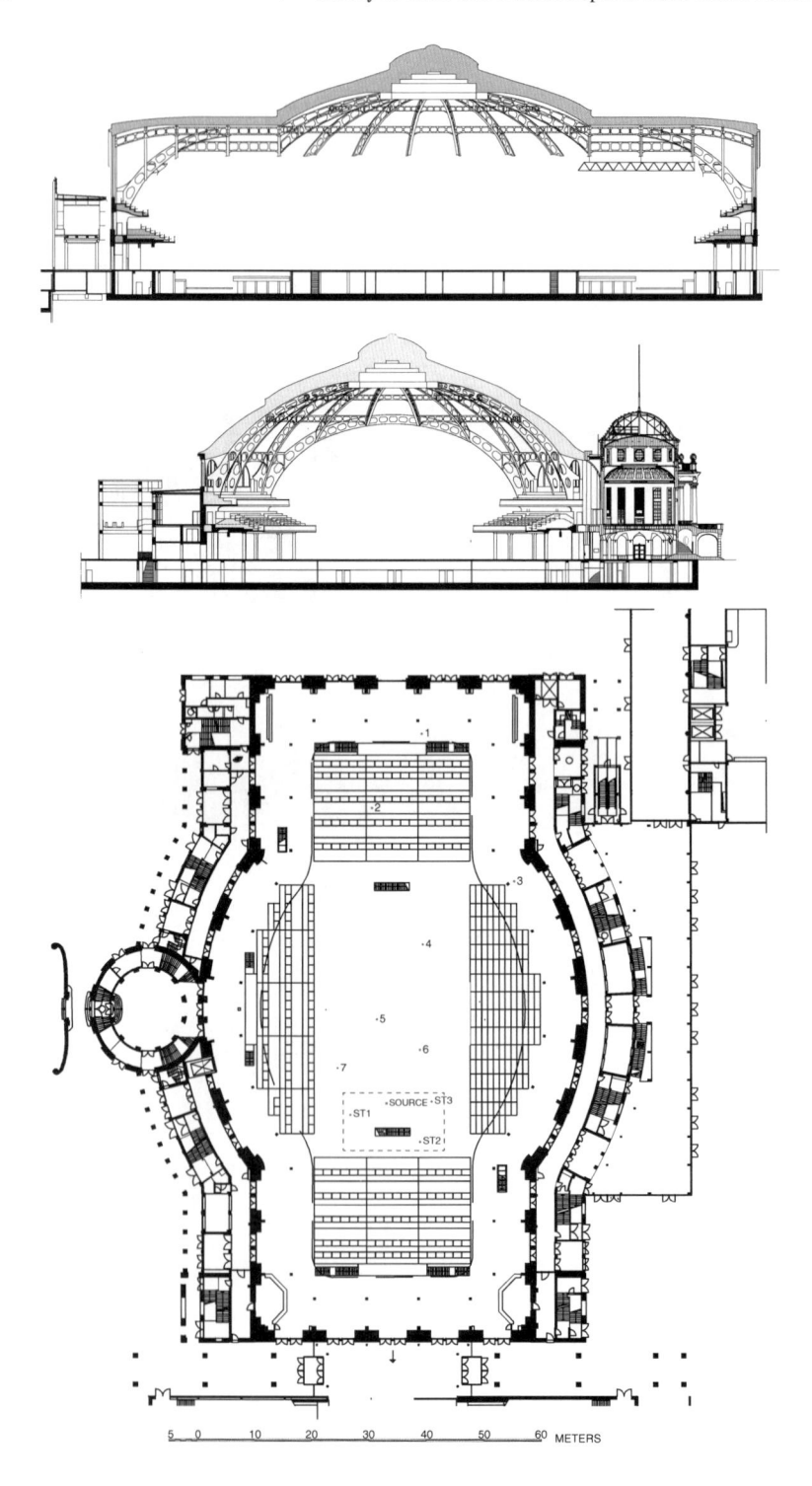

228 7 Gallery of Halls that Present Pop and Rock Music Concerts

Geometrical data	
Volume	110,000 m^3
$L \times W \times H_{(max)}$	$109 \times 64 \times 29$ m

Acoustical data	
Audience area	
$T_{30, 125-2k}$	4.15
EDT_{125-2k}	4.28
$C_{80, 125-2k}$	−4.06
BR_{63} versus 0.5–1k	0.74
BR_{125} versus 0.5–1k	0.91
Stage area	
EDT_{125-2k}	2.97
$D_{50, 125-2k}$	0.44
BR_{63} versus 0.5–1k	0.6
BR_{125} versus 0.5–1k	0.74

Materials Used

Audience Area

Floor: Concrete.
Ceiling: Masonry. Glazing in dome.
Walls: Masonry on concrete.

Stage Area

No stage.

State of Hall When Measured

Empty; no additional seats mounted.

Festhalle.

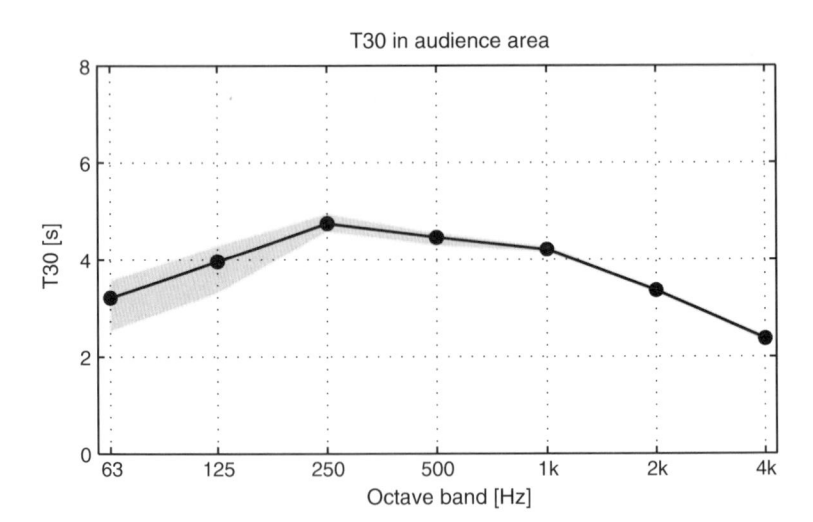

EDT in audience area

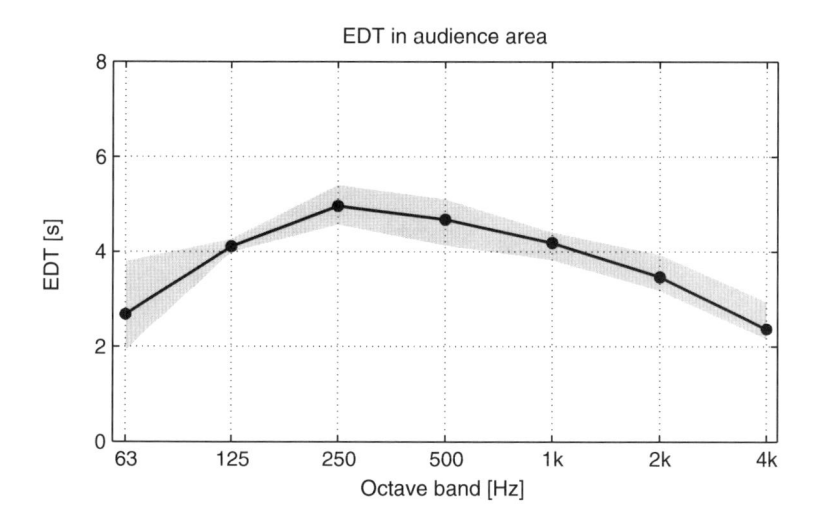

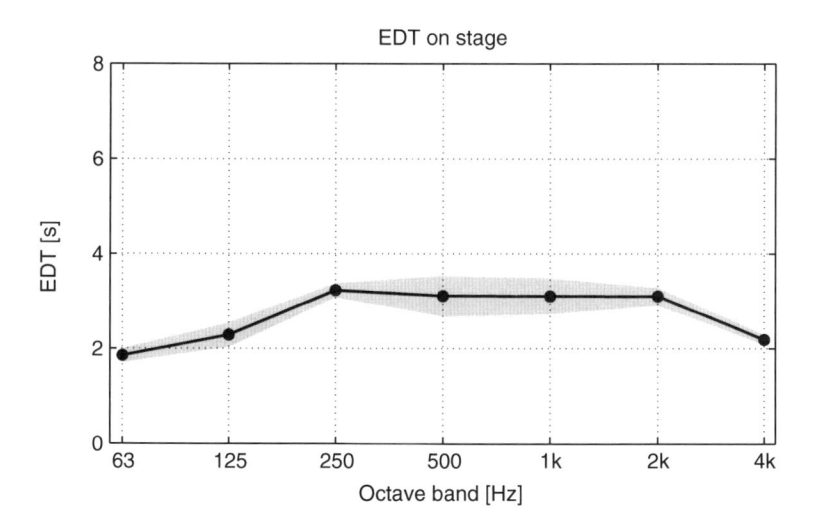

EDT on stage

D50 on stage

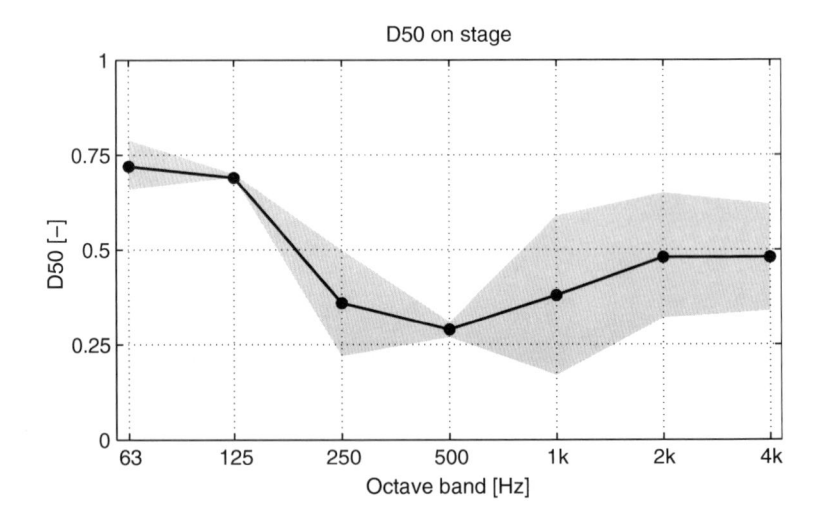

Forest National

Bruxelles
Number of concerts per year: 100
Founded: 1970
Capacity: 10,000
Architect: NA
Acoustician: NA
Owner: Music Hall Group

Since 1970 Forest National has been a venue for major public events such as concerts, musicals, and sports events. The venue was first opened as "Palais des sports," and then commuted into a concert venue. This resulted in a phenomenal list of artists and shows over the years and has put Brussels on the schedule of every international tour manager. The venue has hosted about 3,500 shows attended by over 15 million fans. Yearly, over 500,000 people attend an event in the hall. Forest National is the biggest concert venue in Brussels. Music Hall Group bought the venue in 1995 and renovated it: VIP lounges and VIP boxes were added, and the capacity rated up to 10,000 places. With Maurice Béjart's *Ballets*, the venue's grand opening celebration, the tone was set for quality from the first day onwards and has made it one of the most beloved arenas of the public, the promoters, and the artists.

An impressive list of names of artists and bands of all the styles that contemporary music has to offer have performed on the mythical stage of Forest National. Rock'n'roll legends the Rolling Stones, Queen, or U2; jazz icons Ella Fitzgerald, Count Basie, or Benny Goodman; pop divas Janet Jackson, Diana Ross, or Kylie Minogue; reggae king Bob Marley and the Wailers; French chanson ambassadors Michel Sardou or Johnny Hallyday; New Wavers The Cure, Indochine, or Siouxie and the Banshees; Latin extravaganza Santana, Gloria Estefan, or Enrique Iglesias; hard rock myths Black Sabbath, Iron Maiden, or Metallica; today's hit wonders Katy Perry, The National, Editors, Tokio Hotel, Nelly Furtado, or 50 Cent, and last but not least Belgium's finest artists dEUS, Hooverphonic, Ozark Henry, or Axelle Red, they all had their moment of glory, their standing ovation in Forest National.

In addition to the glitter and the glamour, the venue has also been the home of spectacular operas such as *Aïda*, *Nabucco*, and *Carmen* and moving musicals such as *Notre Dame de Paris*, *Jesus Christ Superstar*, and *Mamma Mia!*.

Other events could be hosted thanks to the magnificent round architecture of the arena. Basketball exhibitions by the Harlem Globe Trotters, equestrian shows of the Spanish Riding School of Vienna, wrestling matches, tennis tournaments, gymnastic championships, and ice-skating family happenings like *Disney On Ice* were a delight for children and adults.

In 2009 Dries Sel was appointed as new CEO with the mission of turning the mythical rock temple into a venue with a legendary future. Forest National started a diversification of its offer, welcoming genres as different as rock, R&B, rap, chanson française, family shows, circus, and so on. A first edition of *KDO!*, a show created by Franco Dragone (Cirque du Soleil's former artistic director) especially for Forest National, sold over 60,000 tickets in the winter of 2009–2010.

234   7 Gallery of Halls that Present Pop and Rock Music Concerts

At the same time the new management team started a series of actions in order to reduce any possible negative impact of the venue on its direct neighborhood and the environment, addressing mobility and parking issues, sound emissions, energy consumption, and so on. The purpose was to create within a few years a brand new Forest National, ensuring its visitors an exceptional experience from ticket buying to the long-time strength of the venue, fabulous shows and concerts, to going home.

If walls could speak ... Since 1970 most great rock stars have passed through this hall.

Large hallways of concrete behind the seats add reverb to the reverberant space.

Forest National

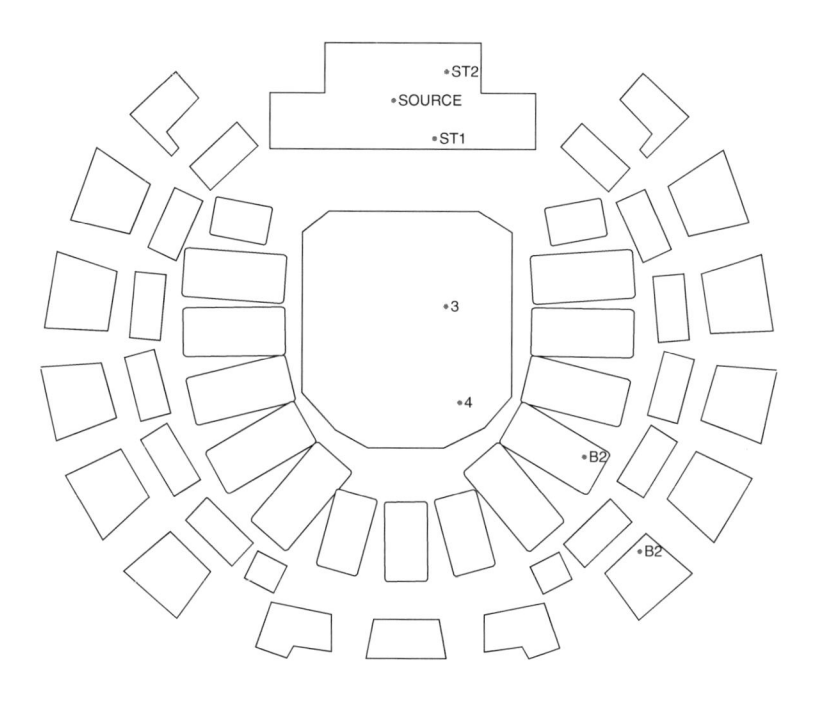

NTS

Geometrical data	
Volume	App. 150,000 m^3
Height	20.5 m

Acoustical data	
Audience area	
T_{empty}	
T_{full}	
D_{50}	
EDT	
BR$_{rock}$	
Stage area	
T_{empty}	
D_{50}	
EDT	
BR$_{rock}$	

Materials Used

Audience Area

Floor: Concrete.
Ceiling: Thin nonperforated trapezoid steel.
Walls: Concrete.
Seats: Upholstered on back and seat.

State of Hall When Measured

Empty; some seats installed on floor. Backdrop.

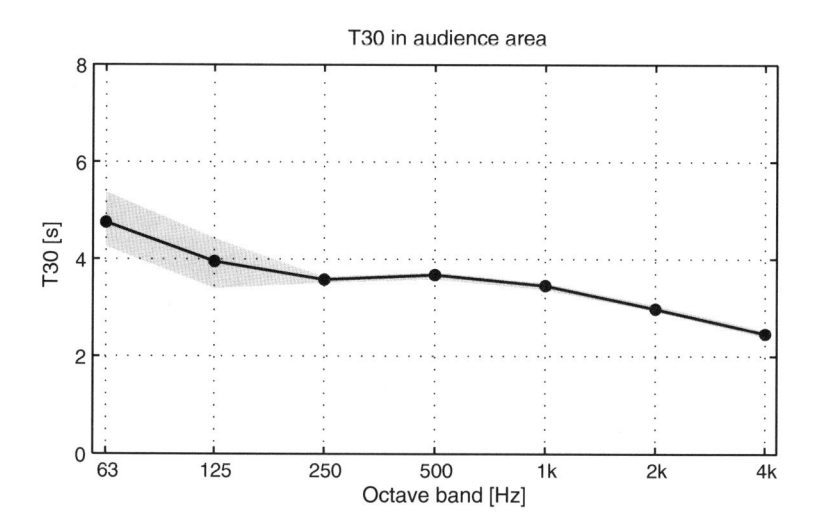

State of Hall When Measured

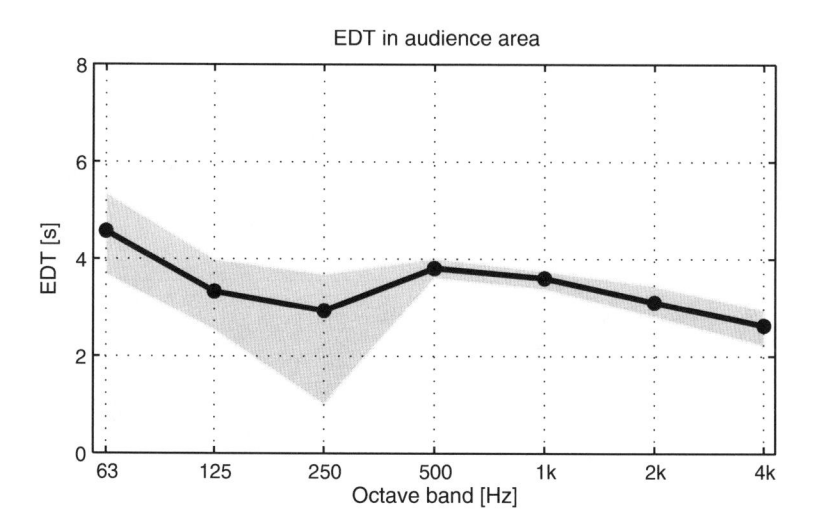

T30 in audience area

EDT in audience area

Globen

Stockholm
Number of concerts per year: 50
Founded: 1989
Capacity: 16,000
Architect: Berg Arkitektkontor AB
Acoustician: Tunemalm Akustik AB, Svante Berg, Esbjörn Adamson, and Lasse Vretblad

The construction work of this colossal hall began in 1986 and ended 2½ years later in 1989. It is the biggest spherical building in the world and has become a symbol for Stockholm and Sweden. The building has a horizontal diameter of 110 m, an inside height of 85 m, and has a volume of 600,000 m^3.

With a volume of 600,000 m^3 the Globe Arena is the largest of all halls in this book.

Globen 239

Curtains can be drawn in front of the windows in order to avoid echoes.

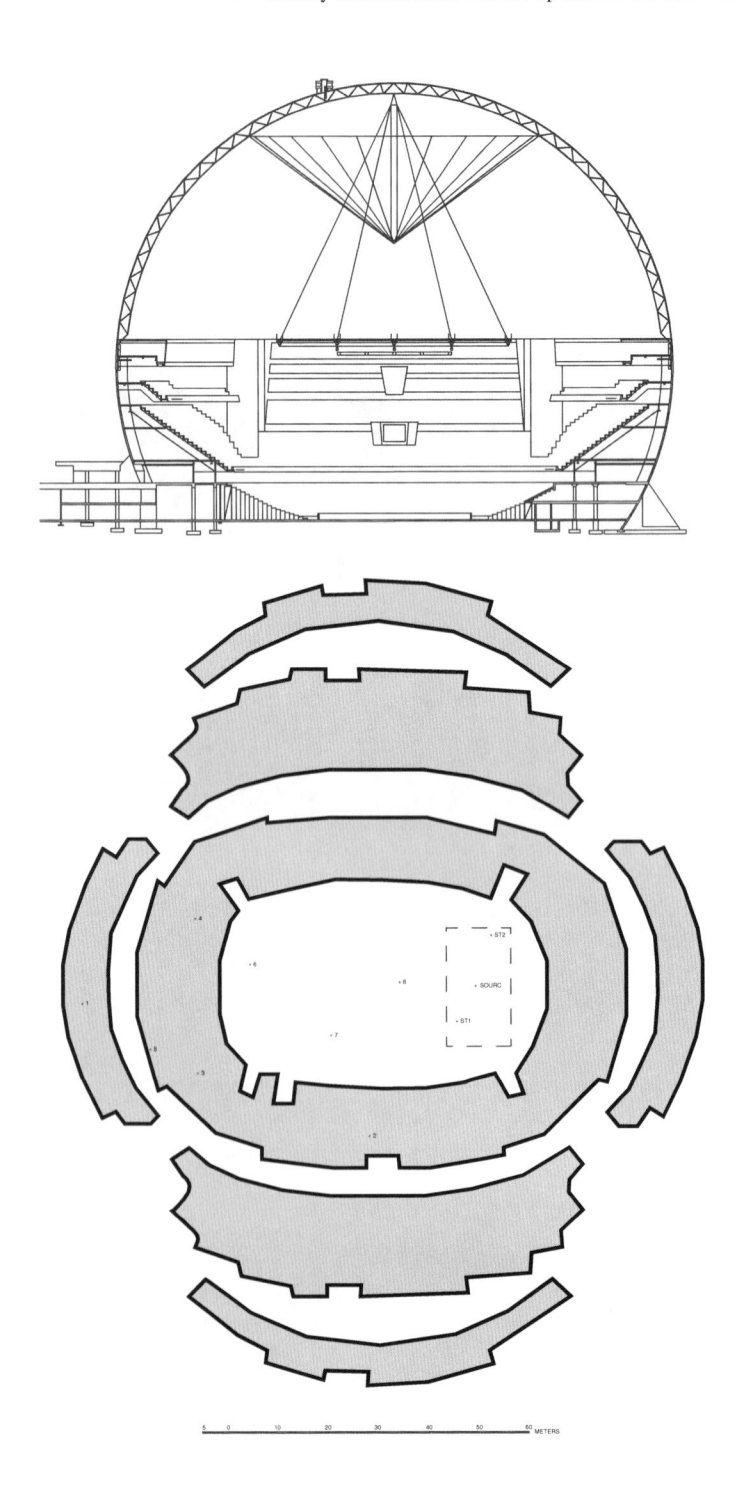

240   7   Gallery of Halls that Present Pop and Rock Music Concerts

Geometrical data	
Volume	600,000 m^3
Height, audience area	85 m
Outer horizontal diameter (max)	110 m

Acoustical data	
Audience area	
$T_{30, 125-2k}$	3.71
EDT_{125-2k}	3.22
$C_{80, 125-2k}$	0.27
BR_{63} versus 0.5–1k	1.89
BR_{125} versus 0.5–1k	1.67

Materials Used

Audience Area

Floor: Concrete.
Ceiling: Mineral wool product.
Seats: Upholstered on seat and back. Reflective when not in use.

Stage Area: NA

State of Hall When Measured

Empty; no additional seats mounted. One fourth of lower seats covered with curtain.

T30 in audience area

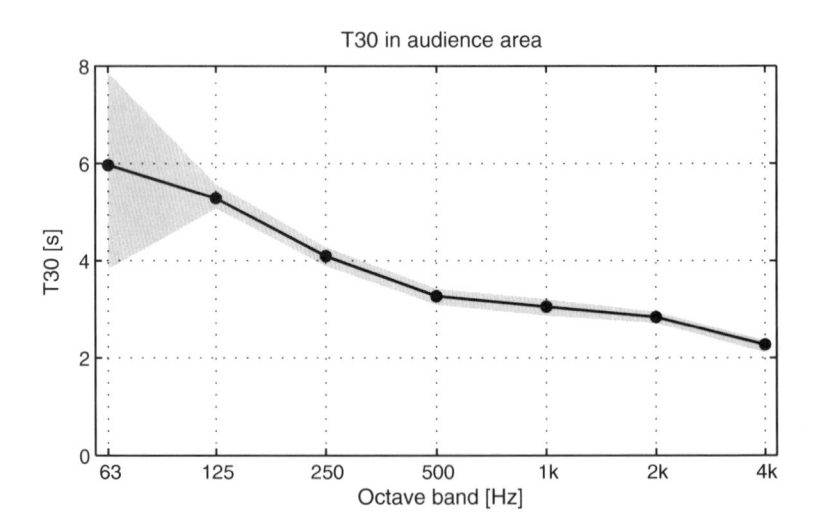

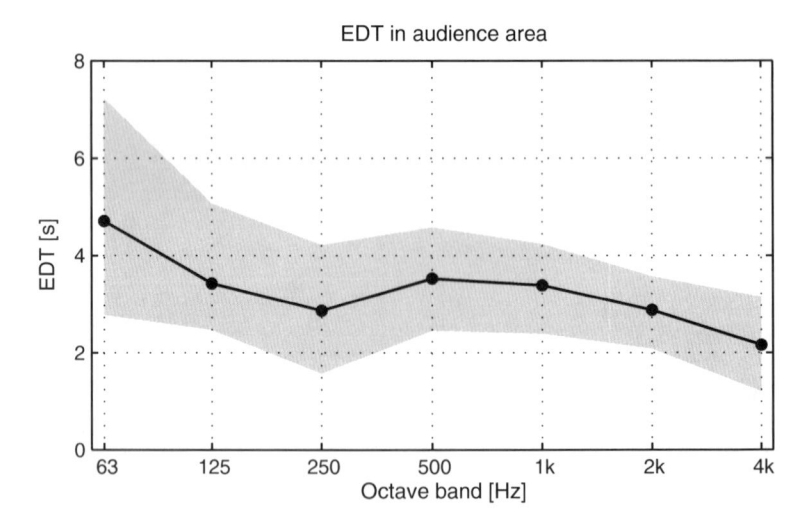

EDT in audience area

Grosse Freiheit

Hamburg
Number of concerts per year: 75
Founded: 1985. Building constructed in 1958
Capacity: 1,250
Architect: Erwin Nagel
Acoustician: N/A

Upstairs from Kaiser Keller where the Beatles played for several months before breaking internationally is the much larger Grosse Freiheit. Considering its size, amazing artists have played here during the past decades, including Johnny Winter and Band, Stevie Ray Vaughn, Sly Dunnbar, Robby Shakespeare, The The, The Ramones, Die Ärzte, Public Enemy, INXS, Cruzados, Meat Loaf, Wet Wet Wet, Faith No More, Manu Dibango, Crowded House, Nick Cave And The Bad Seeds, Sugarcubes, Living Colour, R.E.M., Underworld, John Cale, Texas, Neil Young, George Clinton, Deep Purple, Youssou N'Dour, Leningrad Cowboys, Jonathan Butler, Chris Isaak, Robert Palmer, Maceo Parker, The Jets, Seal, Tower Of Power, Fehlfarben, Curtis Stigers, Pearl Jam, Mothers Finest, Blur, Stereo Mc's, Duran Duran, Einstürzende Neubauten, Ace of Base, Michael McDonald, Björk, Sheryl Crow, Brand New Heavies, Nina Hagen, Huey Lewis, Pantera, Cranberries, Bob Geldof, Jamiroquai, Portishead, Mike and The Mechanics, Joan Armatrading, Suzanne Vega, Kid Kreole and The Coconuts, Robben Ford, Marcus Miller, Crash test Dummies, Rammstein, Bon Jovi, Miriam Makeba, Kool and The Gang, Daft Punk, The Corrs, Willie Nelson, Nena, Runrig, Saga, Macy Gray, Lou Reed, Muse, D´Angelo, dEUS, Arctic Monkeys, Travis, Heather Nova, Amy Mc Donald, and Sterephonics.

This is the kind of venue that was probably built unintentionally of fairly suitable materials for rock concerts.

With a height of less than 6 m there is enough room for a balcony level. And with a carefully adjusted PA system the mid–high frequency reverberation can possibly be tamed by the presence of a packed audience without any (other) porous absorbers in the room.

Grosse Freiheit 245

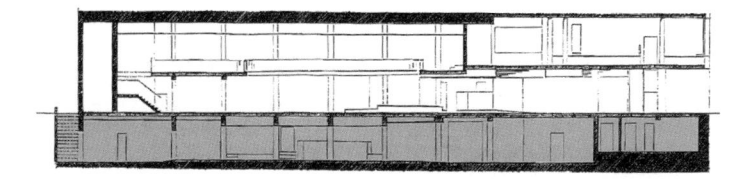

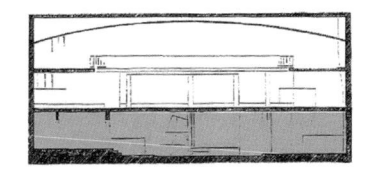

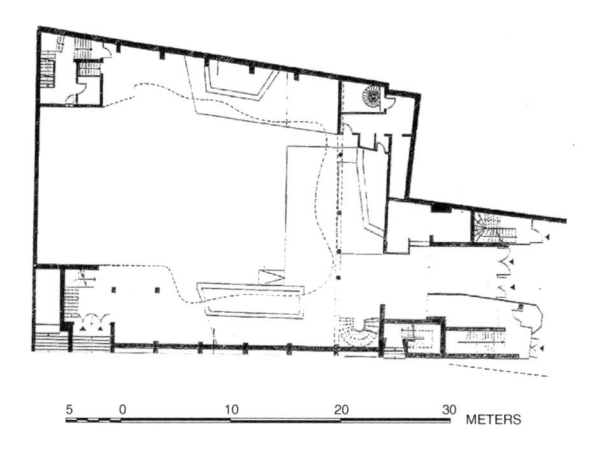

Geometrical data	
Volume	4,200 m^3
$L \times W \times H$	31.5 × 24.5 × 6

Acoustical data	
Audience area	
$T_{30, 125-2k}$	1.48
EDT_{125-2k}	1.52
$C_{80, 125-2k}$	1.75
BR_{63} versus 0.5–1k	0.86
BR_{125} versus 0.5–1k	0.75
Stage area	
EDT_{125-2k}	0.99
$D_{50, 125-2k}$	0.56
BR_{63} versus 0.5–1k	0.64
BR_{125} versus 0.5–1k	0.97

Materials Used

Audience Area

Floor: Concrete; approximately 15 % is covered by wooden audience risers (in the back of the room).
Ceiling: Thin plate with cavity behind.
Walls: Side walls are concrete, rear wall is plate on cavity.

Stage Area

Floor: Vinyl on wood on cavity on concrete.
Ceiling: Thin plate with cavity behind.
Walls: Curtains.

State of Hall When Measured

Empty; a couple of flight cases and a DJ set-up on stage.

T30 in audience area

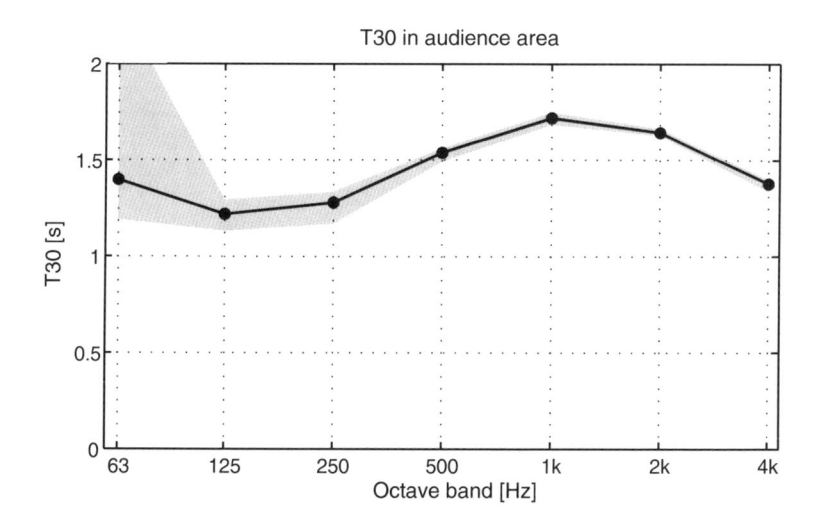

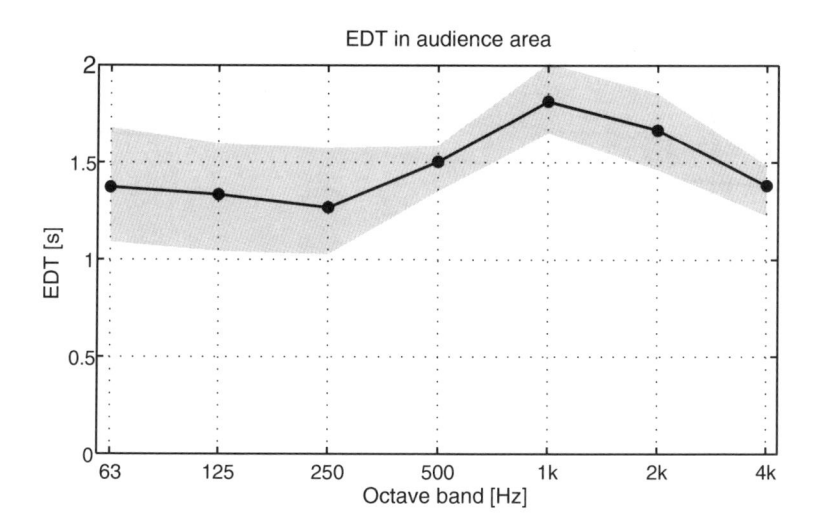

EDT in audience area

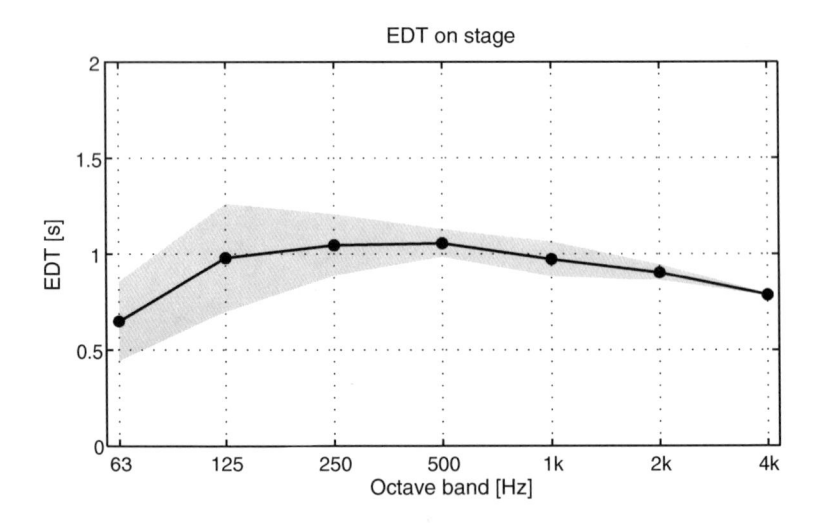

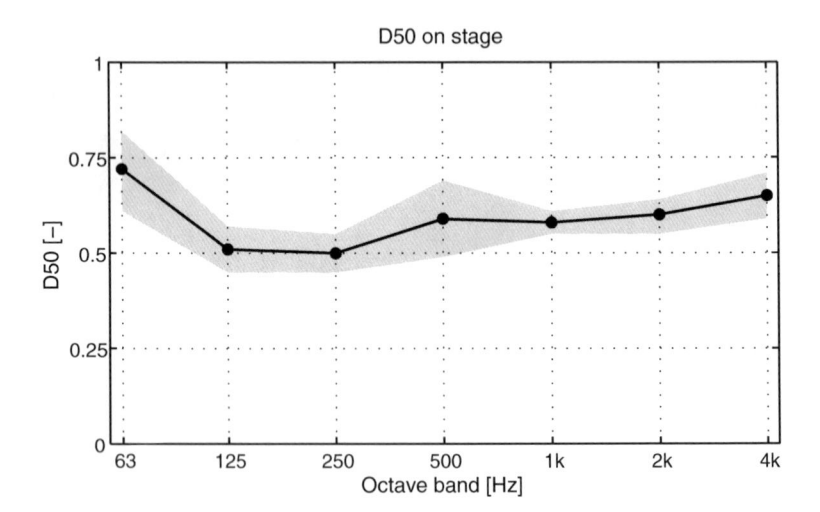

Hallenstadion

Zürich
Number of concerts per year: 70
Founded: 1939; rebuilt: 2005
Capacity: 13,000
Architect: Karl Egender and Wilhelm Müller
Acousticians of rebuild: Kopitsis Bauphysik AG, Wichser Akustik and Bauphysik AG
Owner: AG Hallenstadion

Bis zu 13,000 Besucher haben Platz in der Hallenstadion ARENA. Von der Eisfläche über die Konzertbühne, den Tenniscourt oder den Pferdeparcours bis zum steinig-erdigen Untergrund, auf dem sich Monster Trucks bewegen, bietet die ARENA eine enorme Bandbreite an Möglichkeiten. Auch für Corporate Events ist sie eine beeindruckende Kulisse und ideal nutzbar für Generalversammlungen, Kongresse oder Ausstellungen.

Mit dem FORUM und dem CLUB bietet das Hallenstadion zudem zwei Layouts, die speziell für Kongresse und Business Events von 600 bis 3,000 Personen beziehungsweise Konzerte für bis zu 4,500 Zuschauer konzipiert sind. Das Raumkonzept basiert auf einem standardisierten Layout, das dank diverser Vorinstallationen und der flexiblen Funktionalität sehr schnelle Umbauten und damit kostengünstigere Produktionen ermöglicht.

Hallen Stadion is an up-to-date modern facility with carefully designed acoustics.

The hall also works well acoustically for sports games. All time tennis great Roger Federer was to play a couple of days after this measurement took place.

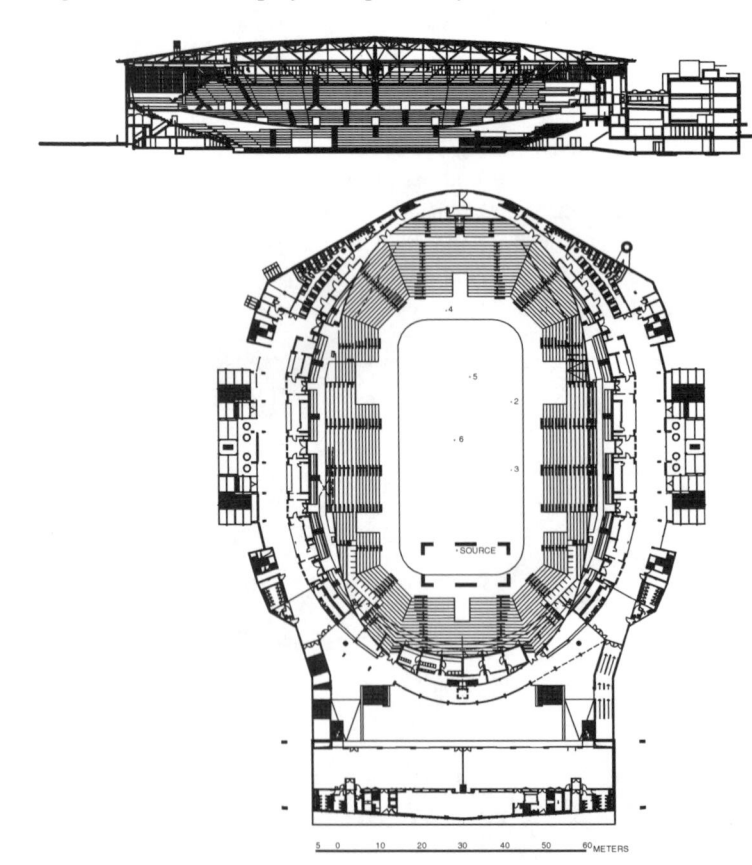

Geometrical data	
Volume	120,000 m^3
$L \times W \times H_{(max)}$	115 × 100 × 25 m

Acoustical data	
Audience area	
$T_{30, 125-2k}$	2.48
EDT_{125-2k}	2.07
$C_{80, 125-2k}$	0.58
BR_{63} versus 0.5–1k	1.46
BR_{125} versus 0.5–1k	1.1

Materials Used

Audience Area

Floor: Concrete
Ceiling: Two layers of 50-mm mineral wool slabs with a 100-mm air space in between.
Walls: Porous absorbers and perforated thick plates. Curtains.
Seats: Upholstered also underneath.

Stage Area

No stage measurement.

State of Hall When Measured

Empty; no additional seats mounted on floor area.

T30 in audience area

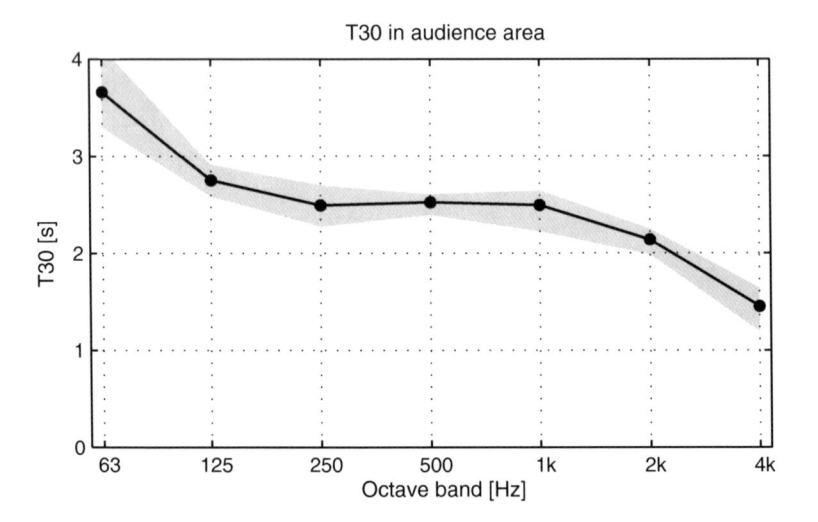

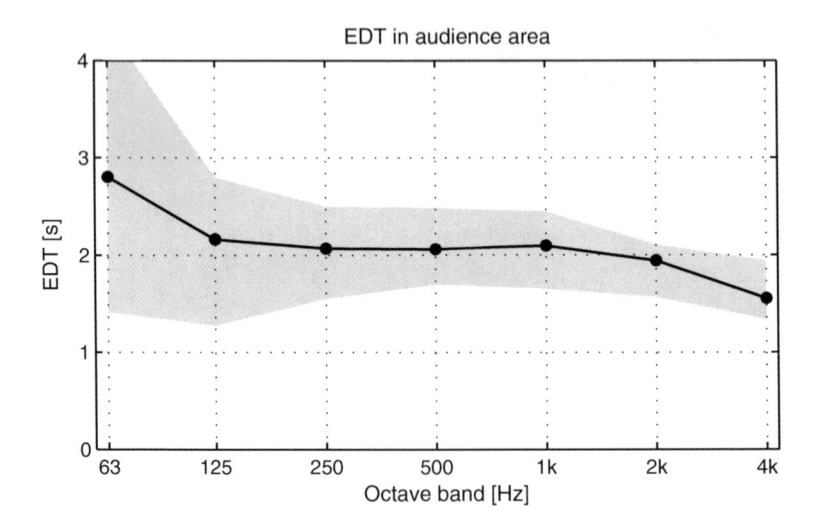

EDT in audience area

HMV Hammersmith Apollo

London
Number of concerts per year: 120
Founded: 1932
Capacity: 5,039
Architect: Robert Cromie
Acoustician: N/A

HMV Hammersmith Apollo is one of London's major live entertainment venues. It is located in Hammersmith, West London and is one of the UK's largest and best-preserved original theatres. It opened on the 28th of March, 1932 as the Gaumont Palace cinema, designed in the Art Deco style by renowned theatre architect Robert Cromie, who also designed the Prince of Wales Theatre in Central London. It was designed on behalf of a joint collaboration between exhibitor Israel Davis and the Gaumont British Theatres chain. It had 3,487 seats and the opening program was Tom Walls, *A Night Like This*, and Helen Twelvetrees in *Bad Company*. It had a large 35-foot deep stage, an excellent fan-shaped auditorium (which, despite its enormous 192 foot width allows remarkable intimacy and excellent sightlines from all parts of the house), 20 dressing rooms, a Compton Manual/15 Ranks theatre organ, and a café/restaurant located on the balcony/foyer area.

HMV Hammersmith Apollo was renamed the Hammersmith Odeon in 1962 and started playing host to many legendary acts of the day, including the Beatles, the Rolling Stones, and Bob Marley. It screened its last regular film in 1984, *Blue Thunder*. Following a sponsorship deal, it was later refurbished and renamed the Labatt's Apollo. During his 1992 sold-out tour, Michael Ball, the musical theater star, best known for his roles in *Les Miserables*, *Phantom of the Opera*, and *Hairspray*, was the last person to play the venue when it was named "Odeon" and the first person to play after it was renamed "Apollo." The venue continued to host long-running shows and musicals such as *Riverdance* and *Dr Doolittle*.

In the early 1990s it reverted back to the Hammersmith Apollo. In 1990, it was designated a Grade II listed building by English Heritage and was upgraded to Grade II* status in 2005. 2003 saw the venue renamed as the Carling Apollo Hammersmith, after another brewery entered into a sponsorship deal with the then-owners, Clear Channel Entertainment, a US-based company (which then spun off as Live Nation UK). Major alterations enabled the stalls to be removable, allowing for both standing and fully seated events. Capacity became 5,039 (standing) and 3,632 (sitting). In 2006, the venue reverted to its former name, the Hammersmith Apollo. The owners were encouraged by Hammersmith & Fulham Council and the Cinema Theatre Association to reinstate the original Compton organ console which had been removed from the building and put into storage in the 1990s. The organ chambers were retained in the building and with its console connected up again, the huge Apollo auditorium is now filled with its sound after 25 years of silence.

The venue changed hands once again in June 2007 when it was bought by MAMA Group, a UK-based entertainment company who own a number of music venues and festivals artist management companies and other music-related businesses such as the UK's most widely circulated music magazine, *The Fly*. In 2009, it was announced that MAMA Group had entered into a joint venture with HMV to jointly run 11 live music venues across the United Kingdom, including the Hammersmith Apollo, the Kentish Town Forum, the Jazz Cafe, and London Garage. Hence, the venue is now known as the HMV Hammersmith Apollo.

A glance at the list of bands having played the venue is almost a lesson in European pop and rock history. In the early 1960s, many of the top American stars performed at the Odeon, including Tony Bennett, with Count Basie, Ella Fitzgerald with Duke Ellington, Louis Armstrong, and Woody Herman and the Herd. However, in late 1964 and early 1965, the Beatles played 38 shows over 21 nights. Special guests on the bill included the original Yardbirds, featuring Eric Clapton. In 1966, Johnny Cash performed at the venue. In the 1970s stars and bands including Bowie, Slade, Bruce Springsteen, Kiss, Thin Lizzy, Sweet, Rush, and Frank Zappa played there. In December 1979, Queen played several further concerts. The Hammersmith Odeon hosted the four-night Concerts for the People of Kampuchea, a benefit concert to raise money for Cambodian residents, who were victims of the tyrannical reign of dictator Pol Pot, of which Queen played the first night.

In the 1980s Blondie, Heroes, Cher, Van Halen, and Osbourne headlined, and in 1981, Motörhead's live album, *No Sleep 'til Hammersmith*, brought the Odeon to the international stage, becoming widely recognized. Duran Duran and Depeche Mode then both recorded live albums entitled after the Hammersmith. Iron Maiden and AC/DC played four consecutive nights. Elton John, Phil Collins, Boy George, and U2 performed three shows there in 1983 on their War Tour. David Gilmore. On February 6–7, 1984, Soft Cell played their last two shows. On June 1st, 1984, Venom accidentally burned Hammersmith's ceiling during a performance, which event can be clearly seen in the *7 Dates of Hell* concert video (during *Countess Bathory*). As a result, Venom was banned from the Hammersmith Apollo for a year. In 1986 A-Ha played six consecutive shows there. On June 9th, 1988, Dire Straits (and Eric Clapton on rhythm guitar) performed a second "warm-up" show at Hammersmith leading up to the *Nelson Mandela 70th Birthday Tribute* to be held on 11 June 11th, 1988 at Wembley Stadium, London. On October 9–11, 1988, Metallica returned to perform, on three consecutive nights, during their Damaged Justice Tour. During the 1990s the hall hosted a number of stage productions but bands such as Pantera and Megadeath found their way to the concert venue. In recent years, performers have included Prince, Oasis, REM, Stereophonics, Kylie, Elton John, Peter Kay, and Paul Weller to name but a few.

HMV Hammersmith Apollo

This hall is lively and performers enjoy good contact visually as well as acoustically with the audience.

There is no porous absorption apart from the upholstered seats.

The fan-shaped hall, in this case very pronounced, seems to work well for amplified music. Mid- to high-frequency sound is beautifully diffused giving an airy and lively atmosphere.

HMV Hammersmith Apollo

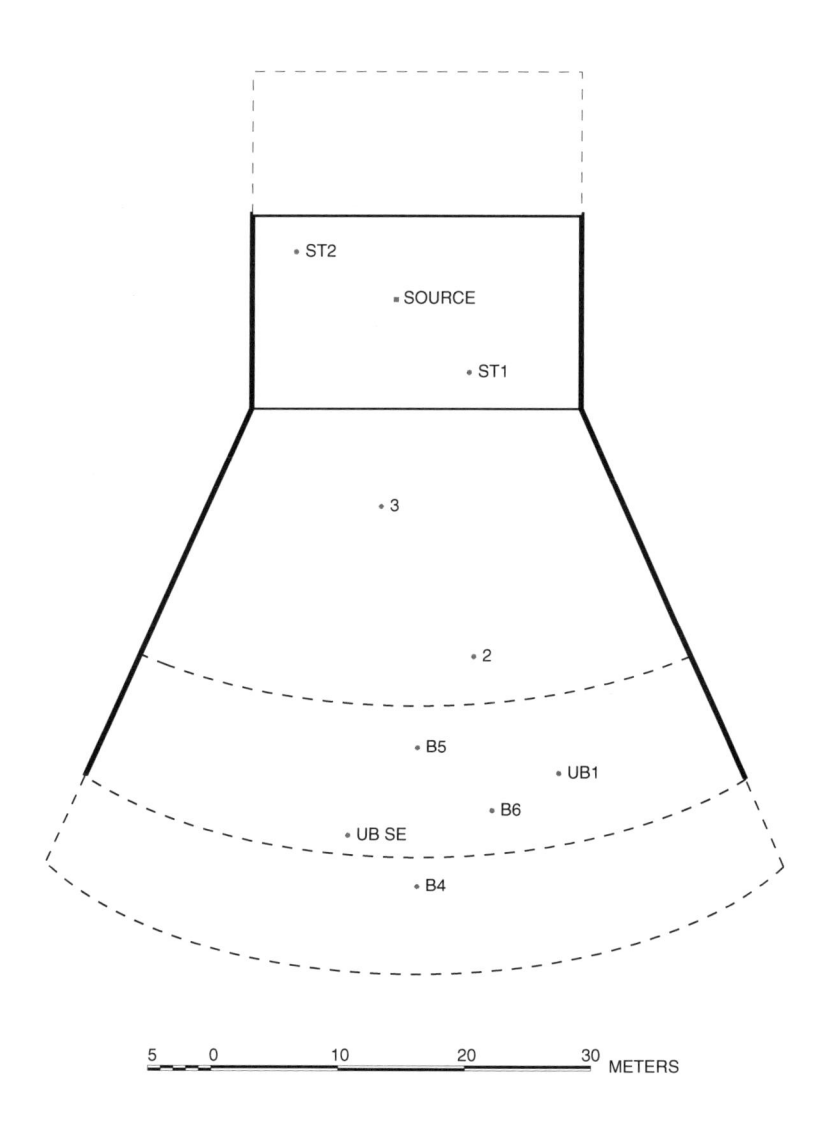

257

- ST2
- SOURCE
- ST1
- 3
- 2
- B5
- UB1
- UB SE
- B6
- B4

5 0 10 20 30 METERS

Geometrical data	
Volume	Approximately 20,000 m^3
Height, audience area	14.4 m
$L \times W \times H$	$58 \times 26 \times 14.4$

Acoustical data	
Audience area	
$T_{30, 125-2k}$	2.35
EDT_{125-2k}	2.12
$C_{80, 125-2k}$	−0.65
BR_{63} versus 0.5–1k	0.98
BR_{125} versus 0.5–1k	1.03
Stage area	
EDT_{125-2k}	2.09
$D_{50, 125-2k}$	0.6
BR_{63} versus 0.5–1k	1.11
BR_{125} versus 0.5–1k	0.96

Materials Used

Audience Area

Floor: Hard plate on concrete.
Ceiling: Under balcony: painted plaster on masonry. Above balcony, rear: porous masonry; in front: painted ornamented plaster or wood direct on painted masonry.
Walls: painted, ornamented masonry.

Stage Area

Floor: Vinyl on wood direct on concrete.
Ceiling: Masonry.
Walls: Masonry.

State of Hall When Measured

Empty; no seats mounted on lower floor. All seats mounted on balcony.

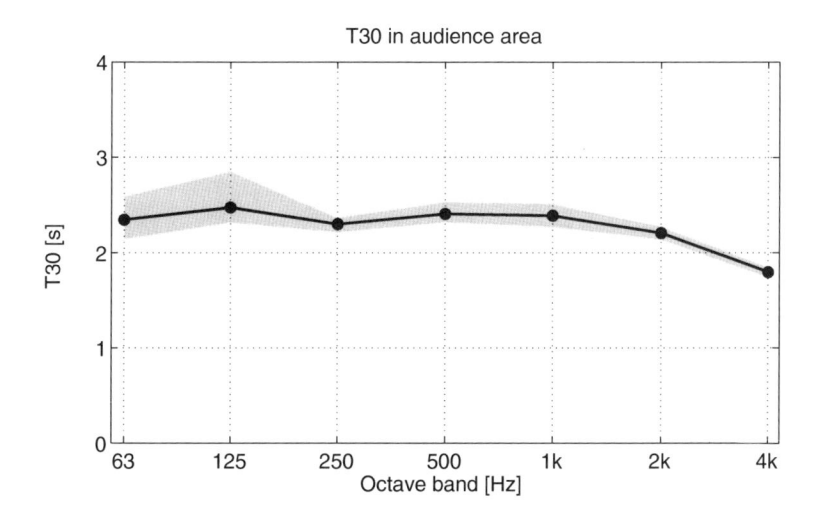

State of Hall When Measured

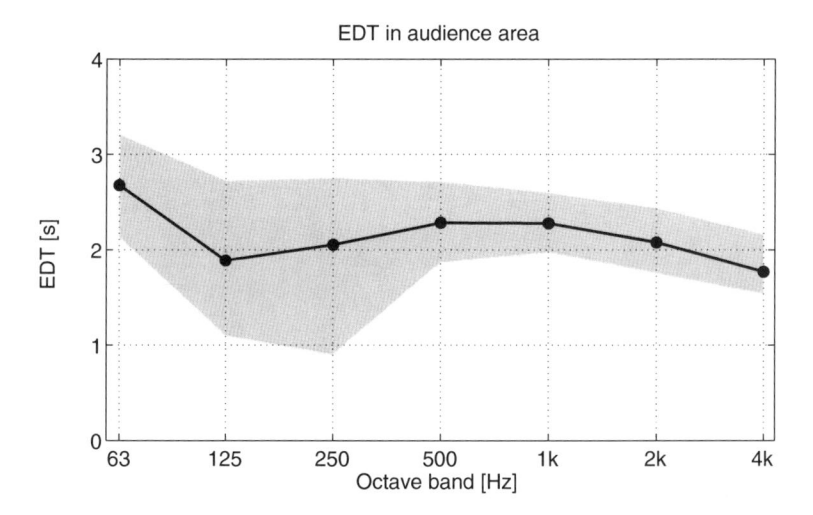

T30 in audience area

EDT in audience area

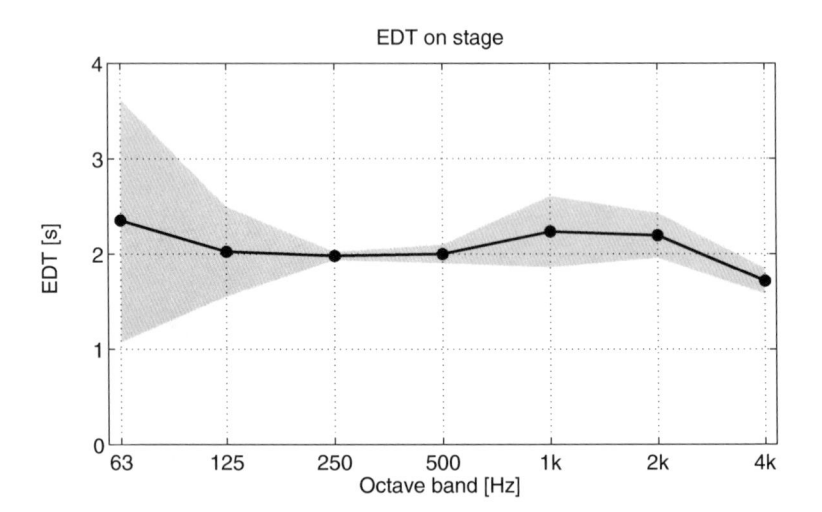

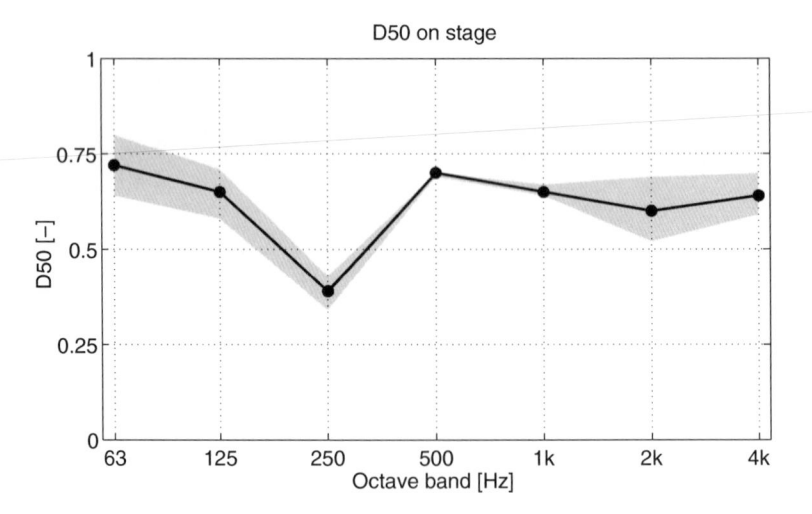

260   7 Gallery of Halls that Present Pop and Rock Music Concerts

Heineken Music Hall

Amsterdam
Number of concerts per year: 50–100
Founded: 2001
Capacity: 5,500
Architect: Frits van Dongen, De Architecten Cie
Acoustician: Rob Metkemeijer, Peutz Acoustics

The founders of the Heineken Music Hall in Amsterdam (NL) had one clear vision: to create a modern multifunctional venue for the main purpose of high-quality (pop) concerts for up to 5,500 visitors. The venue opened its doors in 2001 and is to this day (according to the venue) still the only Dutch multifunctional venue that has been designed both logistically and acoustically for amplified music events.

Erica Bakker of the Heineken Music Hall explains, the starting point was a simple audience surface with balconies, the so-called Black Box which is the larger of two stages in the facility. At the time, many venues seemed to cope with the same problem: lack of sound absorption under 200 Hz. Sound echoes from front to back when not absorbed; this was the one thing that the Heineken Music Hall wanted to prevent. Not only are frequencies transported loud and clear throughout the entire main hall, but the quality of the low frequencies sounds is also higher and more defined. The venue is built according the "box in box" principle; hardly any of the walls touch the outer walls. After concerts people can stay and dance some more in the Beat Box hall located on the first floor of the venue.

With an average of 137 events a year (both concerts and corporate events) and multiple nominations for the Pollstar Award, the venue remains a much-loved location for music lovers. With performers including Joe Cocker, John Mayer, Simply Red, Pink, Lady Gaga, Lionel Richie, Coldplay, Van Morrison, Elvis Costello, Status Quo, and others, the Heineken Music Hall has become a reknowned location within its own league.

Bakker continues:

On a show date, the very first thing a crew can expect in this venue is a warm welcome; a nice cup of coffee and a hot shower. The local crew are always willing to help the production where and when they can. But do keep in mind that they want their venue to be treated the same way they treat their guests; with the utmost respect. One band decided to take being "rock 'n roll" to the extreme and trashed up the entire dressing room. It goes without saying that the venue wasn't happy about this and in their turn decided to let the band clean up their own mess. And they did… Obviously under slight protest, because after all, that is the true nature of "rock 'n roll … !"

The Heineken Music Hall is extreme acoustical engineering in a very delicate architectural design. With a volume of 50,000 m^3 a reverberation time of merely 1 s is indeed an extraordinary achievement.

All wall surfaces are perforated metal plates in front of porous absorption and membranes. Except for the reflective (when not in use) seats and the floor the Heineken Music Hall can be regarded as an anechoic chamber with two bars. Acoustician's paradise.

Heineken Music Hall

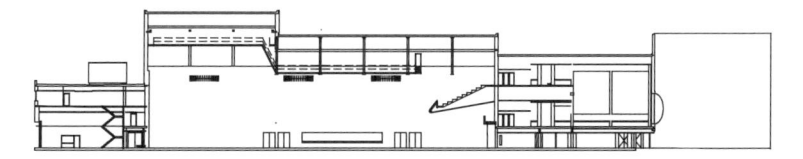

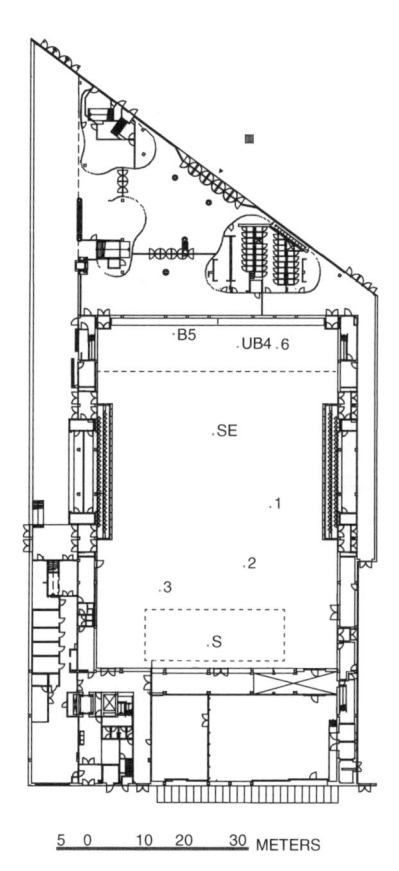

Geometrical data	
Volume	50,000 m^3
Height, audience area	20 m
$L \times W \times H$	$60 \times 43 \times 20$

Acoustical data	
Audience area	
$T_{30, 125-2k}$	1.17
EDT_{125-2k}	1.11
$C_{80, 125-2k}$	3.15
BR_{63} versus 0.5–1k	1.89
BR_{125} versus 0.5–1k	0.95

Materials Used

Audience Area

Floor: Concrete with a polyurethane coating.
Ceiling: 10-cm mineral wool and 20-cm cavity.
Walls: 30-cm layer of absorption material, consisting from 10-cm mineral wool, layers of foil and 20 cm of air.

Stage Area

As audience area.

State of Hall When Measured

Empty; seats mounted in the rear on floor and balcony.

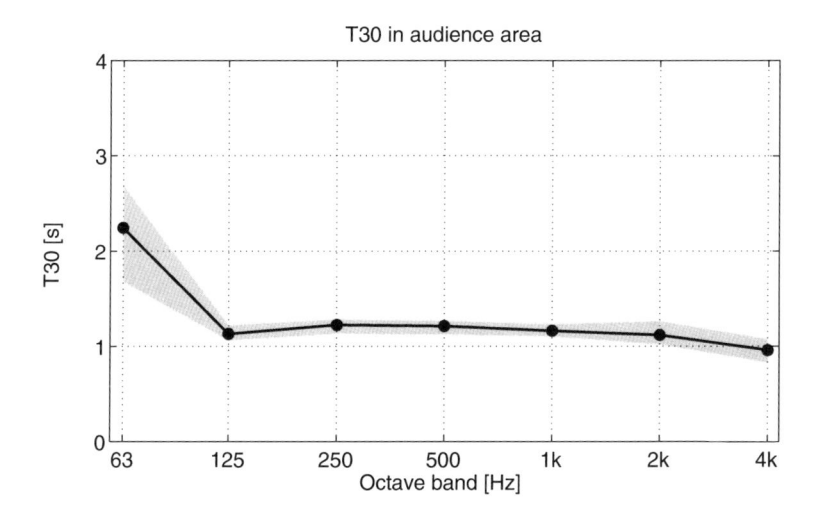

T30 in audience area

EDT in audience area

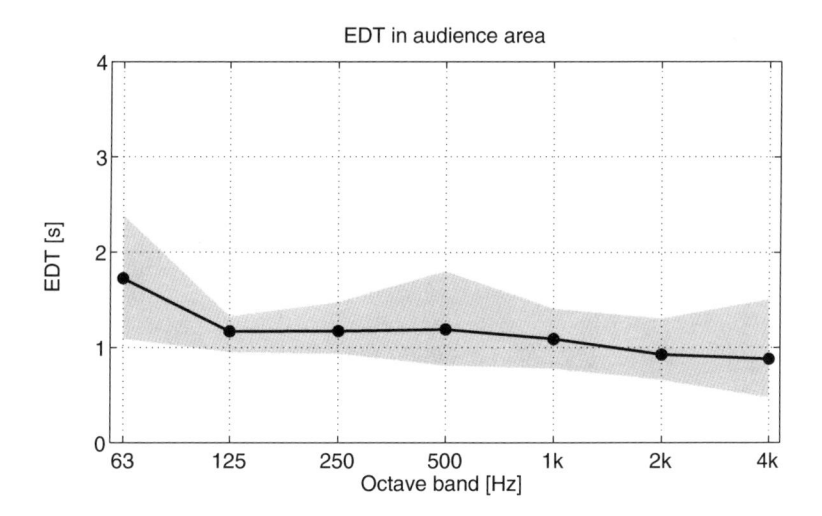

Hanns-Martin-Schleyer-Halle

Stuttgart
Number of concerts per year: 65
Founded: 1983
Capacity: 15,500
Architect: ASP Architekten
Acoustician: NA

The arena is part of a sport complex that includes the adjacent Mercedes-Benz Arena and Porsche Arena. Hanns-Martin-Schleyer-Halle is an indoor sporting arena located in Stuttgart, Germany. The capacity of the arena is 15,500 people. The hall was built in 1983 and is named for Hanns Martin Schleyer, a German employer representative, killed by the terrorist Red Army Faction. It has a 265-m (869-ft) track made of wood. The arena hosted the final phase of the 1985 European basketball championship [1]. It also hosted the Stuttgart Masters when it was a ATP Super 9 event between 1996 and 2001. The arena is also used as a velodrome and was used as the host for the 2003 UCI Track Cycling World Championships.

NeckarPark Stuttgart is one of the biggest and most attractive event sites in Europe. Five state-of-the-art event locations for top international sport, cultural, business, and political events line the Mercedesstrasse in the district of Bad Cannstatt: the Gottlieb-Daimler-Stadium, the Carl Benz Center, the Mercedes Benz Museum, and the Hanns-Martin-Schleyer-Halle and Porsche-Arena hall duo. Extensively modernized and enlarged in 2006, the 15,500 capacity Hanns-Martin-Schleyer-Halle is the largest indoor arena in south Germany. When officially opened in 1983 it was Europe's first multifunctional hall and together with the Porsche-Arena, which opened in May 2006, it forms a unique hall duo in the whole of Europe.

Take a closer look and immerse yourself in a hall complex that is unique and doubly good. A light-flooded and airy lobby unites both halls. The elongated Porsche-Arena is elegantly connected to the glass construction through which people stream into both halls. Flexibility is the key and this is also mirrored in the diversity of the events. From superstars on the national and international music scene to sports events and big show productions, the program of events is as star-studded as it is emotional. More than 14 million visitors to the Hanns-Martin-Schleyer-Halle are clear proof of its attractiveness.

The diversity of events can be enjoyed as a double pack in the Hanns-Martin-Schleyer-Halle and the Porsche-Arena whereby the prerequisite is a perfect and professional organizational structure. Working behind the scenes, it ensures major performances go off smoothly. This applies to a special degree to company presentations, congresses, annual general meetings, and party conferences. An outstanding example is the Porsche Annual General Meeting, which, combined with a big presentation in the Schleyer-Halle, celebrated its premiere in the Porsche-Arena. The hall duo functions in a variety of ways. With a spotlight on for a concert in the Schleyer-Halle, at the same time a first-class handball or ice-hockey match is being played in the Porsche-Arena.

Schleyer Halle.

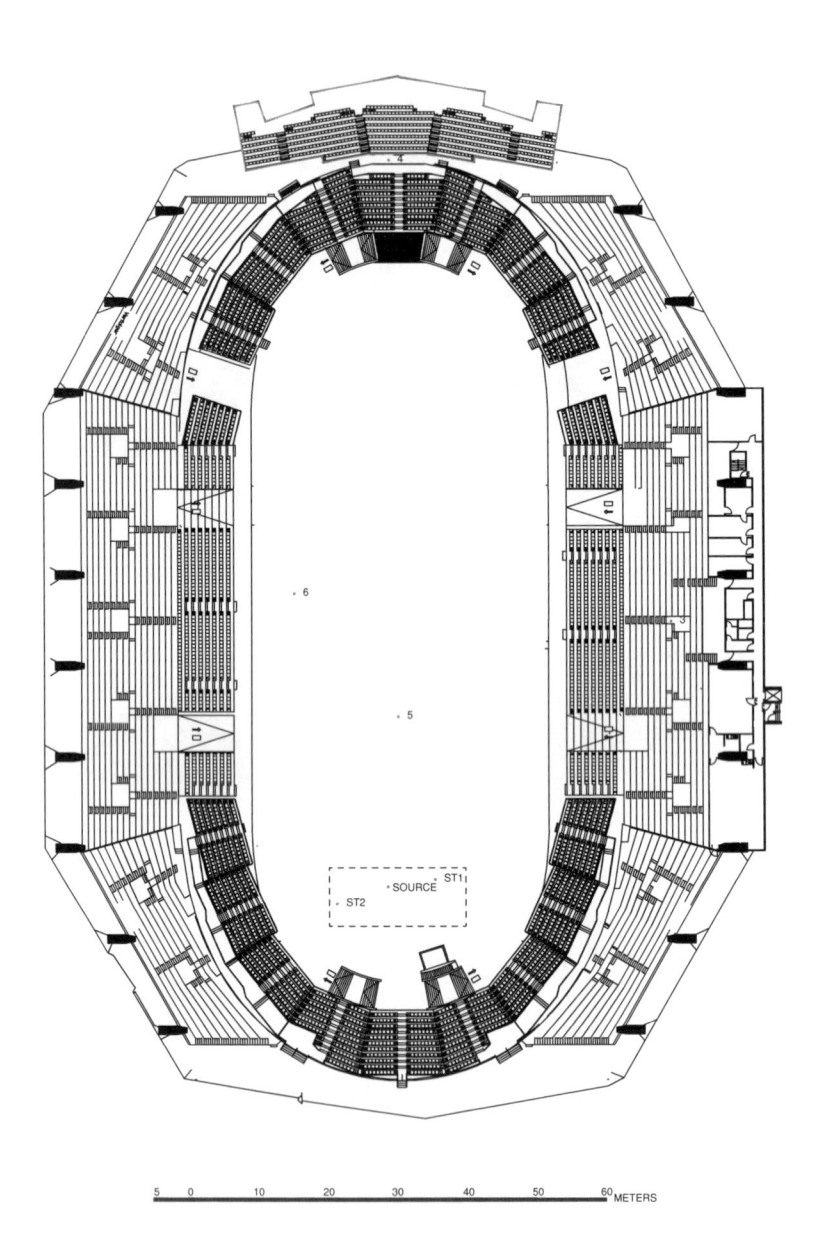

268 7 Gallery of Halls that Present Pop and Rock Music Concerts

5 0   10   20   30   40   50   60 METERS

Geometrical data	
Volume	Approximately 200,000 m^3
Height, audience area	12 m clear height, upper beam 18 m

Acoustical data	
Audience area	
T_{empty}	
T_{full}	
D_{50}	
EDT	
BR_{rock}	
Stage area	
T_{empty}	
D_{50}	
EDT	
BR_{rock}	

Materials Used

Audience Area

Floor: Wood.
Ceiling: Thin trapezoid metal without perforation.
Walls: Concrete thin metal with and without perforation/concrete.
Seats upholstered on seat not on back and perforated underneath seats (see photo).

State of Hall When Measured

Empty: some additional seats mounted, curtains behind stage area and at the rear of the hall but these were only elevated 1 m from the floor.

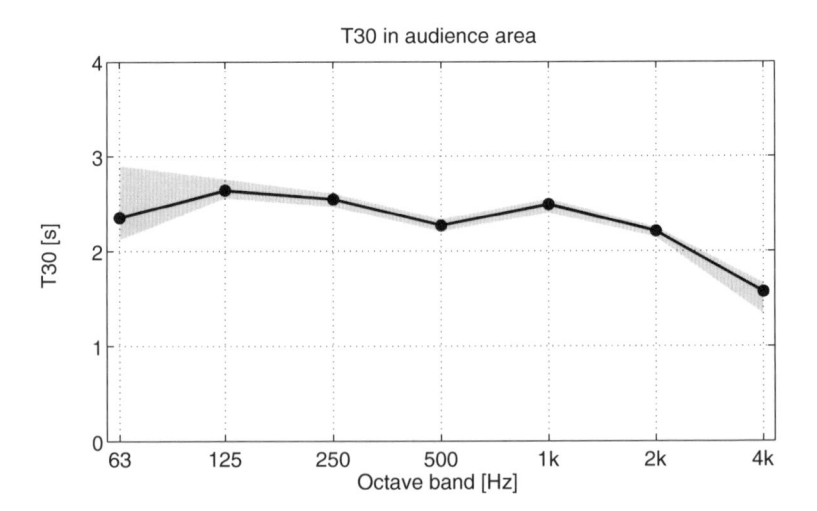

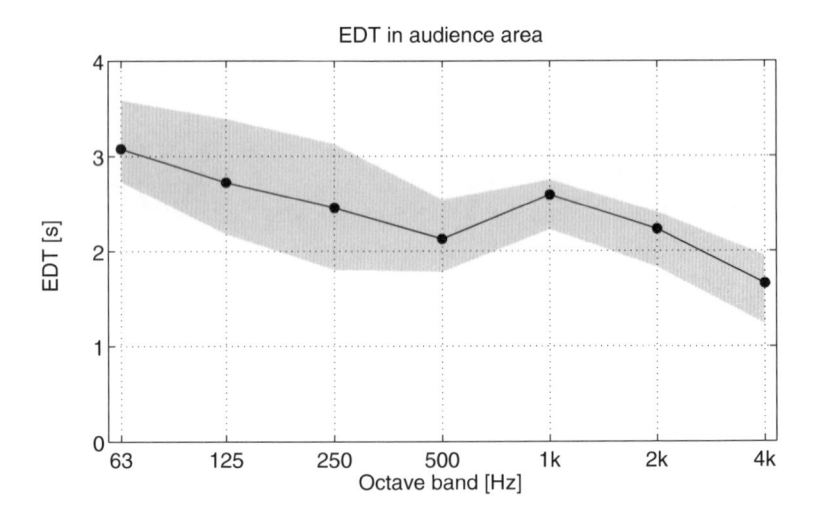

Jyske Bank BOXEN

Herning
Number of concerts per year: N/A
Founded: 2010
Capacity: 15,000
Architect: Årstiderne Arkitekter
Acoustician: Eddy Bøgh Brixen

Jyske Bank BOXEN is Denmark's first multipurpose arena with seating for 12–15,000 people. The multipurpose arena has been specially designed to host a wide range of events and houses numerous service and VIP facilities. Jyske Bank BOXEN is therefore able to stage both national and international events such as sporting fixtures, concerts, shows, and other entertainment. Jyske Bank BOXEN was conceived by the Danish company MCH in close dialogue with sporting associations, concert organizers, business partners, and experts to ensure a state-of-the-art and future-proof venue that lives up to international standards. MCH owns and runs the multipurpose arena.

The first event in Jyske Bank BOXEN took place on October 20th, 2010 when the American pop phenomenon Lady Gaga performed at the venue, which is also known as Denmark's national arena. Just two days later, Prince visited Jyske Bank BOXEN, and since then Linkin Park and Elton John with Ray Cooper have staged shows there. In December 2010, Jyske Bank BOXEN served as the venue for the European Women's Handball Championships, where the tournament's only intermediate round and the finals were held in Herning.

The string of international concerts, shows, and sporting events continued in 2011. For example, the world première of Kylie Minogue's Aphrodite Tour was scheduled to take place at the arena, in addition to the Queen musical *We Will Rock You*, a concert by teen idol Justin Bieber, Roger Waters' The Wall, and a performance by R&B star Rihanna.

Jyske Bank BOXEN is part of "Vision 2025," an ambitious and long-term plan for the future expansion of MCH's physical facilities. The plan was presented on November 1st, 2000, and covers infrastructure/motorway projects, the multipurpose arena (Jyske Bank BOXEN), a stadium, a drive-in cinema, MCH Time World, and an aerial railway.

The first stage of the plan was completed in 2004 with the construction of Denmark's largest column-free hall and MCH Arena, home ground for the professional Danish football club FC Midtjylland. Since then, the infrastructure around Herning has been extended and improved, and most recently Jyske Bank BOXEN has opened. Jyske Bank BOXEN is part of MCH, one of Scandinavia's most flexible experience centers with 15 exhibition halls, a congress center, a football stadium, and now also a multipurpose arena. MCH organizes trade fairs and exhibitions, concerts, sporting events, conferences, meetings, and parties. It is situated in Herning (Denmark) a city with approximately 84,000 citizens and a population of approximately 2.6 million people within a two-hour drive.

JBB arena has a no-nonsense Nordic interior design with wooden chairs.

Apart from porous absorption in the ceiling, membrane absorbers have been used on wall areas. This hall is known in Denmark for its good acoustics for amplified concerts.

Jyske Bank BOXEN

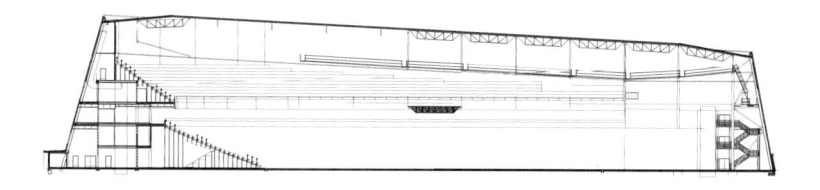

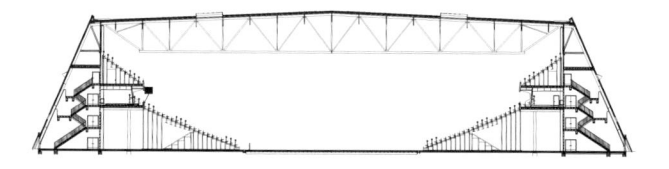

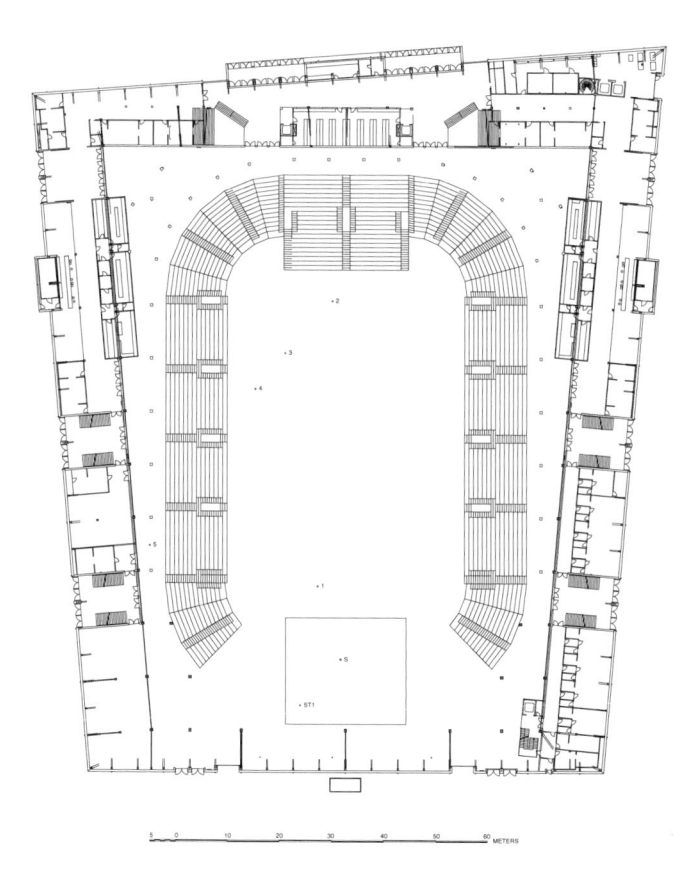

Geometrical data	
Volume	230,000 m³
$L \times W \times H_{(max)}$	115 × 80 × 30 m

Acoustical data	
Audience area	
$T_{30, 125-2k}$	2.81
EDT_{125-2k}	2.92
$C_{80, 125-2k}$	1.35
$BR_{63 \text{ versus } 0.5-1k}$	0.98
$BR_{125 \text{ versus } 0.5-1k}$	0.99

Materials Used

Audience Area

Floor: Concrete.
Ceiling: Steel trapezoid profile with perforation on cavity with mineral wool on top.
Walls: Upper wall areas: 100-mm mineral wool on concrete; lower wall areas: single layer gypsum board on cavity with mineral wool.
Seats are not upholstered.

State of Hall When Measured

Empty; no additional seats mounted.

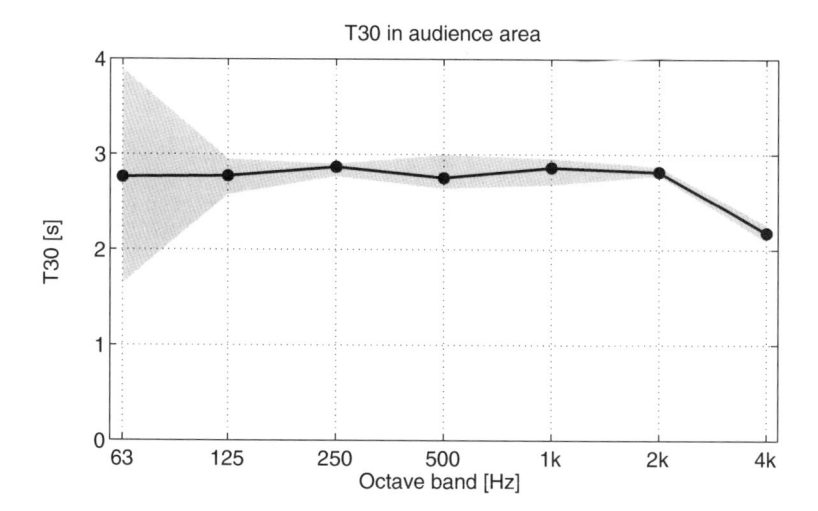

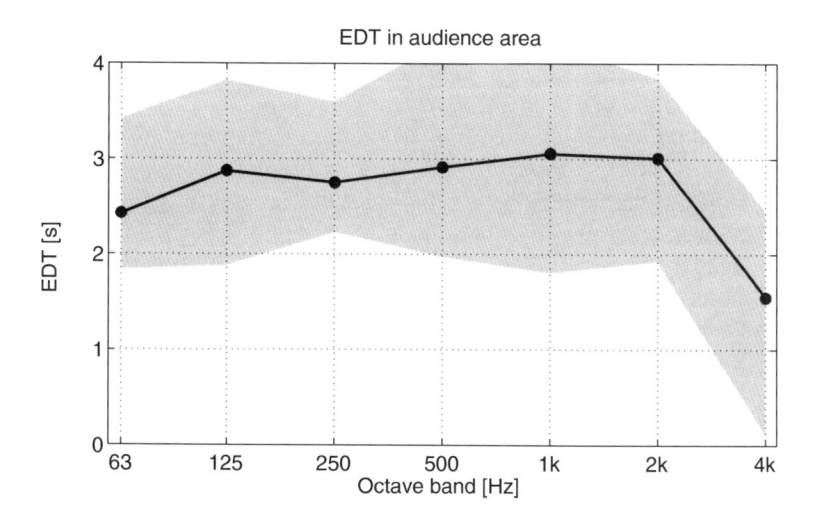

Kaiser Keller

Hamburg
Number of concerts per year: 55
Founded: 1959. Building constructed in 1957/58
Capacity: Approximately 400
Architect: Erwin Nagel
Acoustician: NA

The Beatles residency in Hamburg, the German city where John Lennon, Paul McCartney, George Harrison, Stuart Sutcliffe, and Pete Best regularly performed at a series of four different clubs during the period August 1960 to December 1962, was a chapter in the group's history which honed their performance skills, widened their reputation, and led to their first recording, which brought them to the attention of Brian Epstein. The Beatles' booking agent, Allan Williams, decided to send the group to Hamburg when another group he managed, Derry and the Seniors, proved successful there. Having no permanent drummer at the time, they recruited Best a few days before their departure.

The Beatles arrived very early in the morning on August 17th, 1960, but had no trouble finding the St. Pauli area of Hamburg, as it was so well known. Unfortunately the Indra Club (placed at 58 Grosse Freiheit) was closed, so a manager from a neighboring club found someone to open it up, and the group slept on the red leather seats in the alcoves. The group first played at the club the same night, but were told they had to sleep in a small cinema's storeroom, which was cold and noisy, being directly behind the screen of the cinema, the Bambi Kino. McCartney later said, "We lived backstage in the Bambi Kino, next to the toilets, and you could always smell them. The room had been an old storeroom, and there were just concrete walls and nothing else. No heat, no wallpaper, not a lick of paint; and two sets of bunk beds, with not very much covers Union Jack flags—we were frozen." Lennon put it, "We were put in this pigsty. We were living in a toilet, like right next to the ladies' toilet. We'd go to bed late and be woken up next day by the sound of the cinema show and old German *fraus* [women] pissing next door." After having been awakened in this fashion, the group members were then obliged to use cold water from the urinals for washing and shaving.

Harrison remembered the Reeperbahn and Grosse Freiheit as the best thing the group had ever seen, as it had so many clubs, neon lights, and restaurants, although also saying: "The whole area was full of transvestites and prostitutes and gangsters, but I couldn't say that they were the audience. Hamburg was really like our apprenticeship, learning how to play in front of people." Best remembered the Indra as being a depressing place that was filled with a few tourists, and having heavy, old, red curtains that made it seem shabby compared to the larger Kaiserkeller, a club also owned by Koschmider and located nearby at 36 Grosse Freiheit. After the closure of the Indra because of complaints about the noise, the Beatles played in the Kaiserkeller starting on October 4th, 1960. [From Wikipedia, *"The Beatles in Hamburg"*]

Kaiser Keller

Kaiser Keller is the basement club where the Beatles grew by playing several sets per day for months. Kaiser Keller is still used today for upcoming bands.

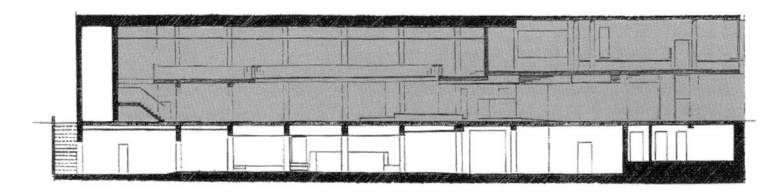

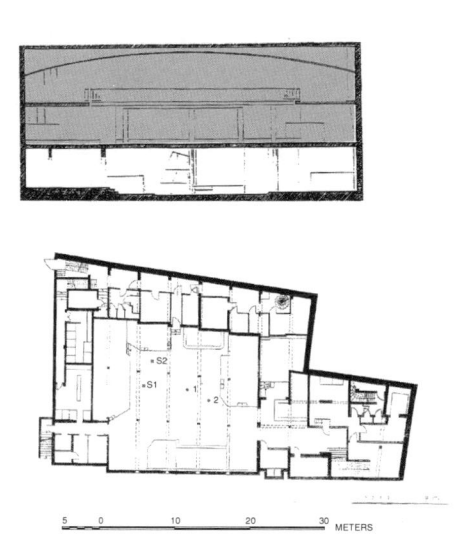

Geometrical data	
Volume	1,200 m^3
Height, audience area	3.1 m
Height of stage	0.4 m
$L \times W \times H$	$21.9 \times 17.8 \times 3.1$

Acoustical data	
Audience area	
$T_{30, 125-2k}$	0.99
EDT_{125-2k}	1.04
$C_{80, 125-2k}$	4.78
BR_{63} versus 0.5–1k	0.64
BR_{125} versus 0.5–1k	0.6

Materials Used

Audience Area

Floor: Tiles on concrete; approximately 20 % is covered by wooden audience risers.
Ceiling: Thin plate with cavity behind.
Walls: Concrete.
Bars made of wood on cavities.

Stage Area

Floor: Wood on cavity.
Ceiling: Thin plate with cavity behind.
Walls: Curtains.

State of Hall When Measured

Empty: chairs and sofas on stage.

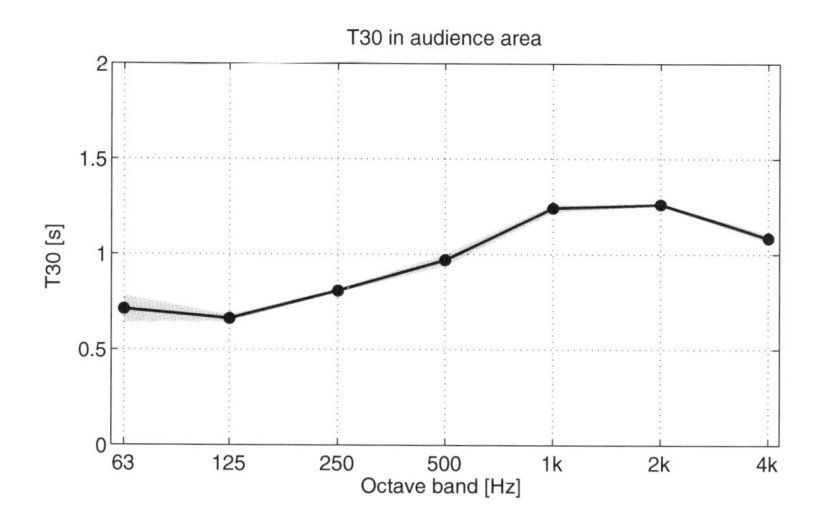

State of Hall When Measured

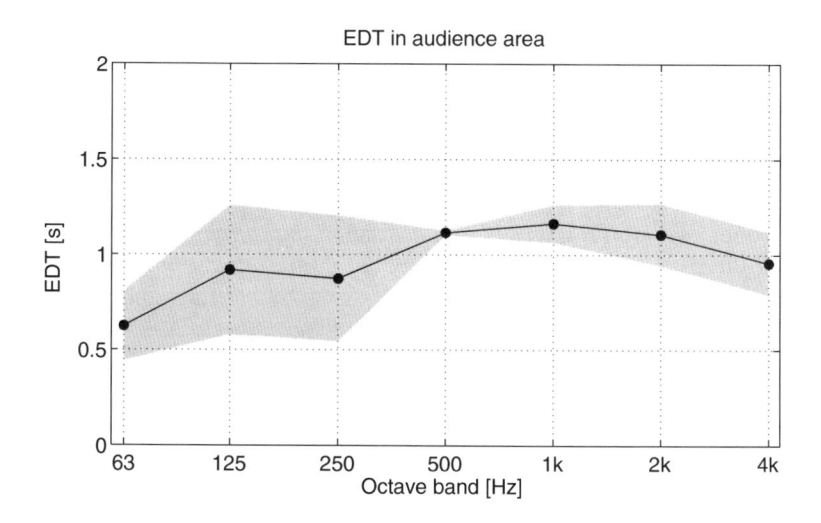

T30 in audience area

EDT in audience area

Live Music Club

Trezzo sull'Adda (Milan)
Number of concerts per year:
Founded: 1997/2007
Capacity: 1,500
Architect: Baruffi Valeria/Studio Achea
Acoustician: N/A

The Live Club was founded in 1997 in Trezzo sull'Adda as a entertainment alternative in the province of Milan. The music programming was soon given exclusive tribute nights with the most important names of the Italian indie scene. Soon artists on the international scene began to play on the live club stage, and causing the choice of a new location. The transfer took place in June, 2007. A new structure was designed and prepared to become one of the best clubs in Italy for live music, theater, events, and conventions. Inside the new venue the Live Club Restaurant and offices for the Live Club staff treat every aspect of production of an event, from the technical to the creative, administrative, promotional. The Live Club position at the crossroads between the provinces of Milan, Bergamo, and Lecco makes very easy to reach from the entire Lombardy region.

This is Live Club's fourth season of live events and DJ sets in its new structure and has already hosted big national and international artists (Afterhours, Caparezza, Vibrazioni, Joe Satriani, Buddy Guy, Misfits, Marillion, Saxon, Gotthard and many more).

For its fourth season too, national and international artists together with classic tribute bands play on Friday and Saturday nights. The weekly concerts are dedicated to alternative artists on the international music scene events ranging from good old rock to Italian and international reggae, without forgetting metal, electronic music, and hip hop. Before and after the shows the music/images experiment continues thanks to DJ and VJ performances.

Going beyond traditional expectations, Live Club has also hosted in its brand new structure dance events (such as the 12th Adda Danza International Modern Dance Show) and theater events in cooperation with the city of Trezzo sull'Adda, that used Live Club's stage to organize several initiatives.

Live Music Club

Several balcony levels enhance the audience dynamics.

Before sound check: tonight's band enjoying brunch "on the house" while rigging is in process on stage. Large sheets of foam suspended from walls to ceiling create a somewhat silent atmosphere in the bar.

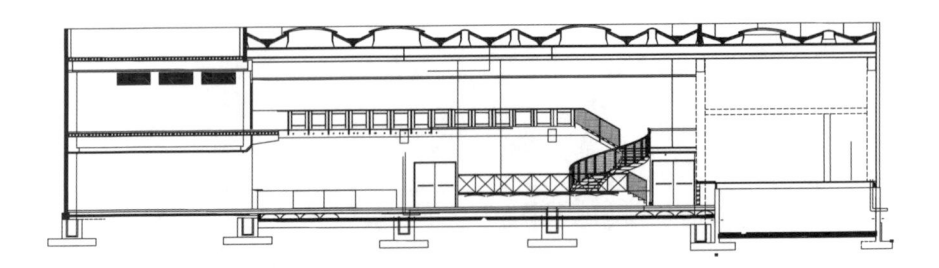

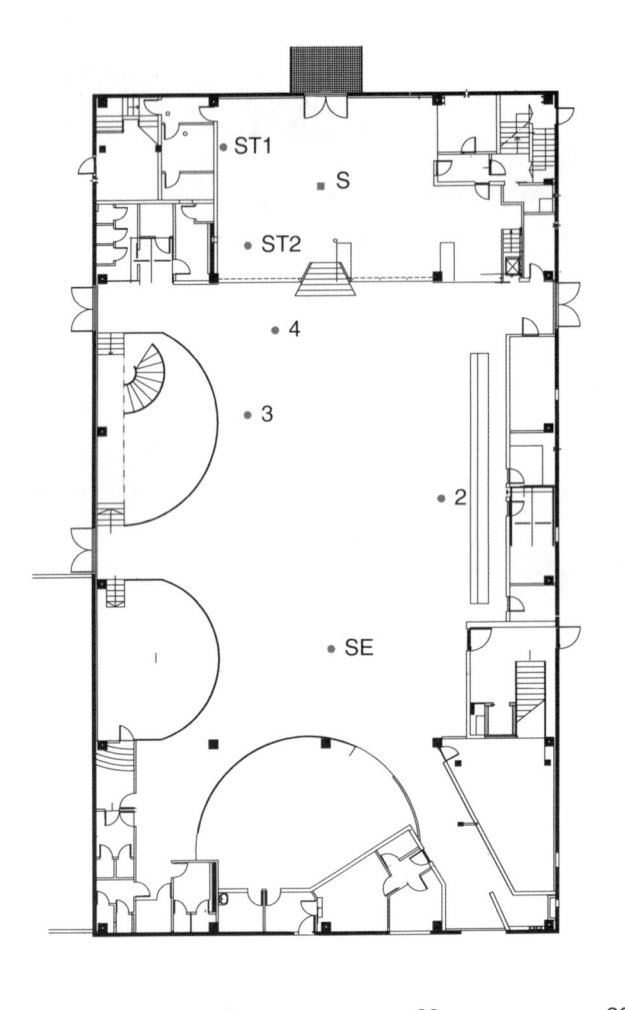

282   7 Gallery of Halls that Present Pop and Rock Music Concerts

Geometrical data	
Volume	8,000 m³
Height, audience area	8.7 m
Surface area of stage	270 m²
Height of stage	1.2 m
$L \times W \times H$	$43 \times 23 \times 8.7$

Acoustical data	
Audience area	
$T_{30, 125-2k}$	1.17
EDT_{125-2k}	1.15
$C_{80, 125-2k}$	2.70
$BR_{63 \text{ versus } 0.5-1k}$	0.75
$BR_{125 \text{ versus } 0.5-1k}$	0.96
Stage area	
EDT_{125-2k}	0.73
$D_{50, 125-2k}$	0.62
$BR_{63 \text{ versus } 0.5-1k}$	0.66
$BR_{125 \text{ versus } 0.5-1k}$	0.67

Materials Used

Audience Area

Floor: Concrete.
Ceiling: Suspended mineral wool with foam product on top.
Walls: Concrete. Over the bar a large area is covered with approximately 5-cm–thick foam slabs with a large distance to wall.

Stage Area

Floor: Vinyl on wood direct on concrete.
Ceiling: Suspended mineral wool with foam product on top.
Walls: Backdrops on back wall and side walls.

State of Hall When Measured

Empty: lighting rigs on stage.

284　　　　　　　　7　Gallery of Halls that Present Pop and Rock Music Concerts

Live performance at the Live Music Club.

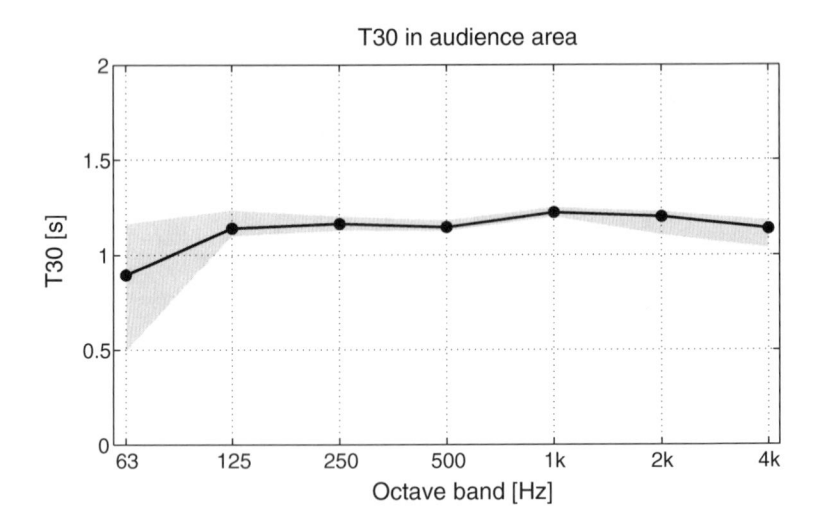

EDT in audience area

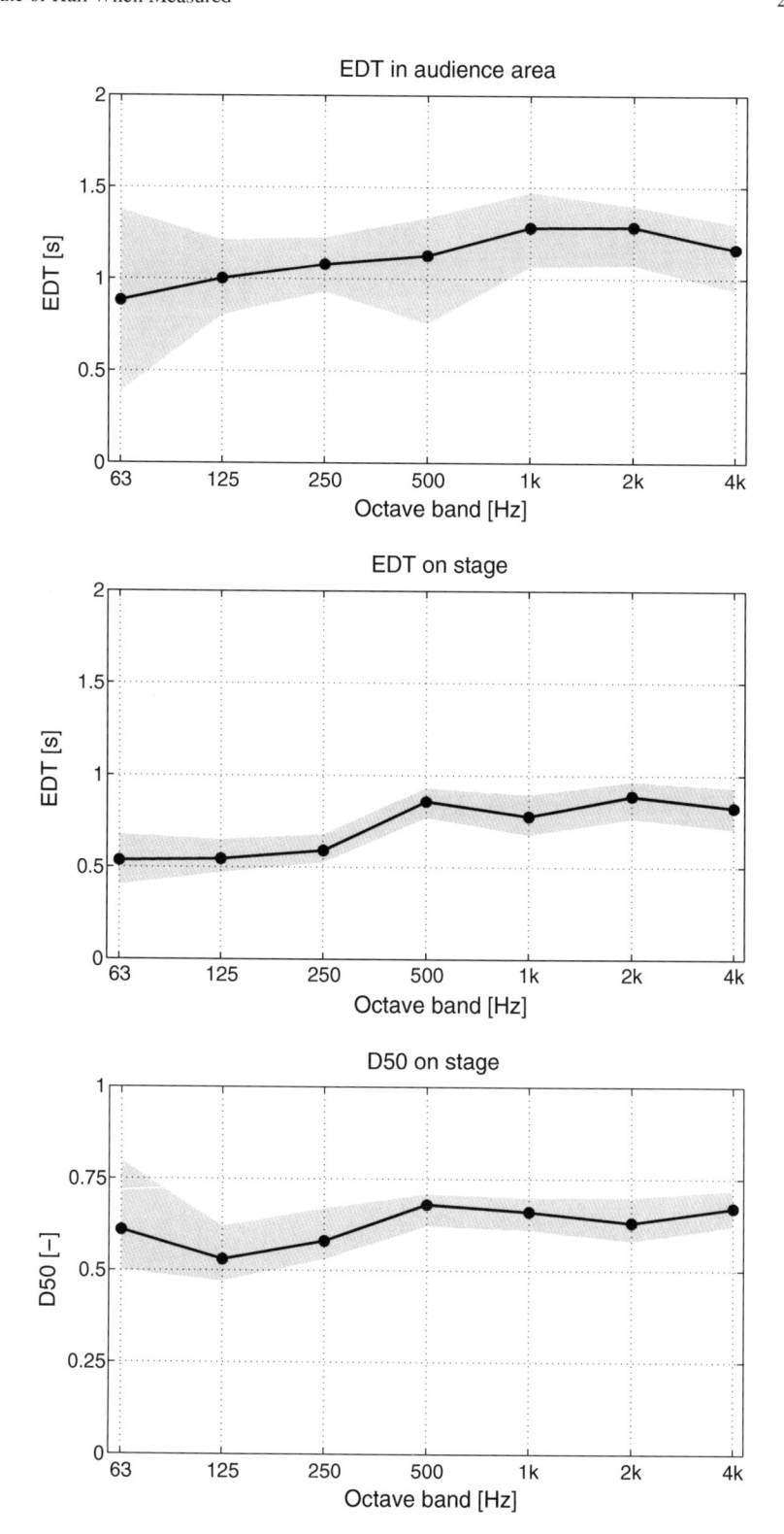

EDT on stage

D50 on stage

LKA/Langhorn

Stuttgart
Number of concerts per year: 100–120
Founded: 1984
Capacity: 1,500
Architect: N/A
Acoustician: N/A

Er hat hatte sie alle: Nirvana, Sheryl Crow, Nina Hagen, Die Ärzte, Rammstein, Guildo Horn, Nickelback, The Black Eyed Peas, Die Schürzenjäger, Truck Stop, Jeanette Biedermann und und und. Im Liveclub LKA in Stuttgart-Wangen gaben und geben sich die Stars und solche, die es werden wollen, die Klinke in die Hand. Und das seit mehr als 25 Jahren. Das LKA ist zur Institution der Konzerthallen in der Landeshauptstadt geworden.

Damit hatte 1984 keiner gerechnet. Thomas Müller, damals Geschäftsführer der Diskothek Oz, lebte mit seiner amerikanischen Freundin in den Patch Barracks. Da bekam er mit, dass die GIs einen Countryclub vermissten. Im Industriegenbiet von Stuttgart- Wangen wurde er fündig. Die Halle eines insolventen Unternehmens erschien geeignet. Wo bislang Durchlauferhitzer gelagert waren, eröffneten sie den Countryclub Longhorn. Mitbewerber beim Insolvenzverwalter damals war übrigens Werner Schretzmeier, der für sein Theaterhaus eine Heimat suchte. Das Longhorn erhielt den Zuschlag—zur Freude der GIs, die schnell den Club bevölkerten. Aber auch Deutsche zählten zu den Kunden, die zu den Klängen des DJs und der Countrybands, die live spielten, tanzten. Das Longhorn wurde zum größten Country—and Westernclub außerhalb der USA, schrieb die US-Zeitung „Stars and Stripes".

Das erste Rockkonzert sorgte am 14. Dezember 1987 für ein volles Haus. Konzertveranstalter Henning Tögel und seine Moderne Welt suchten für den Auftritt von Nina Hagen eine preisgünstige Location. Das Longhorn, das 1,500 Zuschauern Platz bietet, zeigte sich bereit, Nina Hagen zu empfangen. Der Club platzte aus allen Nähten. „Es war gnadenlos voll", erinnert sich Thomas Müller. „Die Kellner kamen nicht mehr durch." Die Saat war gelegt, Konzerte das zweite Standbein im Longhorn. Auf Nina Hagen folgten am 10. März 1988 Bobby Womack und weitere 40 Bands und Künstler. Innerhalb von neun Tagen gastierten The Pogues („Die haben sich in der Garderobe geprügelt"), The Exploited („Das suchte der Kassierer mit den Einnahmen das Weite"), Linton Kwesi Johnson, Savoy Brown und Truck Stop.

Eine weitere Änderung erfolgte 1993: Die GIs waren abgezogen worden, aus dem Countryclub wurde nach zwei monatigem Umbau das LKA, Longhorn-Kultur-Austausch, mit Livekonzerten, Rockdisco und Nachwuchsförderung. Die Countryutensilien verschwanden, Andy Blattner, der schon Gitarren von Prince besprayen durfte, zauberte Motive aus der Sixtinischen Kapelle („das finde ich neutral, spielen bei uns doch Bands unterschiedlicher Stile") an die Wände. Am 3.

September 1993 eröffnete das LKA wieder seine Pforten. Weit mehr als 1,000 Nachwuchsbands bot das LKA seitdem Bühne und Plattform, sich unter professionellen Bedingungen Gehör zu verschaffen.

Weiterhin werden namhafte Bands begrüßt, die auf dem Weg in die großen Hallen und Stadien erst einmal das LKA Longhorn bespielten: Die Ärzte, Rammstein, Eminem, Nickelback, Korn, The Black Eyed Peas, Sheryl Crow. 1991 spielten Nirvana im Vorprogramm von Sonic Youth. Die Resonanz war eher bescheiden. Kurz darauf erschien „Nevermind" mit „Smells like teen spirit". Der Rest ist Musikgeschichte.

Das Bemühen um den Nachwuchs und das Engagement blieb nicht ohne Folgen: 2004 und 2005 wurde das LKA als bester nicht-geförderter Club ausgezeichnet. 2006 gab es den DASDING-Publikumspreis. Und 2009 erhielt das LKA den Gaston, „Gastro-Award Bester Club in Baden-Württemberg 2009". Aktuell wurde das LKA 2011 mit dem MARS "Music Award Region Stuttgart" unter der Rubrik "Best Live Location 1000" ausgezeichnet. Keine Frage, das LKA Longhorn ist Institution unter den Konzerthallen in Stuttgart.

LKA is a legendary club driven by sheer passion.

288   7 Gallery of Halls that Present Pop and Rock Music Concerts

A state-of-the-art sound system is united with a traditional style rock club. A formula of the success of the LKA.

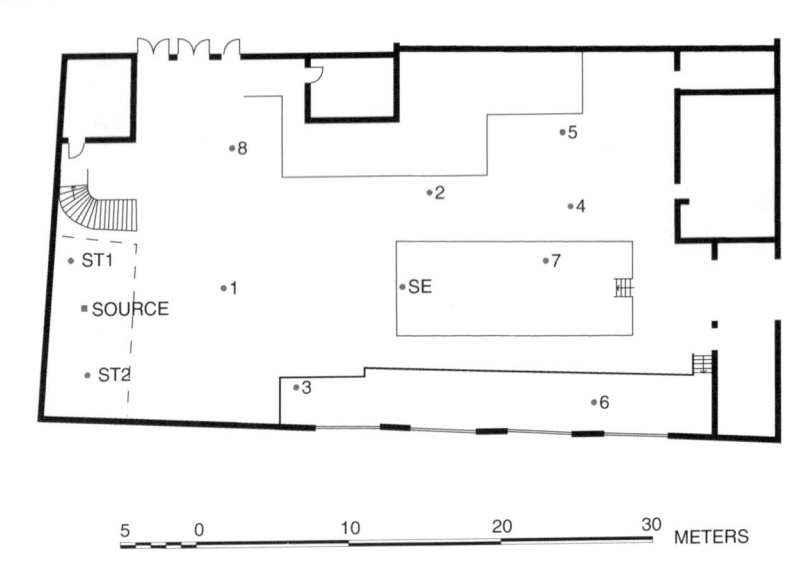

Geometrical data	
Volume	Approximately 5,000 m^3
Height, audience area	6–8 m
$L \times W \times H$	$40 \times 22 \times 6$

Acoustical data	
Audience area	
$T_{30, 125-2k}$	1.15
EDT_{125-2k}	1.13
$C_{80, 125-2k}$	3.06
BR_{63} versus 0.5–1k	1.42
BR_{125} versus 0.5–1k	1.33
Stage area	
EDT_{125-2k}	0.47
$D_{50, 125-2k}$	0.86
BR_{63} versus 0.5–1k	0.66
BR_{125} versus 0.5–1k	0.89

Materials Used

Audience Area

Floor: Concrete; wood direct on concrete on the dance floor in front of the stage.
Ceiling: 100-mm mineral wool with a 200-mm air cavity behind.
Walls: Concrete; upper half: 10 cm of mineral wool in linen direct on concrete.

Stage area

Floor: Wooden floor on concrete.
Ceiling: 100-mm mineral wool with a 200-mm air cavity behind.
Walls: 5-cm-thick wood fiber slabs with a 3-cm cavity behind.

State of Hall When Measured

Empty; chairs and tables are always mounted in the rear half of the room.

The snooker room with posters from famous visiting bands.

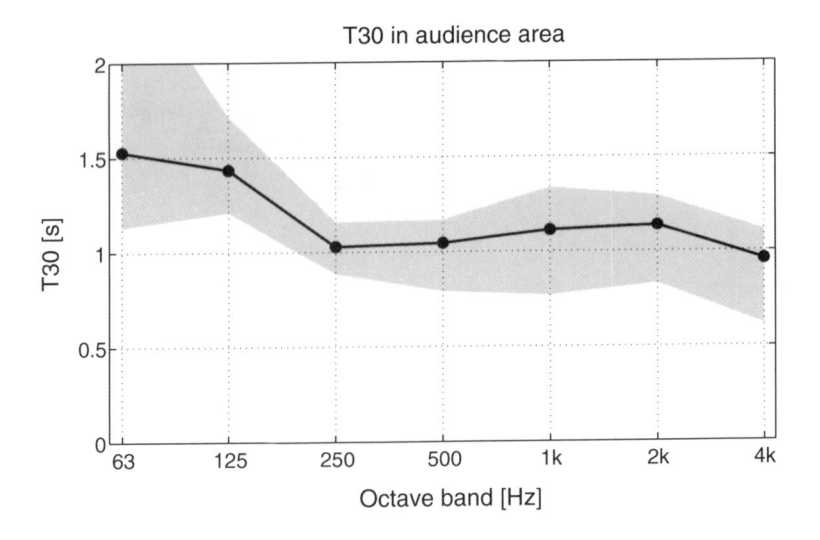

State of Hall When Measured

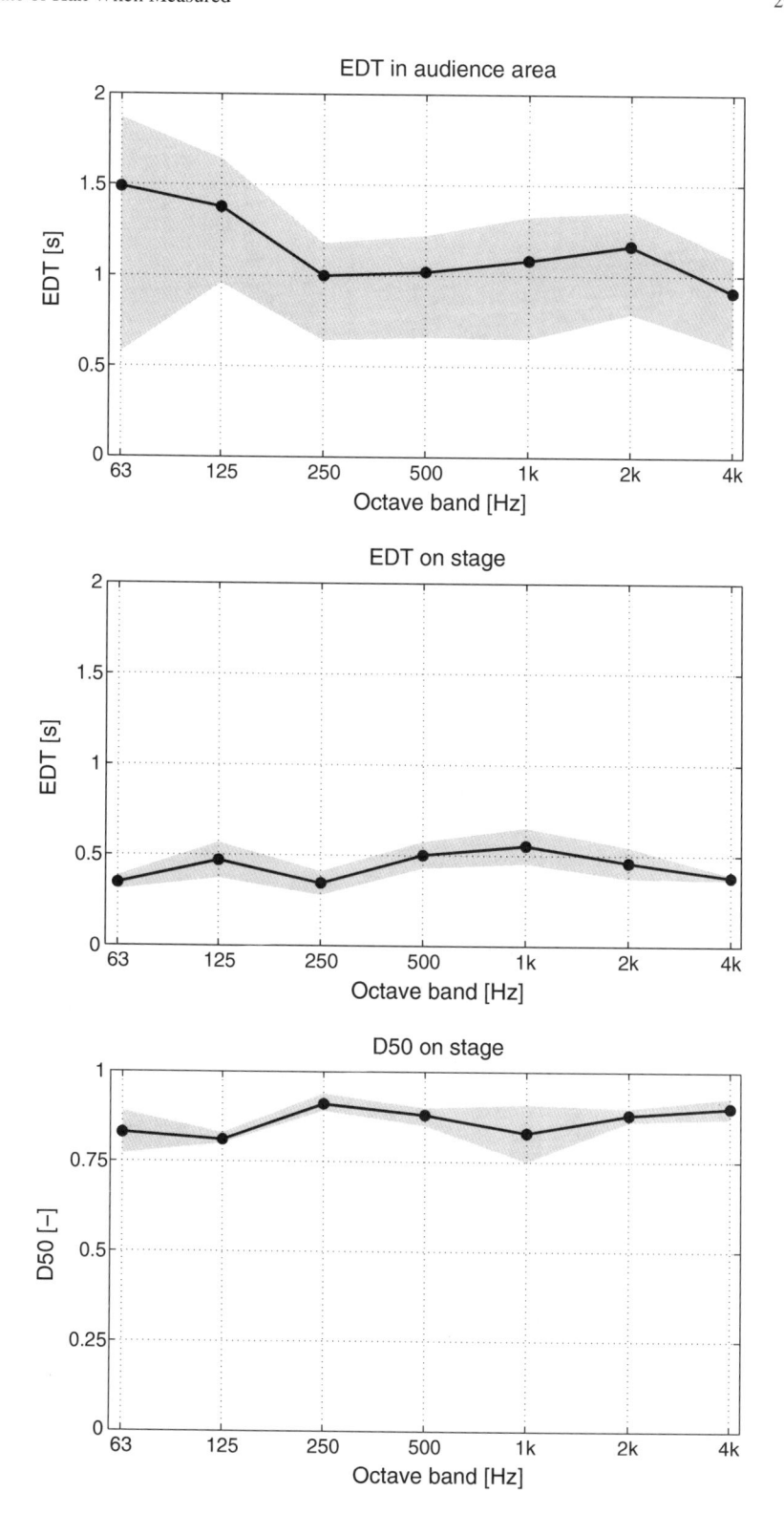

Mediolanum Forum

Milan
Number of concerts per year: N/A Total in venue: N/A
Founded: 1990
Capacity: 11,000
Architect: N/A
Acoustician: N/A

The Forum of Milan, which is today called Mediolanum Forum, opened in 1990 and is the main covered polyfunctional venue in the north of Italy.

Along with the PalaLottomatica in Rome, which is also managed by the ForumNet Group, the Mediolanum Forum is the only Italian structure to be included in the prestigious European Arenas Association (EAA), which features important European arenas. The building is arranged over three floors and has a total area of 40,000 m^2. Thanks to its modularity and suitability for any type of event, over the years it has become a point of reference for the biggest international events, including concerts with the most famous artists, sporting events of primary importance, shows, conventions, fairs, exhibitions, competitions, and television productions as well as smaller events, such as gala dinners and pre- or after-show activities. The Mediolanum Forum offers various internal spaces with different dimensions and characteristics. These can be used simultaneously or independently, as required by the type of event. As well as the Central Arena, the venue offers the Premium, Gallery, and Gold Halls, the Quota Otto spaces, and various external spaces. The Arena's upper tier (Quota Ventuno) hosts the "Sky Belvedere," a refined environment that directly overlooks the parterre and can be exclusively reserved for press conferences, meetings, and corporate hospitality activities. The Mediolanum Forum has ample dedicated parking facilities and will soon have a dedicated stop on Line 2 of the Milanese Underground.

The Central Arena is the heart of the Mediolanum Forum. It is the largest space within the venue and is situated at 4.96 m above the external ground level.

As well as hosting international music star concerts, large shows, and sporting events with different space configurations and areas, the Central Arena also hosts conventions, gala dinners and receptions, competitions, and fairs. Different space configurations can in fact be offered through the use of the building's long or short sides, the stalls, a modular use of the stands, and so on. A dimming system for the stands enables the capacity to be reduced from 11,000 to 3,500 places, and the parterre area alone can host more than 2,000 seated people. Along the two sides of the Arena there are 12 changing rooms of different sizes (that are available as service spaces) and loading and unloading take place through a roomy freight elevator and two vehicle entrances that lead straight to the parterre area. The versatility of the Central Arena is completed by the vertical height of the space which leaves plenty of room for all types of equipment to be suspended.

Mediolanum Forum is a traditional sports arena that changes its name from time to time.

A young and helpful crew meets those visiting the venue during office hours.

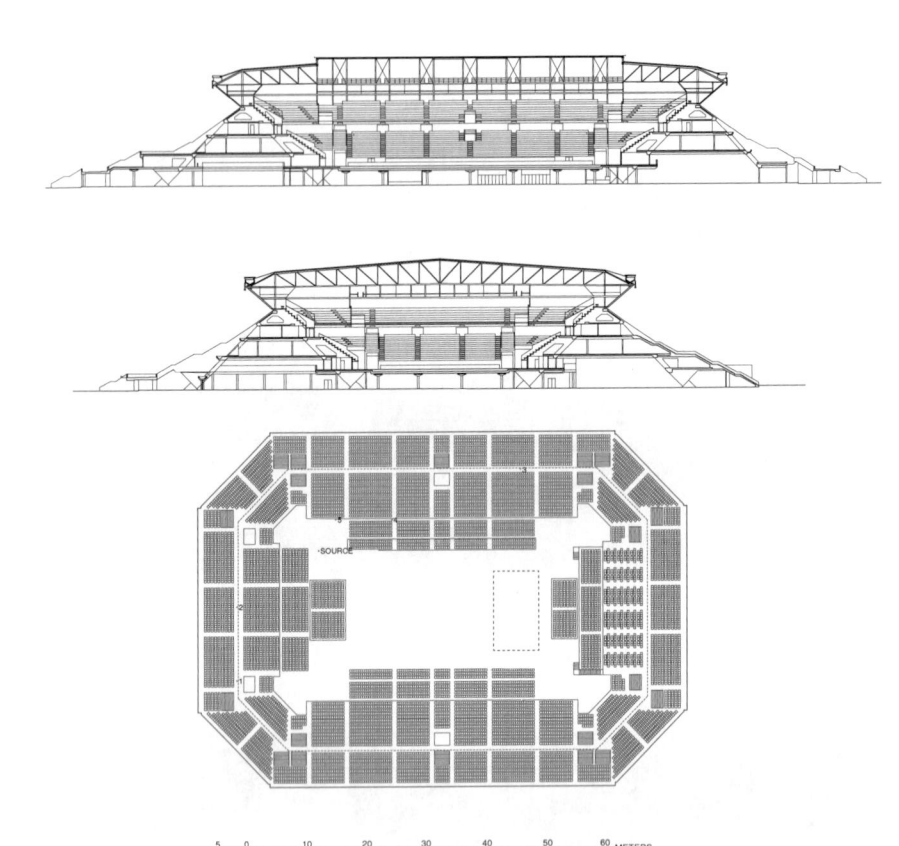

294   7 Gallery of Halls that Present Pop and Rock Music Concerts

Geometrical data	
Volume	Approximately 150,000 m^3
$L \times W \times H_{(max)}$	Approximately. 117 × 81 × 26 m

Acoustical data	
Audience area	
$T_{30, 125-2k}$	2.46
EDT_{125-2k}	2.56
$C_{80, 125-2k}$	−8.98
BR_{63} versus 0.5–1k	1.26
BR_{125} versus 0.5–1k	1.05

Materials Used

Audience Area

Floor: Concrete.
Ceiling: Perforated metal.
Walls: Concrete.
Seats are made of unupholstered plastic.

State of Hall When Measured

Empty; no additional seats mounted. Curtains drawn in front of the upper balcony level as for the 7,000 person configuration. Floor covered with rubber layer for the installment of the ice hockey field. Speakers were put in a 90-degree angle unlike measurements in other venues because of work on the floor.

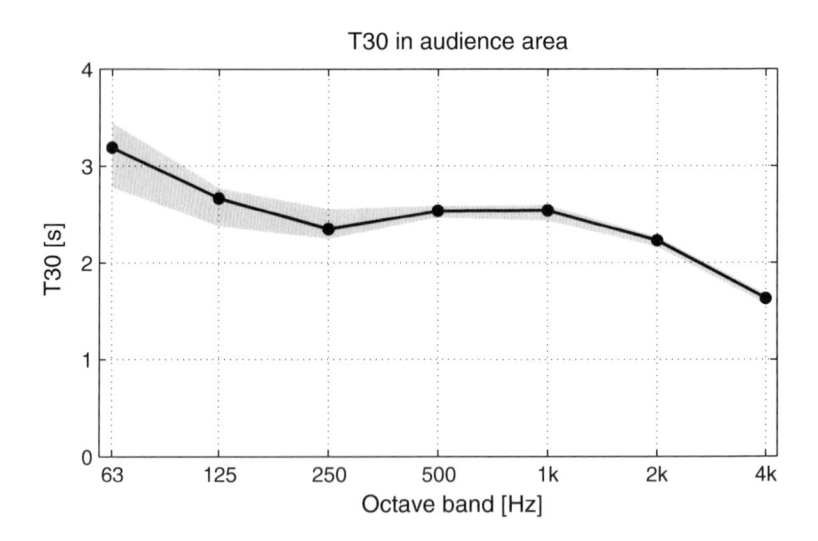

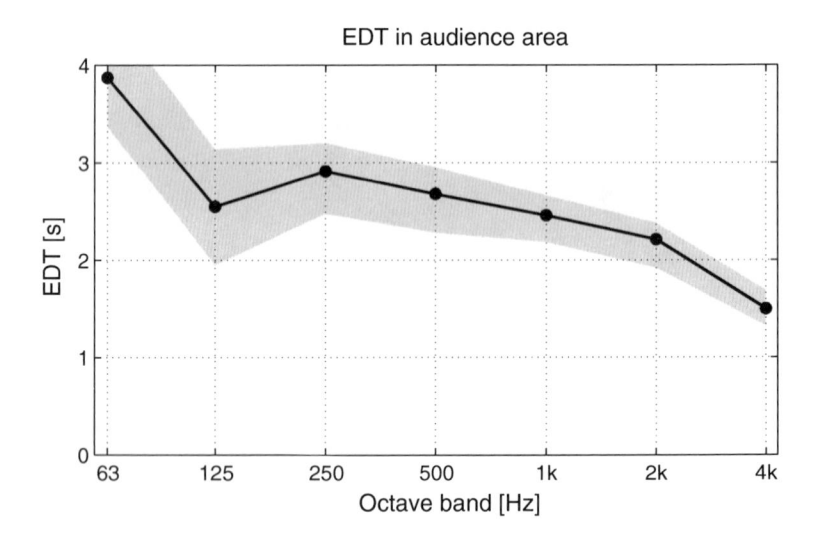

7 Gallery of Halls that Present Pop and Rock Music Concerts

Melkweg—The Max

Amsterdam
Number of concerts per year: 150. In both halls in the venue: 400
Founded: 1970 (Melkweg)/1995 (this hall, The Max)
Capacity: 1,500
Architect: Jim Klinkhamer (Jonkman & Klinkhamer Architects)
Acoustician: N/A
Owner: The building is owned by Municipality of Amsterdam. Melkweg is a foundation

The Melkweg (English translation: "Milky Way") is a popular music venue and cultural center in Amsterdam, the Netherlands. It is located on the Lijnbaansgracht, near the Leidseplein, a prime nightlife center of Amsterdam. It is housed in a former factory and warehouse and is divided into a number of spaces of varying sizes. In addition to a large hall for rock and pop music concerts (The Max), there's a second space for live music and there are also spaces for dance/theater, cinema, photography, and media-art. The Melkweg is run by a nonprofit organization that has existed since 1970. The building itself dates back to 1898.

In 40 years, Melkweg developed from a social cultural meeting place to a professional cultural center, attracting over 400,000 visitors a year. Many of the big stars of pop and world music have performed here in their early days, including the Grateful Dead, U2 (first show outside Northern Ireland), Youssou N'Dour, Nirvana, Prince, Lady Gaga, and so on. The Melkweg originally had a concert hall with a capacity of 700, and extended it by creating a second, larger space for live music in 1995: The Max. This hall was enlarged from 1,000 to 1,500 capacity in 2007. Since 2009 Melkweg also programs concerts in the new Rabozaal (1,400 capacity), created together with their neighbors: the City Theatre. Photos © DigiDaan/Melkweg.

Melkweg is a legendary Amsterdam rock club.

Acousticians have obtained a high degree of absorption even at low frequencies in the ceiling.

Melkweg—The Max

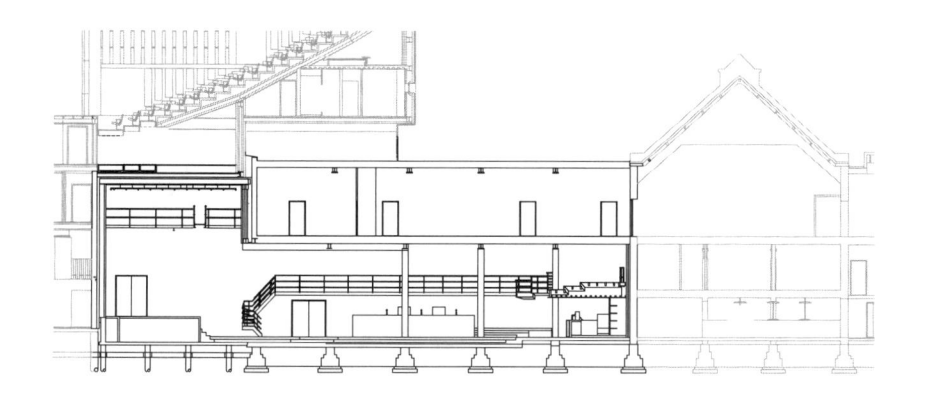

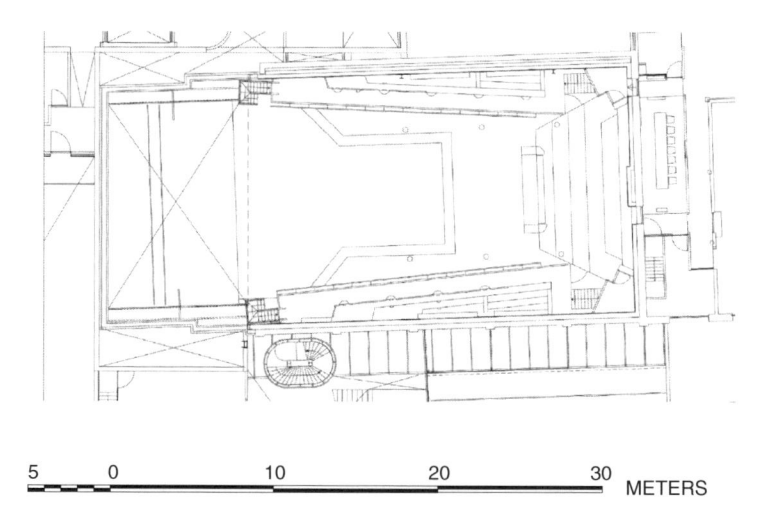

Geometrical data	
Volume	2,600 m^3
$L \times W \times H$	32 × 15 × 5.6 m
Height of stage	1.7 m
Stage opening	14.5 × 4 m

Acoustical data	
Audience area	
$T_{30,125-2k}$	0.81
EDT_{125-2k}	0.78
$C_{80,125-2k}$	6.03
BR_{63} versus 0.5–1k	0.76
BR_{125} versus 0.5–1k	0.78
Stage area	
EDT_{125-2k}	0.32
$D_{50,125-2k}$	0.9
BR_{63} versus 0.5–1k	1.12
BR_{125} versus 0.5–1k	0.83

Materials Used

Audience Area

Floor: Rubber on concrete.
Ceiling: Suspended mineral wool with large cavity above.
Walls: Layers of gypsum board on cavity. End wall: wooden fiber slabs.

Stage Area

Floor: Wood on cavity.
Ceiling: Suspended mineral wool with large cavity above.
Walls: Layers of gypsum board on cavity. End wall: wooden fiber slabs. Backdrop.

State of Hall When Measured

Empty: some equipment on stage.

State of Hall When Measured 301

Melkweg audience.

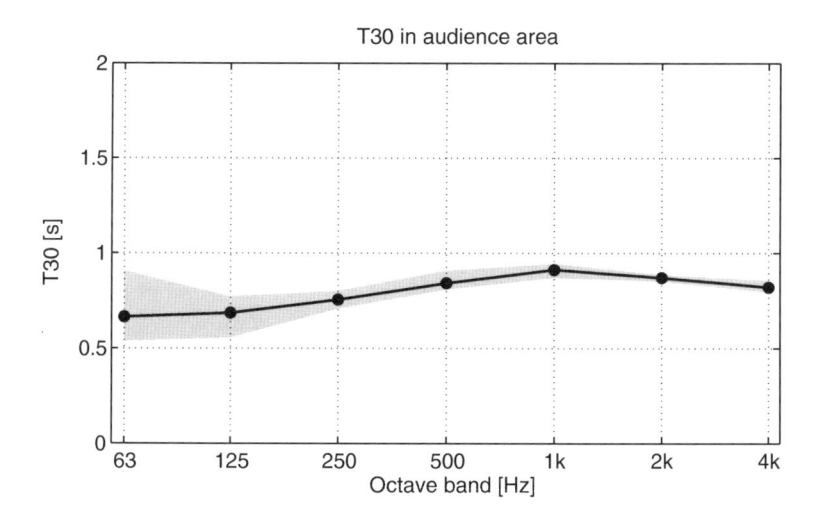

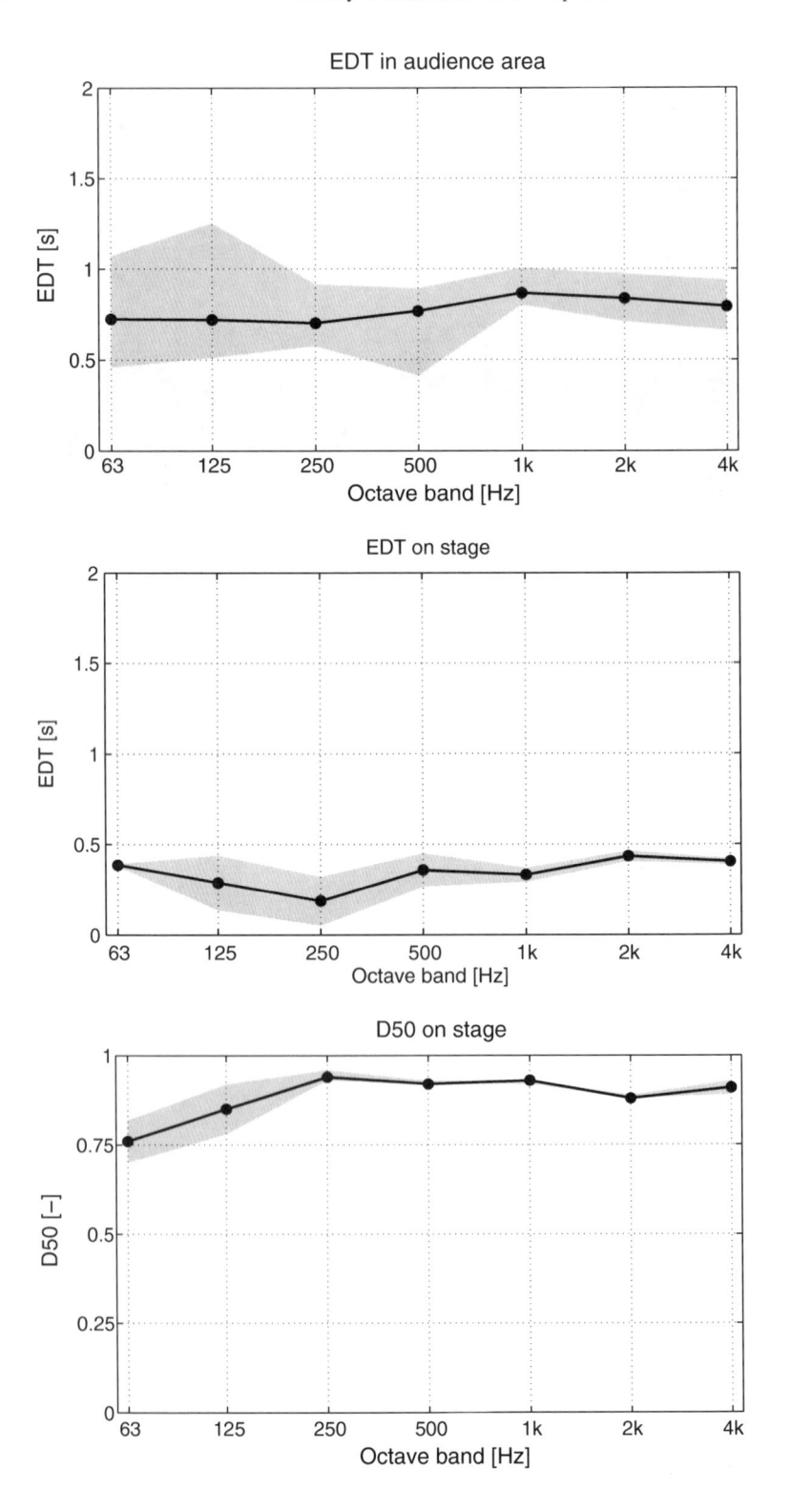

MEN Arena

Manchester
Number of concerts per year: 75+
Founded: 1995
Capacity: 21,000
Architect: DLA/Elerbe Beckett
Acoustician: Ove Arup and Partners

The Manchester Evening News Arena, managed and operated by SMG Europe, is one of the busiest venues in the world and the largest indoor arena in Europe. Opened in 1995, the MEN Arena has staged the biggest names in live entertainment from U2, the Rolling Stones, Madonna, and Pavarotti to the record-breaking 2010/2011 residency by local comedian Peter Kay. Attracting over one million visitors each year, the 21,000 capacity Arena was named "International Venue of the Year" in 2002 by industry publication *Pollstar* and has been nominated an unrivaled 10 consecutive times.

In 1995 the venue was officially opened by Torvill and Dean who broke the UK Box Office record for a single ice performance to over 15,000 fans. In 1998 Ricky "Hitman" Hatton beat Karl Taylor in his debut MEN Arena bout, the first of 14 fights at his hometown venue. 1999 saw the Arena host the homecoming of Manchester United after they won "The Treble." In 2000 Mike Tyson's UK debut fight at the Arena was watched by over 100 million people worldwide. 2001 saw local band James perform their farewell show at the Arena, before returning with a sell-out concert in 2007! The venue also held the 2002 Commonwealth Games' Boxing and Netball tournaments, as well as U2 playing to 19,384 fans, the biggest single audience of their European Elevation Tour. Bono declared the MEN Arena "the best concert venue in the country." In 2004 Madonna returned to Manchester for the first time since her 1980s British debut at the infamous Hacienda to perform two sell-out shows. In 2005 Comedian Lee Evans broke the Guinness world record for a solo act performing to the biggest comedy audience and comedian Peter Kay, who used to work as a MEN Arena Steward, returned with his record-breaking "Mum Wants A Bungalow" Tour. Now an annual event, the first Versus Cancer charity concert was staged by Ex-Smith's member Andy Rourke in 2006. 2007 saw Kylie Minogue crowned as the biggest-selling solo artist in the history of the Arena after her January Showgirl Homecoming dates bring her total Arena performances to 17; and local band Take That became the biggest-selling act of all time with a total 27 MEN Arena performances following their 11 shows in December. In 2008 the Arena became the UK's first music and entertainment venue to host competitive swimming after staging the 9th FINA World Short Course Swimming Championships in April and the Arena became Kylie Minogue's number one venue after official figures reveal the international pop star has performed more shows at the Arena—and to more people—than any other venue in the world. Peter Kay broke box office records in 2009 after all 20 nights for his 2010 spring residency sold out in an hour. The local comedian later announced he would return for an additional eight shows in 2011. In 2009 Westlife announced they would return to the Arena with two 2010 shows making them the biggest selling group (29 performances) in the history of the venue, surpassing the joint record held with Take That. Also in 2009 the Arena's summer family blockbuster, "Walking With Dinosaurs" won "Best Family Show" at the Manchester Evening News Theatre Awards, and the venue smashed attendance figures after welcoming over one and a half million ticketholders. The MEN Arena celebrated its 15th anniversary with a star-studded Birthday Bash featuring Alexandra Burke, Scouting For Girls, Pixie Lott, The Script, Beverley Knight, The Saturdays, and many others. The summer of 2011 saw the Manchester venue stage the world premiere of *Batman Live* (July) and the European premiere of *Glee* (June) as well as Kylie cementing her position as the biggest selling solo artist in the history of the venue after her four "Aphrodite Les Folies" in April shows brought her total Arena performances to 27, and the Children In Need Rocks Manchester in November, featured a star-studded line-up including Gary Barlow, Coldplay, Lady Gaga, JLS and local favorites Elbow.

*Based on concert ticket sales between January 2002 and June 2007, as calculated by *Pollstar*.

MEN in full swing. Photo: Karin Albinsson.

Curtains in front of seats do not have a significant impact on T_{30} when they are upholstered. Curtains can on the other hand be used in front of reflective surfaces to prevent echoes and for lowering T_{30} at higher frequencies.

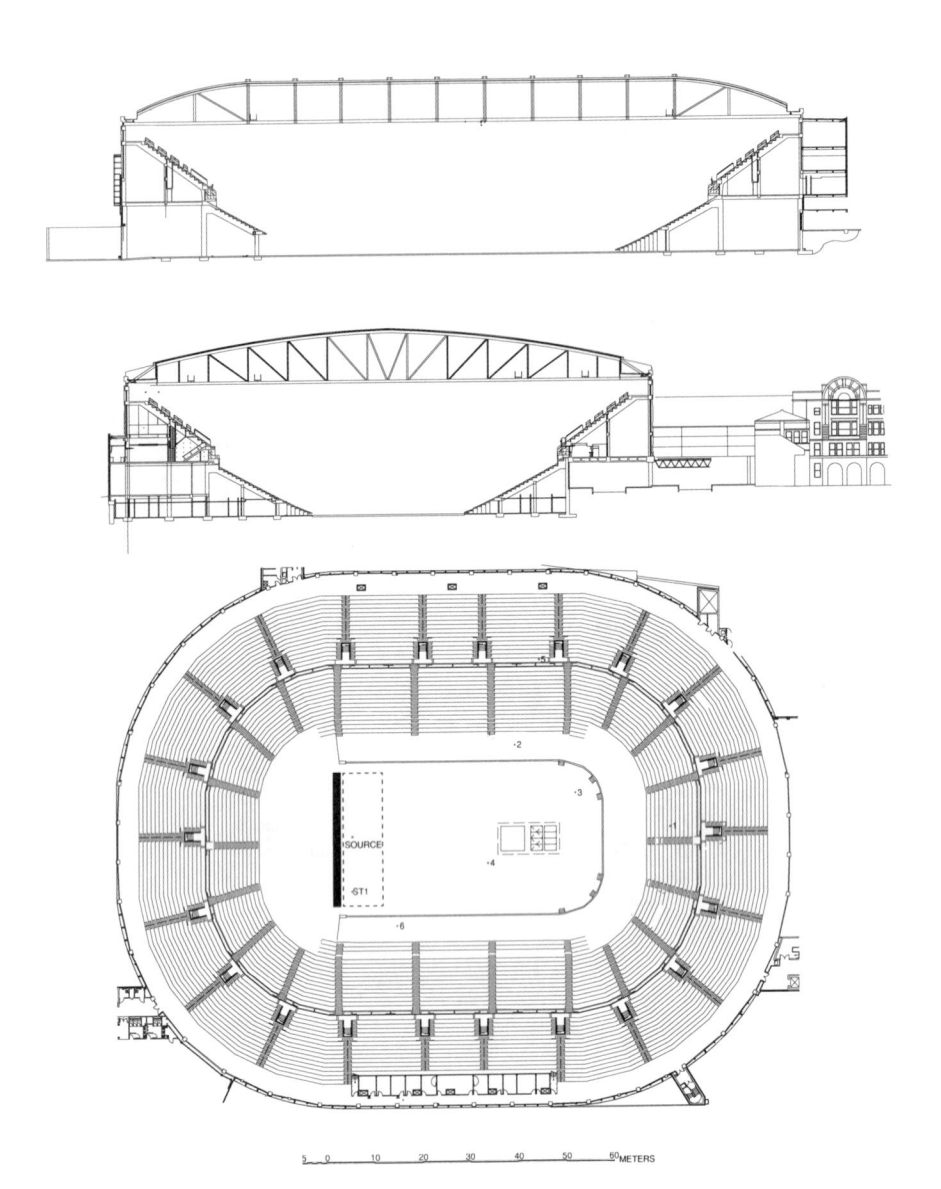

7 Gallery of Halls that Present Pop and Rock Music Concerts

Geometrical data	
Volume	NA
$L \times W \times H$	$130 \times 100 \times 36$

Acoustical data	
Audience area	
$T_{30, 125-2k}$	2.47
EDT_{125-2k}	2.22
$C_{80, 125-2k}$	−0.40
$BR_{63 \text{ versus } 0.5-1k}$	1.46
$BR_{125 \text{ versus } 0.5-1k}$	1.34

Materials Used

Audience Area

Floor: Concrete.
Ceiling: Thin metal trapez, perforated.
Walls: Very little wall area of concrete.
Chairs are upholstered also on rear back.

State of Hall When Measured

Empty; some rigging on floor. Most of the end of the venue was detached by large curtains probably causing only a very slight drop in reverberation time mostly above 1 kHz.

Roger Waters secures the fireworks in the MEN.

T30 in audience area

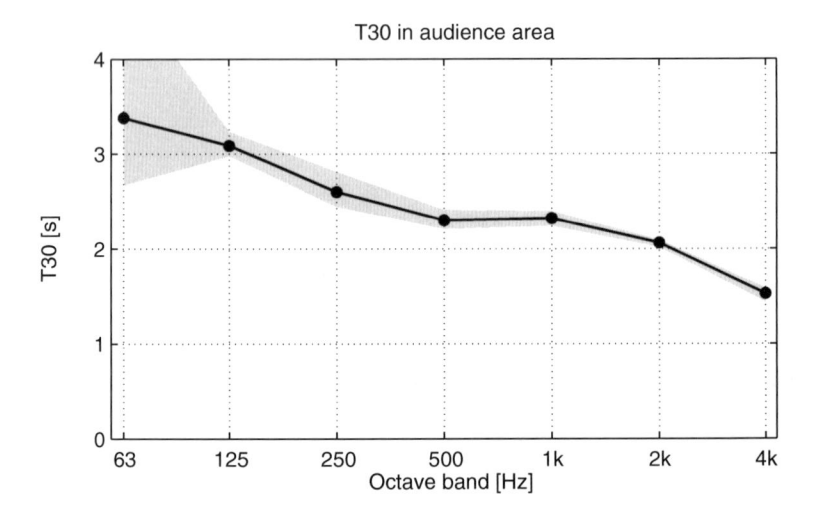

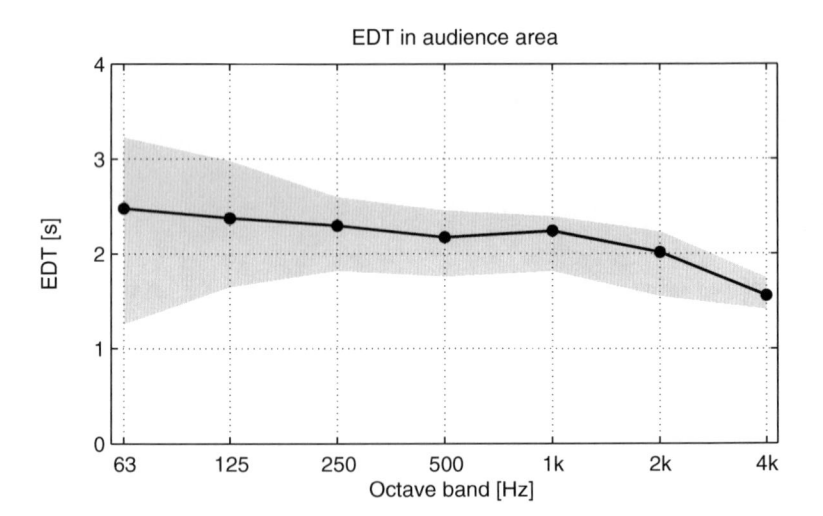

EDT in audience area

Nosturi

Helsinki
Number of concerts per year: 120
Founded: Built in 1957, converted to a concert hall in 1999
Capacity: 900
Architect:

- Original building 1957- Kaj Salenius.
- Conversion 1999- Jan Tromp and Markus Nevalainen, Valvomo Architects.

Acoustician: Akukon/Henrik Möller.

Nosturi, a live music venue with a character as versatile as that of its visitors, sits by the seaside near downtown Helsinki. The shipyard's old warehouse is operated by the Live Music Association (ELMU ry), whose offices can be found within, along with a fully licensed restaurant and rehearsal spaces for 50 bands.

Nosturi takes pride in providing a platform for a broad spectrum of culturally relevant events, including—but not limited to—concerts, raves, youth discos, performing arts, theatre, and art gallery functions. Everything from unknown underground bands to the largest mainstream acts have graced the stage of Nosturi at one time or another.

There are over 120 shows a year in the main hall and over 100 gigs downstairs at the bar.

Nosturi's main hall operates on two floors, which means that the maximum audience capacity is nothing short of 900 people. The floor level in Nosturi is narrower than the balcony level. One end of the space houses the sizeable stage, which can be observed from both floors.

The ground floor is also home to the restaurant, going by the name of Ravintola Alakerta, which serves as an excellent venue for smaller gigs and events. Alakerta can manage up to 120 standing guests, and works ideally for parties of about 70 seated guests. What is more, the outdoor summer terrace at the waterfront can comfortably seat well over 100 people in the sunshine.

As a venue, Nosturi is easily modifiable to suit a particular purpose by altering interiors and fittings. The mezzanine, for example, can seat an audience of 250. Whether the event requires a motorcycle-able catwalk, a dinner table the length of the entire venue, or a retractable roof, everything can be (and has been!) done. Should your guests wish to arrive directly to the venue by boat, this is also possible. Naturally.

310   7   Gallery of Halls that Present Pop and Rock Music Concerts

What it is all about! Photo: Eeka Mäkynen.

Classic rock venues are often found in uninhabited areas. Photo: Eeka Mäkynen.

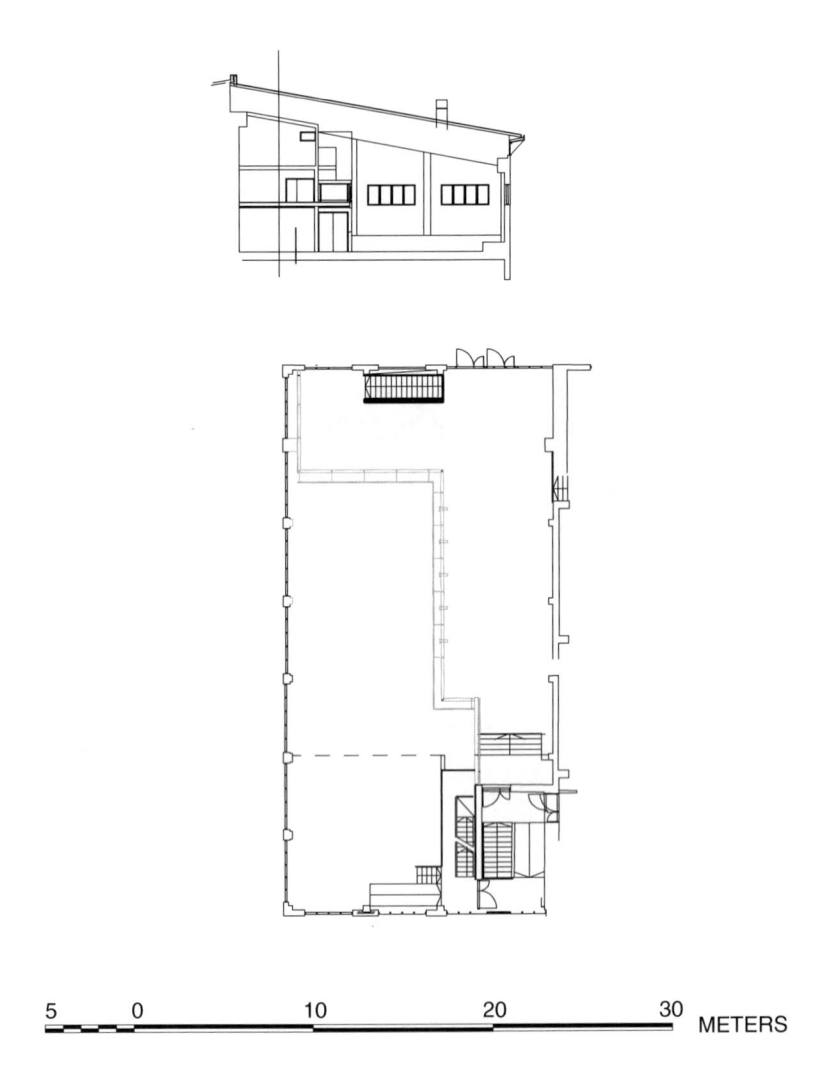

7   Gallery of Halls that Present Pop and Rock Music Concerts

Geometrical data	
Volume	3,000 m^3
$L \times W \times H$	$29 \times 14 \times 6\text{--}10$ m

Acoustical data	
Audience area	
$T_{30,125-2k}$	0.64
EDT_{125-2k}	NA
$C_{80,125-2k}$	NA
BR_{63} versus 0.5–1k	2.48
BR_{125} versus 0.5–1k	1.89

Materials Used

Audience Area

Floor: Mainly concrete; raised platforms are wood on cavity.
Ceiling:

- Main ceiling: Mineral wool on concrete.
- Ceiling under balcony: Porous plate made of wood shavings and cement.

Walls: Brick and concrete, curtains and some absorbtive wall plates directly on walls.

Stage Area

Floor: Plywood.
Ceiling: Mineral wool on concrete.
Walls: Brick and concrete with molton curtains.

State of Hall When Measured

Empty.

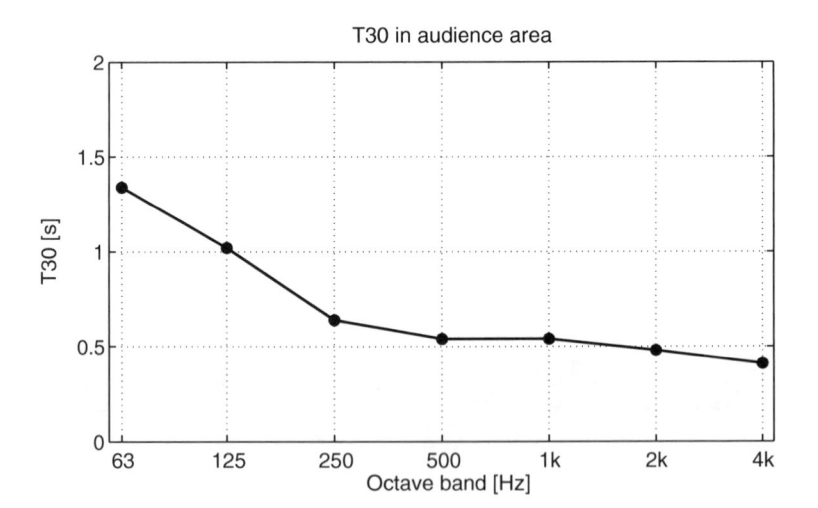

O₂ Berlin

Berlin
Number of concerts per year: 70. In all halls in the venue: 100
Founded: 2008
Capacity: 17,000
Architect: HOK/Kansas City
Acoustician: Wolfgang Ahnert/Acoustic Design Ahnert

Since opening on September 10th, 2008 the O₂ World Berlin has changed Germany's entertainment landscape. Owned and operated by the Anschutz Entertainment Group, one of the leaders in the sport and entertainment business, the O₂ World is widely regarded as one of Europe's most modern arenas. It is the main stage in the German capital for sports and entertainment with a capacity of up to 17,000 fans including club seats, luxury suites, private restaurants, clubs, and hospitality spaces in addition to a full array of the most modern amenities.

The Kansas City based architectural firm HOK was responsible for designing the arena that was constructed in just 727 days. The development process was managed and controlled by ICON Venue Group to complete the project on time and on budget. The O₂ World debuts many never-before experienced technological systems, conveniences, and features including one of the world's largest architectural LED lighting grids. Spanning across the entire south glass façade on the exterior of the new arena, the breathtaking 1,390-m² installation displays motion graphics and video content on a massive scale. Acoustic Design Ahnert (ADA) was contracted to secure the optimal acoustical set-up for the O₂ World. To reduce the reverberation time sound-absorbing material was installed throughout the arena bowl such as ceiling banners and panels on all the walls.

Right from the start O₂ World was put to test with four events in as many days. Opening weekend kicked off with Metallica blasting the inaugural tunes, followed by Herbert Grönemeyer the next day. Music took a break for a hockey game on day three, but was back with a Coldplay concert to complete the premiere. Although the ambitious opening program made the O₂ World an instant success, it did not happen without the expected and unexpected challenges of "breaking in" a brand new arena. Due to a busy PR schedule for the release of their record, *Death Magnetic*, Metallica were a little behind schedule and conducted their sound check while 17,000 anxious fans were already lined up outside the arena. Herbert Grönemeyer, playing a center-stage gig for the first time, performed until way after midnight, thereby challenging crew and production to execute the changeover to the hockey game, scheduled for the next afternoon, in an almost impossible time frame.

Since then O₂ World has become a must-play venue featuring concerts by Paul McCartney, Tina Turner, Pink, Eagles, Eric Clapton, Elton John, Neil Young, Kings of Leon, Supertramp, Alicia Keys, Beyoncé, The Scorpions, Lady Gaga, Joe Cocker, Kiss, Rod Stewart, Sting, Peter Gabriel, Leonard Cohen, and Depeche Mode including several own-promoted events by artists including The Black Eyed

Peas, Paul van Dyk, Die Fantastischen Vier, and Britney Spears. It has also hosted the MTV European Music Awards and the German Music Awards ECHO.

O$_2$ Berlin.

The American way: thick porous banners in the ceiling installed for broadband absorption.

O₂ Berlin

There is 20-cm mineral wool behind the perforated metal walls with double gypsum plates behind. A good combination of absorption and insulation is achieved for the purpose.

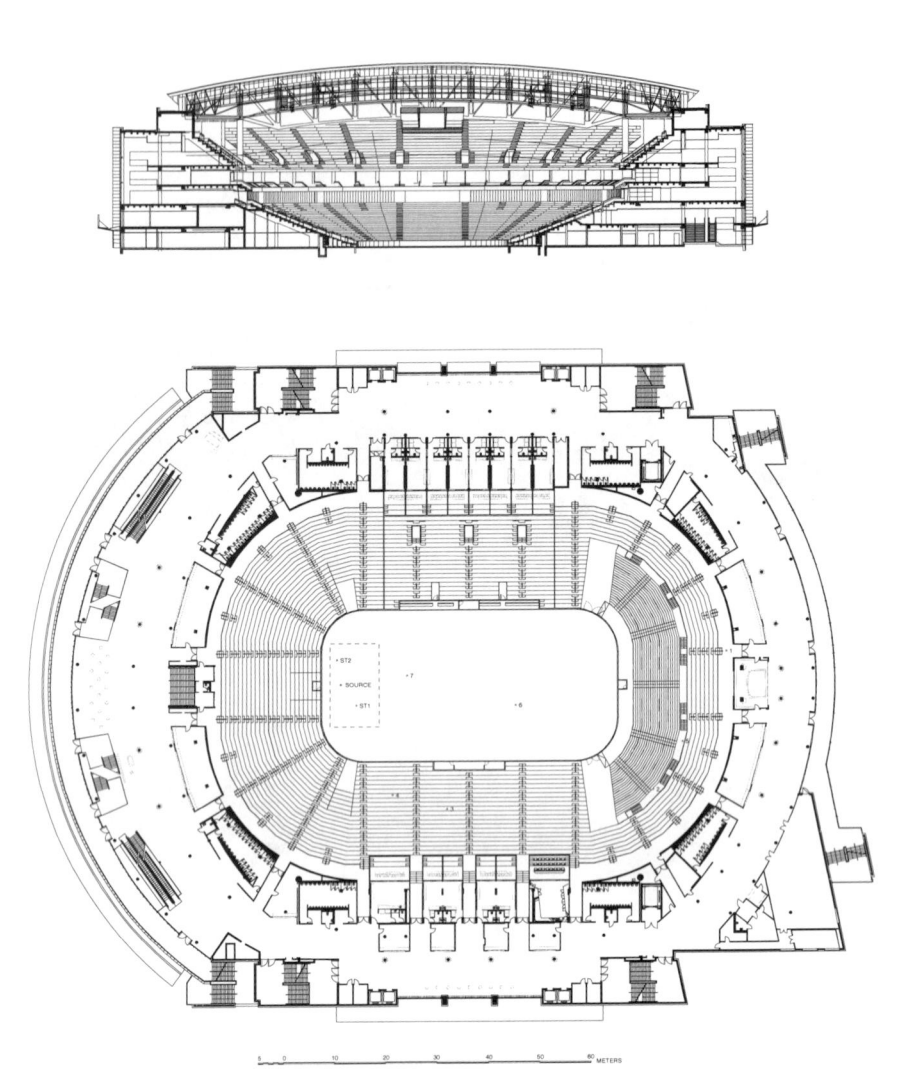

318   7   Gallery of Halls that Present Pop and Rock Music Concerts

Geometrical data	
Volume	280,000 m³
$L \times W \times H_{(max)}$	$100 \times 95 \times 35$ m

Acoustical data	
Audience area	
$T_{30, 125-2k}$	2.44
EDT_{125-2k}	2.03
$C_{80, 125-2k}$	−0.65
BR_{63} versus 0.5–1k	1.7
BR_{125} versus 0.5–1k	1.19

Materials Used

Audience Area

Floor: Vinyl on concrete.
Ceiling: Thin trapezoid metal nonperforated. Suspended approximately 6-cm thick porous absorptive banners at a distance from approximately 0.3–0.6 m from metal.
Walls: Perforated metal then 20-cm mineral wool and then double layer gypsum in front of large cavity.
Seats are upholstered.

State of Hall When Measured

Empty; no seats mounted on floor.

T30 in audience area

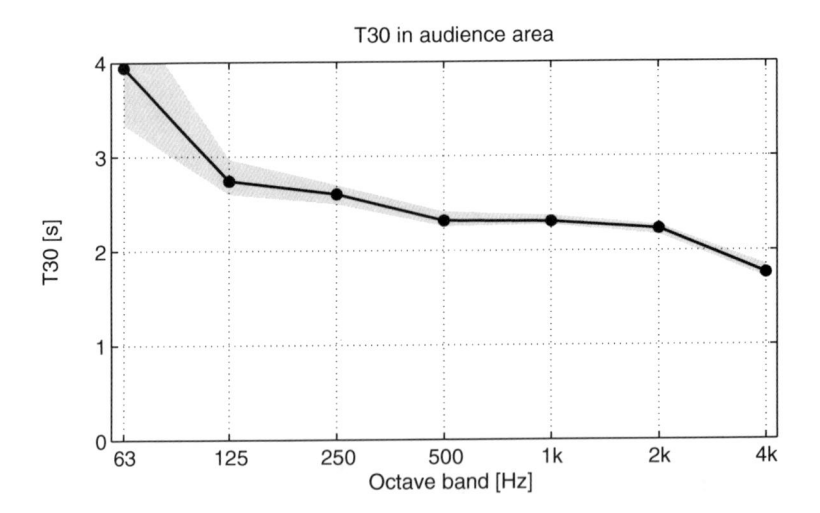

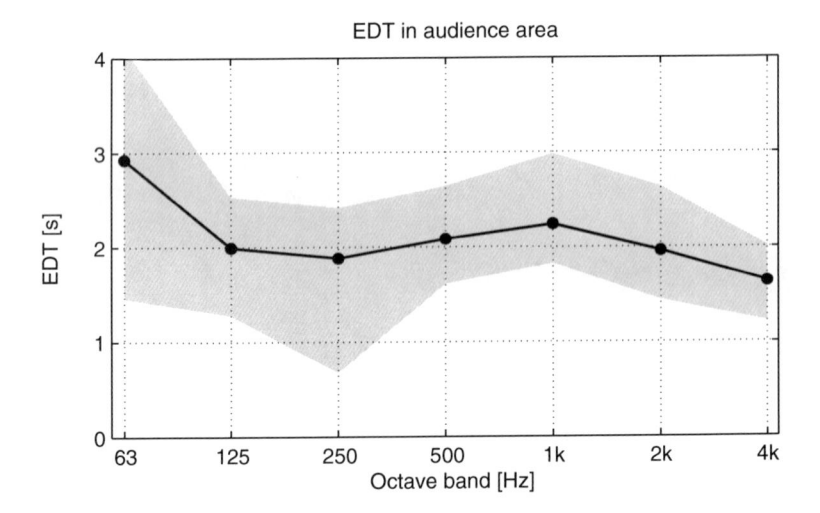

EDT in audience area

O₂ World Hamburg

Hamburg
Number of concerts per year: 50. Total number of events: approximately 130
Founded: 2002
Capacity: 16,000
Architect: EVATA Finland OY (now Pöyry)
Acoustician: EVATA Finland OY (now Pöyry)

On November 8, 2002 the largest multifunctional arena in northern Germany opened its gates. The EVATA Finland OY Company was responsible for the building's architecture and acoustics. The arena can host up to 16,000 people. Since its opening the arena hosted more than 1,100 concerts, shows, and sporting events. Stars such as Alicia Keys, Beyoncé, Bon Jovi, Bruce Springsteen, Paul McCartney, Metallica, Shakira, Take That, and KISS performed in the venue, formerly known as Color Line Arena. Many boxing world champions were crowned and the home teams Hamburg Freezers (hockey) and HSV Handball (handball) celebrated goals and victories. About 130 events take place every year with more than one million visitors attending the shows, concerts, and home games. The arena can be minimized to a 6,500- and even a 4,000-person configuration.

Since fall of 2007 the O₂ World Hamburg is part of the Anschutz Entertainment Group, a subsidiary of the Anschutz Corporation, one of the world's leading sports and entertainment presenters. In April 2010 the arena was renamed O₂ World Hamburg and has since undergone remodeling to make one of Europe's most modern arenas even more attractive.

During the remodeling process a state-of-the-art 360-degree LED board was installed below the upper gallery. Also remodeled were 24 of the suites in the arena. The guests can choose between three different styles: classic, lounge, and high table. A highlight for special occasions is the show suite. Up to 48 people can enjoy an event in the new suite, which also can be separated into two suites with 24 seats each.

Upgrading their premium and suites sections, the O₂ World Hamburg started to build an extension to the existing building in April of 2011. The extension, which will be finished in the spring of 2012, will serve as the new entry for the arena's premium guests and suite-holders. It will offer new ways for hospitality thanks to a state-of-the-art club level and lounge level and it will be home for several new offices for arena employees.

As one of Europe's most modern multifunctional arenas the O₂ World Hamburg is open for almost any idea artists may come up with. Rammstein is known to play with fire. Cirque de Soleil once used a stage that was 40 m long. The Ben Hur show featured a historic Roman setting and live chariot racing. German rock icon Peter Maffay entered the stage on a Harley Davidson motorcycle. The *Dinosaur* show brought the giants from primeval times back to life. And the *Night of the Jumps* featured the world's most exciting motocross stunt show, with the riders almost touching the roof.

The O$_2$ World Hamburg hosts all these concerts, shows, and sporting events for one reason: to entertain Hamburg. And so it does; it had 146 events in 2010 and was thereby ranked number 8 in the annual *Pollstar* magazine ranking of the top 100 concert arenas worldwide in ticket sales.

The O$_2$ World Hamburg is a busy arena that some may know by its former name: Color Line Arena.

The line of color is kept blue.

Geometrical data	
Volume	500,000 m^3
$L \times W \times H_{(max)}$	150 × 110 × 33 m

Acoustical data	
Audience area	
$T_{30, 125-2k}$	2.66
EDT_{125-2k}	2.49
$C_{80, 125-2k}$	−2.69
$BR_{63 \text{ versus } 0.5-1k}$	2.58
$BR_{125 \text{ versus } 0.5-1k}$	1.69

Materials Used

Audience Area

Floor: Concrete.
Ceiling: A layer of trapezoidal sheet metal without perforation, then a layer of solid insulation and roofing felt on top.
Walls: Walls made of lime-sand brick. The walls on the lower level are completely covered in cotton material.

State of Hall When Measured

Empty; no additional seats mounted.

T30 in audience area

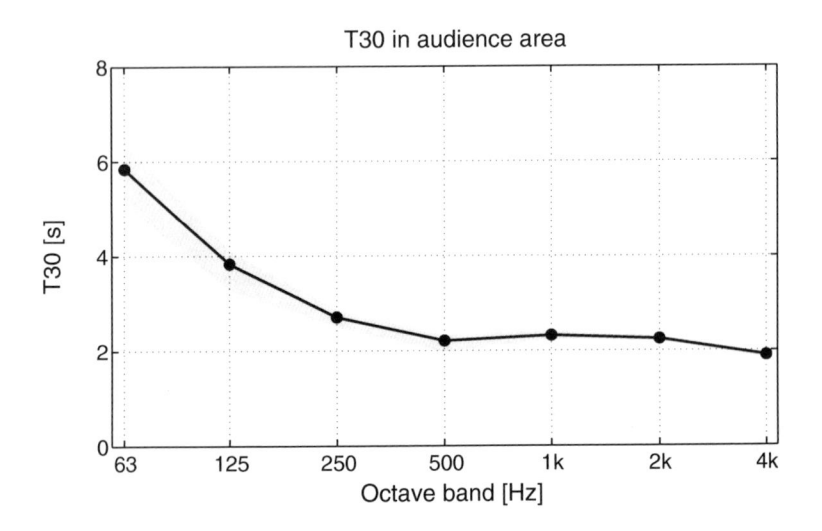

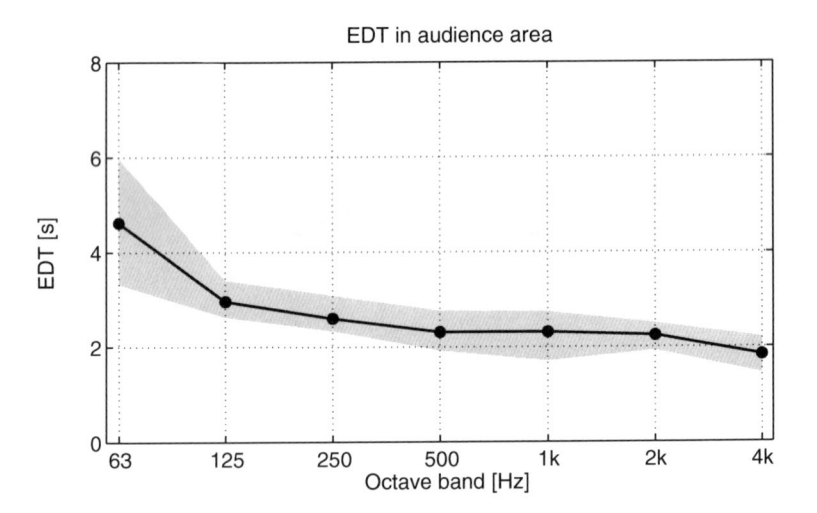

EDT in audience area

O₂ London

London
Number of concerts per year: 60. In all halls in the venue: 120
Founded: 2007
Capacity: 20,000
Architect: N/A
Acoustician: Vanguardia Consulting

The O_2, visually typeset in branding as *The O_2*, is a large entertainment district on the Greenwich peninsula in South East London, England, including an indoor arena, a music club, a Cineworld cinema, an exhibition space, piazzas, bars, and restaurants. It was built largely within the former Millennium Dome, a large dome-shaped building built to house an exhibition celebrating the turn of the third millennium; as such, *The Dome* remains a name in common usage for the venue. Naming rights to the district were purchased by O_2 plc (now Telefónica Europe plc) from its developers, Anschutz Entertainment Group (AEG), during the development of the district. AEG owns the long-term lease on the O_2 Arena and surrounding leisure space.

Construction of the arena started in 2003 and finished in 2007. Owing to the impossibility of using cranes inside the dome structure, the arena's roof was constructed on the ground within the dome and then lifted. The arena building's structure was then built around the roof. The arena building, which houses the arena and the arena concourse, is independent of all other buildings in The O_2 and houses all the arena's facilities. The whole arena building takes up 40 % of the total dome structure. The venue, rebranded the O_2, was reopened to the public on June 24th, 2007 with a concert by Bon Jovi in the arena. The O_2 celebrated its first year with a book, including a double-page picture of Elton John from his September 2007 *Red Piano* show.

The seating arrangement throughout the whole arena can be modified, similar to the Manchester Evening News Arena [4]. The ground surface can also be changed among ice rink, basketball court, exhibition space, conference venue, private hire venue, and concert venue. The arena was built to reduce echoing which has previously been a problem in many London music venues [5]. The sound manager for U2, Joe O'Herlihy, worked with acoustic engineers on acoustics.

Despite The O_2 arena's being open for only seven months of the year, the venue sold over 1.2 million tickets in 2007, making it the third most popular venue in the world for concerts and family shows narrowly behind the MEN Arena (1.25 million) and Madison Square Garden in New York (1.23 million). In 2008, it became the world's busiest venue taking the crown from the MEN Arena with sales of more than two million [6]. The O_2 arena since its opening in 2007 has been host to many concerts, from UK bands and artists to international superstars. The O_2 was named the "World's Best Venue" by *Pollstar* in 2009. *Pollstar* figures for 2010 placed The O_2 as the world's number one music venue with a huge 1,737,654 tickets sold last year, more than any other arena in the world. The *Pollstar* industry listings chart showed The O$_2$ arena's year end ticket sales beat their nearest competitor, Madison Square Garden, by over 50 %.

- Anschutz opened The O$_2$ arena on June 23rd, 2007, with a free event for all of the building's employees billed as the O$_2$ premiere featuring Peter Kay, Tom Jones, Kaiser Chiefs, and Basement Jaxx, with the show hosted by Dermot O'Leary. Snow Patrol then played a concert to an audience made up of sponsors, local residents, local business employees and winners of an online competition.
- David Campbell, President and CEO of AEG Europe, commented: "Prince's 21 nights at The O$_2$ are a testament to London's eagerly awaited new entertainment destination. We're thrilled that a world-class artist like Prince is part of our opening season, and that he'll be breaking a world record in our arena."
- The Spice Girls performed 17 sold-out shows during their 2007/2008 *Return of the Spice Girls* tour. The first date's tickets were sold in 38 s.
- Led Zeppelin performed a one-off reunion concert at the arena on December 10th, 2007. Over one million people applied online for tickets.
- Michael Jackson was preparing for a sell-out series of 50 shows at the O$_2$, due to take place July 13th, 2009–March 6, 2010, when he died much too early of cardiac arrest.
- Legendary heavy metal band native to London, Iron Maiden, was scheduled to perform here on August 5th and 6th, 2011, the last dates of The Final Frontier World Tour.
- Bon Jovi performed 12 nights in June 2010 as part of their *The Circle Tour*. They also became the first -ever artists to perform on the roof of the O$_2$ Arena a few days prior to their 12-night residency.

Packed. The O$_2$ in London was rated best international arena in the world 2011 by Pollstar.

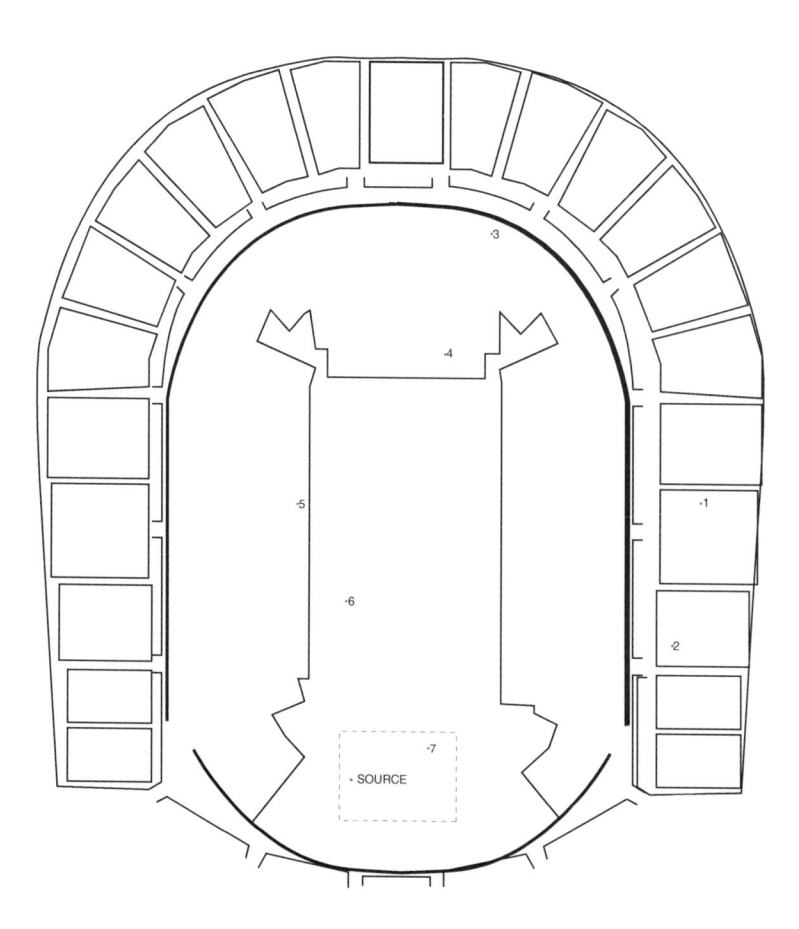

O₂ London

Geometrical data	
Volume	400,000 m³
$L \times W \times H_{(max)}$	approximately 125 × 115 × 43 m

Acoustical data	
Audience area	
$T_{30, 125-2k}$	2.17
EDT_{125-2k}	1.83
$C_{80, 125-2k}$	−2.69
BR_{63} versus 0.5–1k	2.06
BR_{125} versus 0.5–1k	1.38

Materials Used

Audience Area

Floor: Concrete.
Ceiling: Perforated liner over entire ceiling areal with 40-cm cavity with broadband absorption.
Walls: Upper tier rear wall: acoustic panel finish; broadband absorber of 60-cm depth. Lower tier back wall. Acoustic panels 80 % coverage above 1 m. Balcony fronts: Stretch fabric panels with absorbent behind.
Upholstered seats with upholstered underside.

State of Hall When Measured

Empty; no seats on floor.

Spectacular. O$_2$ World, London from the outside.

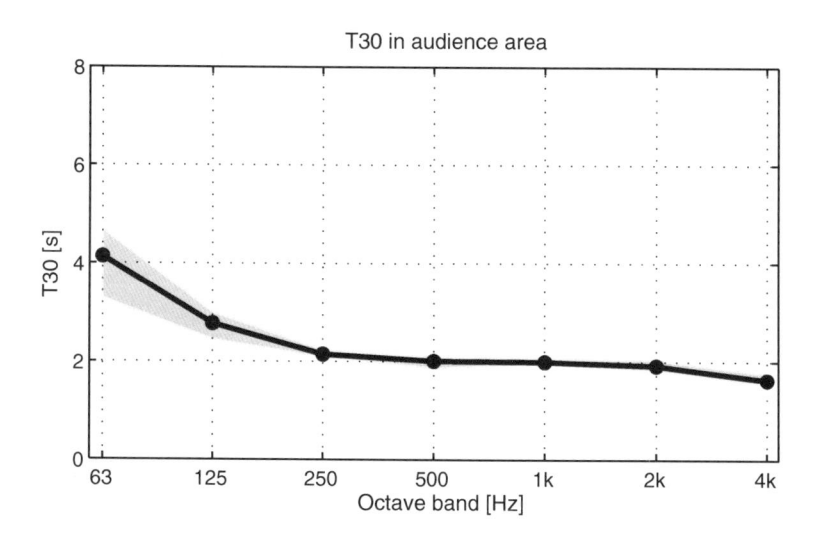

EDT in audience area

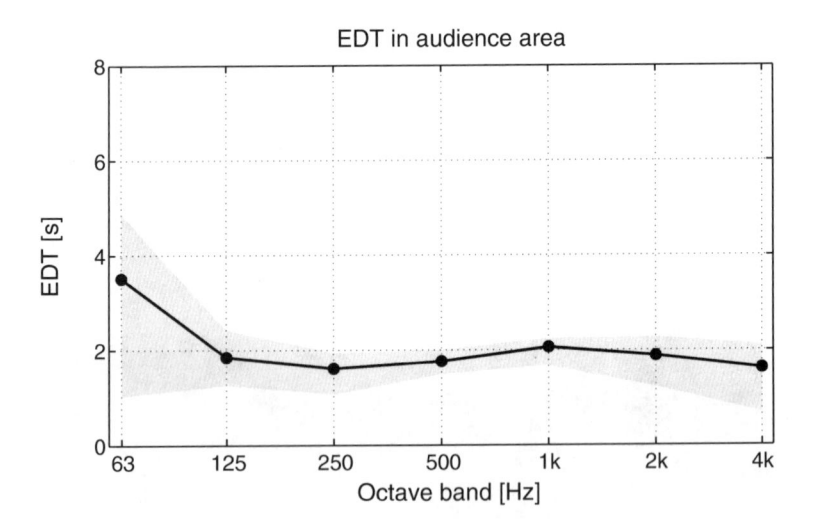

O$_{13}$ Tilburg

Tilburg
Number of concerts per year: 100
Founded: 1998
Capacity: 2,000
Architect: Mels Crouwel, Benthem Crouwel Architects
Acoustician: Peutz
Owner: City of Tilburg

Unforgettable shows, amazing dance events, festivals, furious stand-up comedians, one-off projects, and even fresh young talent got the chance to perform. It is easy to say that O$_{13}$ is an unique venue in the center of Tilburg. O$_{13}$ consists of three halls: large and medium-sized auditoria (Dommelsch Hall and Small Hall) and a smaller space for new/specialist music trends (Stage01). It also contains a studio and several rehearsal rooms. Spatially the building meshes with the Tivoli parking structure as one large built mass that screens off the street and cleanly marks off the park. The expanded metal screen of the parking facility is extended on one side to project in front of the facade of O$_{13}$. Façades and roof are clad in black EPDM rubber filled with glass wool and sporting real CDs on its surface.

In 1998, O$_{13}$ was the first concert hall in The Netherlands which was newly built for the sole purpose of live music. Since then artists such as Robbie Williams, Nine Inch Nails, Muse, Alice Cooper, The Roots, Tool, Blondie, Damien Rice, Kraftwerk, Simple Minds, Snoop Dogg, 30 s To Mars, Slayer, Queens Of The Stoneage, Sigur Ros, Editors, Interpol, Armin van Buuren, Pablo Francisco, Ice Cube, Chuck Berry, and many others have performed in O$_{13}$. O$_{13}$ appeals to all lovers of pop music from The Netherlands, Belgium, Germany, and sometimes even the whole of Europe. Quality, public opinion, and affordability are of great importance. O$_{13}$ organizes about 400 activities with more than 230,000 visitors each year.

Everybody on the sloped audience area enjoys great views and great sound.

No one in the audience of max. 2,000 persons is more than 22 m away from the PA speakers.

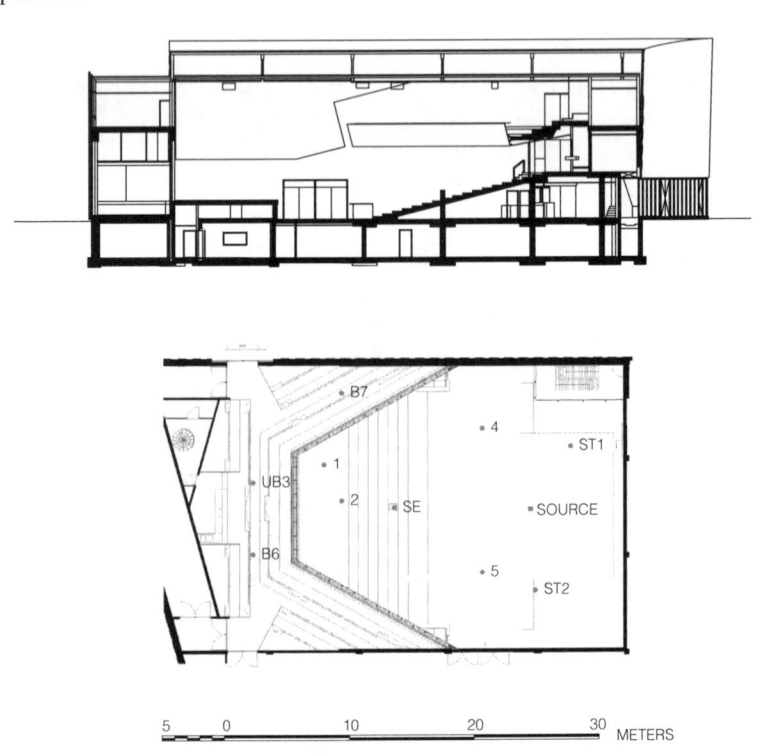

Geometrical data	
Volume	7,500 m^3
$L \times W \times H$	$32 \times 22 \times 11.0$ m
Surface area of stage	8×22.4 m
Height of stage	1.67
Stage opening	

Acoustical data	
Audience area	
$T_{30, 125-2k}$	1.07
EDT_{125-2k}	0.88
$C_{80, 125-2k}$	4.89
$BR_{63 \text{ versus } 0.5-1k}$	1.73
$BR_{125 \text{ versus } 0.5-1k}$	1.01
Stage area	
EDT_{125-2k}	0.51
$D_{50, 125-2k}$	0.89
$BR_{63 \text{ versus } 0.5-1k}$	0.97
$BR_{125 \text{ versus } 0.5-1k}$	0.91

Materials Used

Audience Area

Floor: Concrete.
Ceiling: Suspended mineral wool.
Walls: Concrete. Wood fiber slabs on cavity on side walls in front of the stage near the loudspeakers.

Stage Area

Floor: Nonabsorptive vinyl direct on concrete.
Ceiling: Suspended mineral wool.
Walls: Wood fiber slabs on side walls with cavity for a total of 7-cm depth.
Backdrop: curtains on the back wall.

State of Hall When Measured

Empty.

O₂ Tilburg is indeed a modern facility.

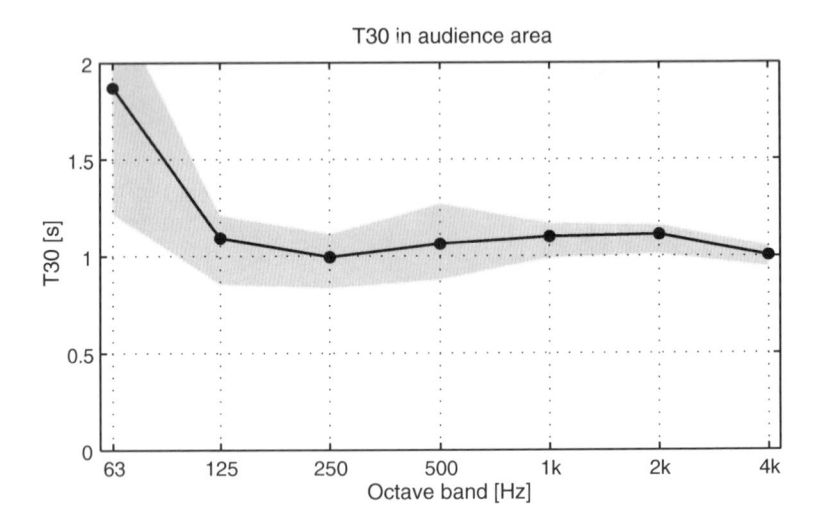

EDT in audience area

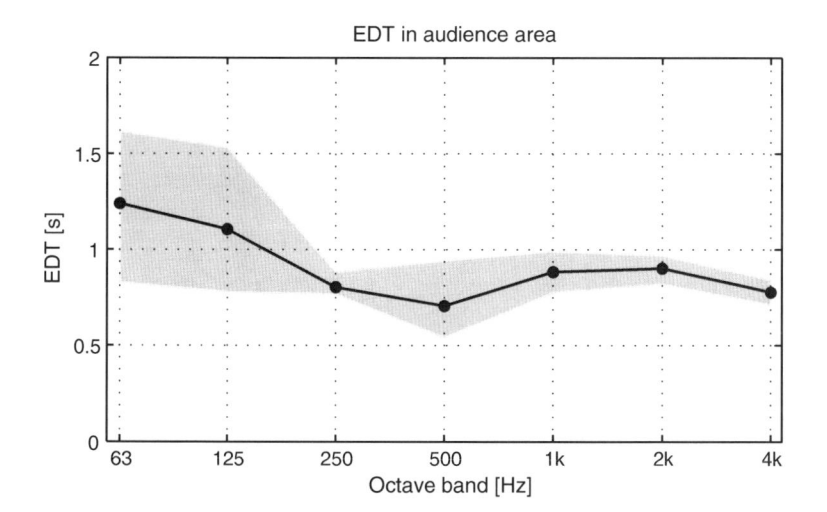

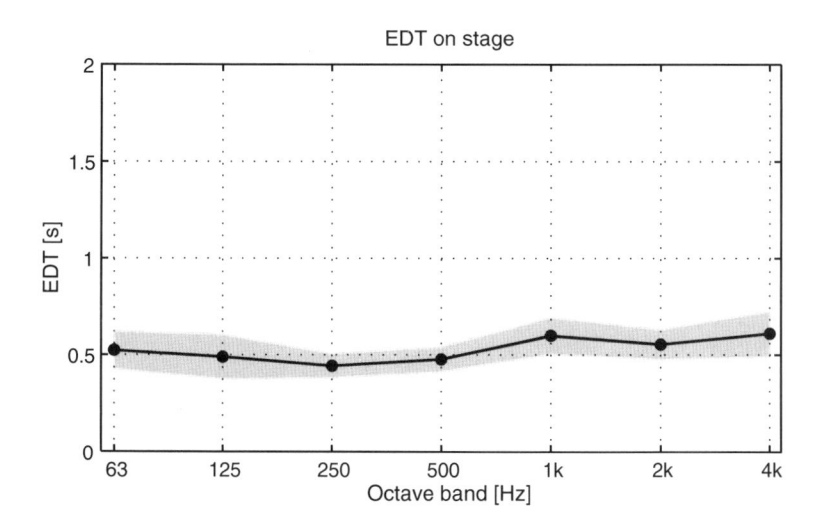

EDT on stage

D50 on stage

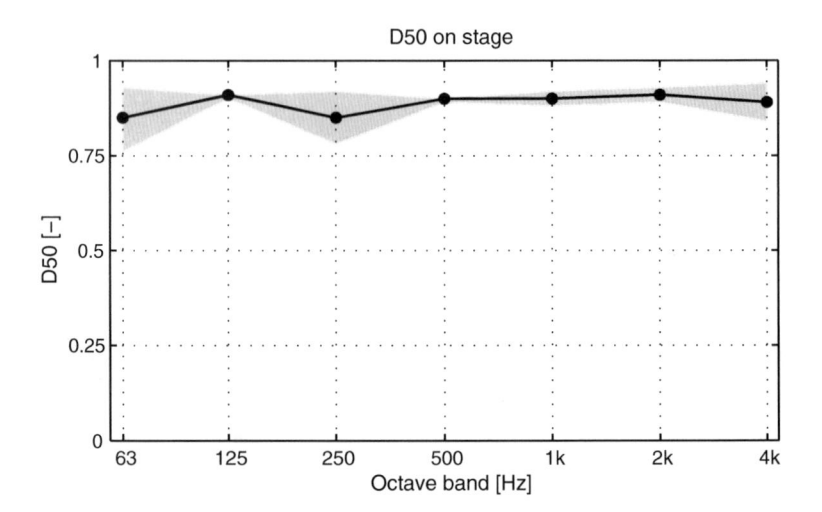

Olympia

Paris
Number of concerts per year: 120. In all halls in the venue: 320
Founded: 1954
Capacity: 2,200
Architect:
Acoustician:
Owner

Founded in 1888, by Joseph Oller, the creator of the Moulin Rouge, today it is easily recognizable by its giant red glowing letters announcing its name. It opened in 1889 as the "Montagnes Russes" but was renamed the Olympia in 1893. In addition to musicians, the Olympia played host to a variety of entertainment including circuses, ballets, and operettas. However, following a steady decline in appearances by the great stars, from 1929 until 1944 it served as a movie theater. It may have opened as a music hall under the German occupation of France during World War II, but certainly in 1945 after the Liberation, it was a music hall free to Allied troops in uniform. Attendees had to listen to the playing of four national anthems before the varied programs that always ended with a spirited French can-can performed by dancers, some of whom were no longer young.

Thereafter, at times it may have reverted to movies again until Bruno Coquatrix revived it as a music hall with a grand reopening in February 1954. After his death, it ultimately went into another decline and was in danger of being torn down and turned into a parking lot but on January 7th, 1993, France's then minister of culture, Jack Lang, issued a preservation order for the Olympia that resulted in two years of construction work to rebuild a perfect replica of the façade and the grandeur of its famous red interior. Édith Piaf achieved great acclaim at the Olympia, giving several series of recitals from January 1955 until October 1962. Jeff Buckley, long an admirer of Piaf, gave what he considered the finest performance of his career there in 1995, which was later released in 2001 on *Live at L'Olympia*. Jacques Brel's 1961 and 1964 concerts at L'Olympia are legendary and preserved to this day on new CD releases. Marlene Dietrich's 1962 Olympia concert was broadcast. On May 3–4, 1972, The Grateful Dead played two concerts here as part of their first major European tour. Both shows were recorded and songs from each were released on their 1972 live album *Europe '72*.

Inaugurated by the biggest star in France at the time, singer/dancer La Goulue, the venue has showcased a wide variety of performers, from French acts including Alan Stivell, Edith Piaf, Charles Aznavour, Adamo, Gilbert Bécaud, Johnny Hallyday, and Mireille Mathieu, to international stars of very different musical genres: Chuck Berry, Joséphine Baker, The Animals, Celine Dion, Cher, Diana Ross and The Supremes, Aretha Franklin, Stevie Wonder, James Brown, Diana Krall, War, Fairuz, Jeff Buckley, Robert Plant and The Strange Sensation, Jimi Hendrix, Judy Garland, Kraftwerk, Nana Mouskouri, Genesis, Wings, Paul McCartney, Paul Simon, Led Zeppelin, David Bowie, Bob Dylan, Black Sabbath, PJ Harvey, Lady Gaga, The Corrs, Luciano Pavarotti, Morrissey, The Shadows, Nelly Furtado, New

Order, Nick Cave, Nina Simone, Patti Smith, Phil Collins, Roger Hodgson, Primal Scream, KISS, Scorpions, Simple Minds, The Beatles, The Cure, The Jackson 5, Otis Redding, Red Hot Chili Peppers, The Rolling Stones, Dave Matthews Band, Björk, Dionne Warwick, Violetta Villas, Velvet Underground, Leonard Cohen, Manu Dibango, Madonna, Katy Perry, Christina Aguilera, and James Brown.

The red seats on the lower level are removable for standing audience pop and rock performances.

Piaf, Beatles, Bowie, Brel, Dylan, Zeppelin, Pavarotti, Jackson 5, Stones, Hendrix, Buckley, and Wonder.

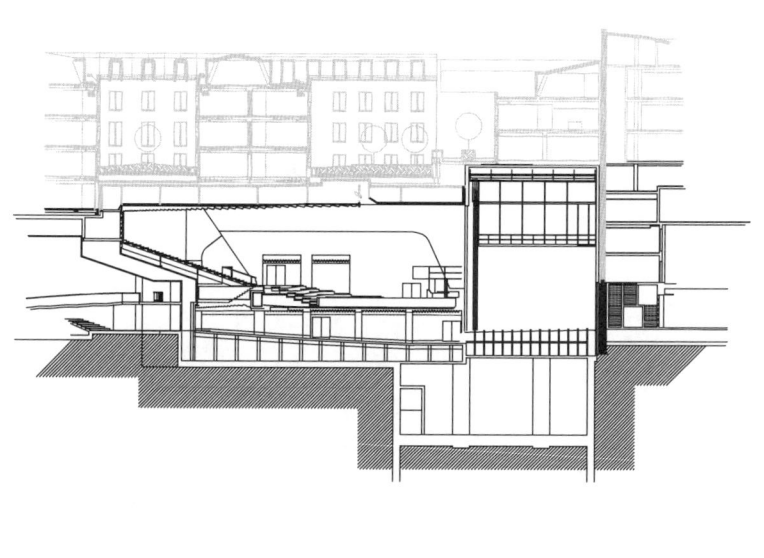

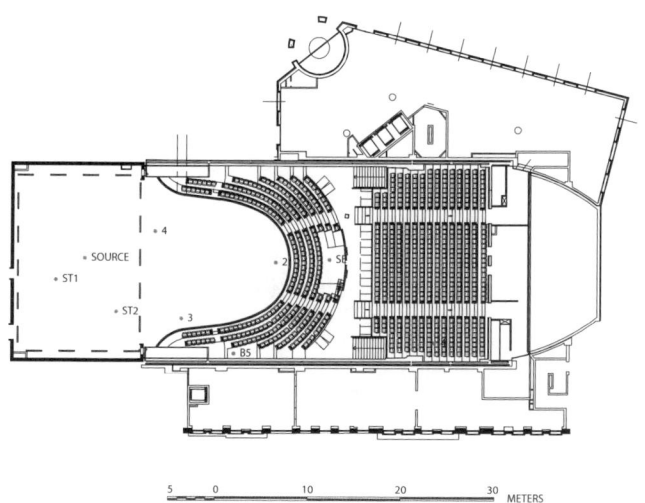

Olympia

Geometrical data	
Volume	13,000 m^3
$L \times W \times H$	$50 \times 20 \times 14.3$ m

Acoustical data	
Audience area	
$T_{30,125-2k}$	1.23
EDT_{125-2k}	1.1
$C_{80,125-2k}$	3.69
BR_{63} versus 0.5–1k	1.63
BR_{125} versus 0.5–1k	1.47
Stage area	
EDT_{125-2k}	0.82
$D_{50,125-2k}$	0.84
BR_{63} versus 0.5–1k	1.02
BR_{125} versus 0.5–1k	1.56

Materials Used

Audience Area

Floor: Concrete with temporary flooring for the mounting of chairs. Carpet on balcony.
Ceiling: Painted concrete. Suspended reflector above balcony.
Walls: Painted plates on cavity. Walls are not vertical but tilted slightly upwards. Back wall of thick perforated panels with cavity behind.

Stage Area

Floor: Vinyl direct on concrete.
Ceiling: Painted concrete.
Walls: 5-cm thick wood fiber panels direct on concrete.

State of Hall When Measured

Empty with seats mounted. All seats on the lower level can be removed for standing audience concerts.

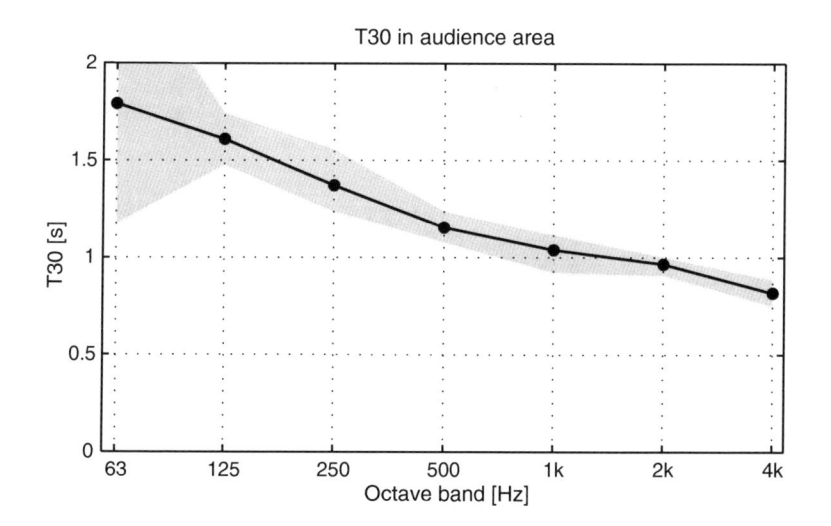

State of Hall When Measured

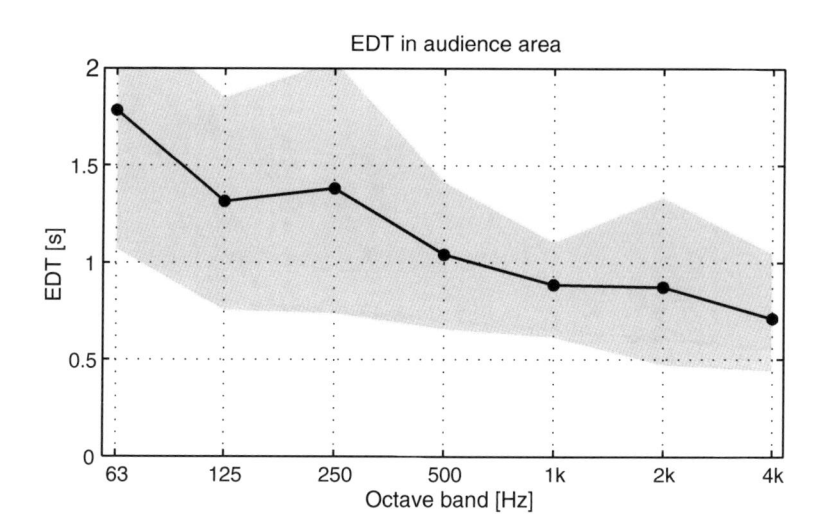

T30 in audience area

EDT in audience area

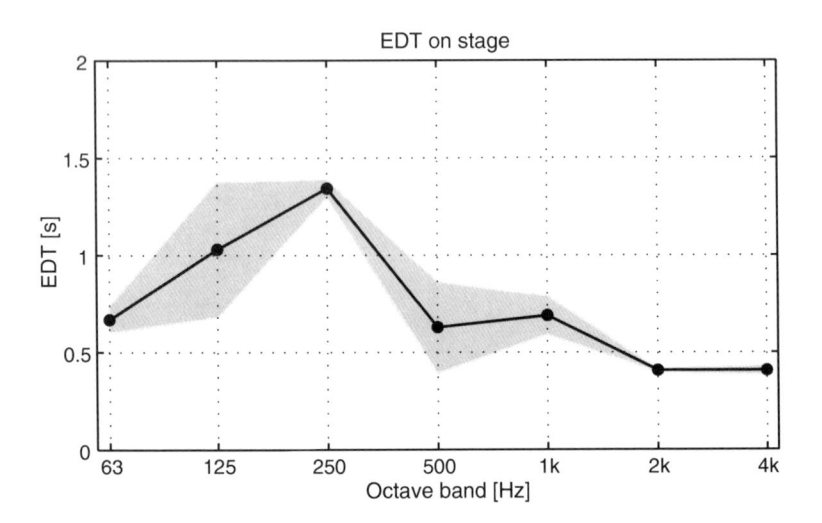

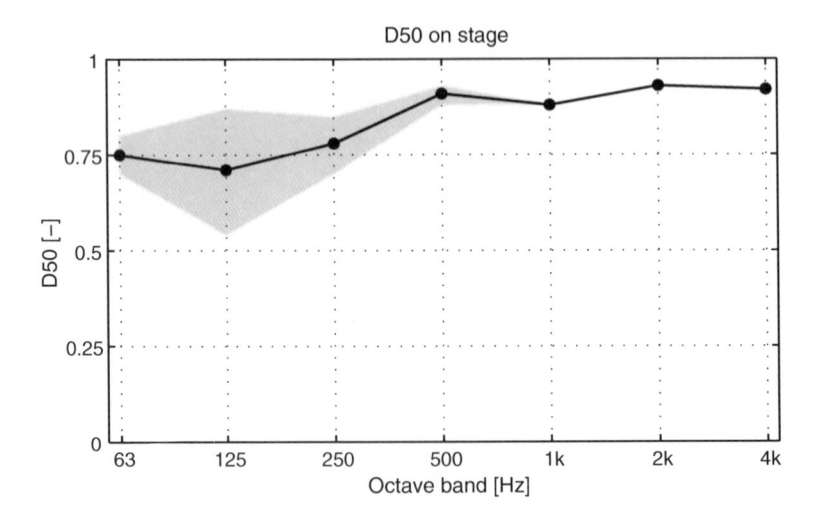

342   7 Gallery of Halls that Present Pop and Rock Music Concerts

Oslo Spektrum Arena

Oslo
Number of concerts per year: 60
Founded: 1990
Capacity: 9,700
Architect: LPO Arkitektkontor AS
Acoustician: Multiconsult and later Brekke and Strand
Owner: Norges Varemesse

Oslo Spektrum was originally built as an indoor multipurpose arena, ideally suited for trade fairs, sports, and concert events. The first event on the program when the doors were opened to the public on December 9th, 1990 was an ice hockey game, closely followed by a rock concert featuring the famous Norwegian group a-ha. In the 20 years that have passed, Oslo Spektrum has undergone continuous improvement making the arena more flexible and better suited to the needs of organizers, audiences, and performers.

The building was designed by architects LPO Arkitektkontor AS in an architecture competition organized by Oslo City Council in connection with the development of a large inner-city area in the 1980s. A centrally located concert arena with a capacity of almost 10,000 people put both Oslo and Norway on the international concert tour map and was designed to attract international performers and events.

The building is constructed in concrete and brick and decorated internally and externally with handmade bricks produced by artist Guttorm Guttormsgaard and ceramicist Søren Ubisch based on print fragments created by artist Rolf Nesch. When audiences enter the main arena, they encounter a space that is decorated in dark tones, with muted lighting and well thought-out practical features, designed to offer the best possible overall experience.

The movable grandstand and a sophisticated system of curtains offers scope for a wide variety of events. The arena can be configured as a full concert hall with general admission and/or seating. The flexible curtain system also enables the creation of a smaller arena that can be extended in response to the needs of the audience.

When the arena was built in 1990, the sound and acoustic conditions were adapted to the wide range of events that would be hosted. Since its inception, the arena has undergone a continuous process of improvement and a number of adjustments have been made in line with the shift towards a greater focus on concert events.

Oslo Spektrum hosts approximately 100 events a year. These include concerts, musicals, trade fairs, ice shows, sporting events, conferences, and banquets. The arena is the venue for the annual Nobel Peace Prize Concert. Elton John, Bob Dylan, Leonard Cohen, Pavarotti, Tina Turner, Frank Sinatra, Iron Maiden, and Bruce Springsteen are just some of the artists who have appeared on the Oslo Spektrum stage.

Oslo Spektrum has undergone radical acoustic improvements just prior to this measurement thanks to a dynamic leadership and correct consulting. The result is outstanding.

A low T30 even at 125 Hz ensures a high value of critical distance. Due to an intelligent shape of the hall only few members of the audiences will suffer from a too-reverberant sound. Here the challenge is speaker coverage.

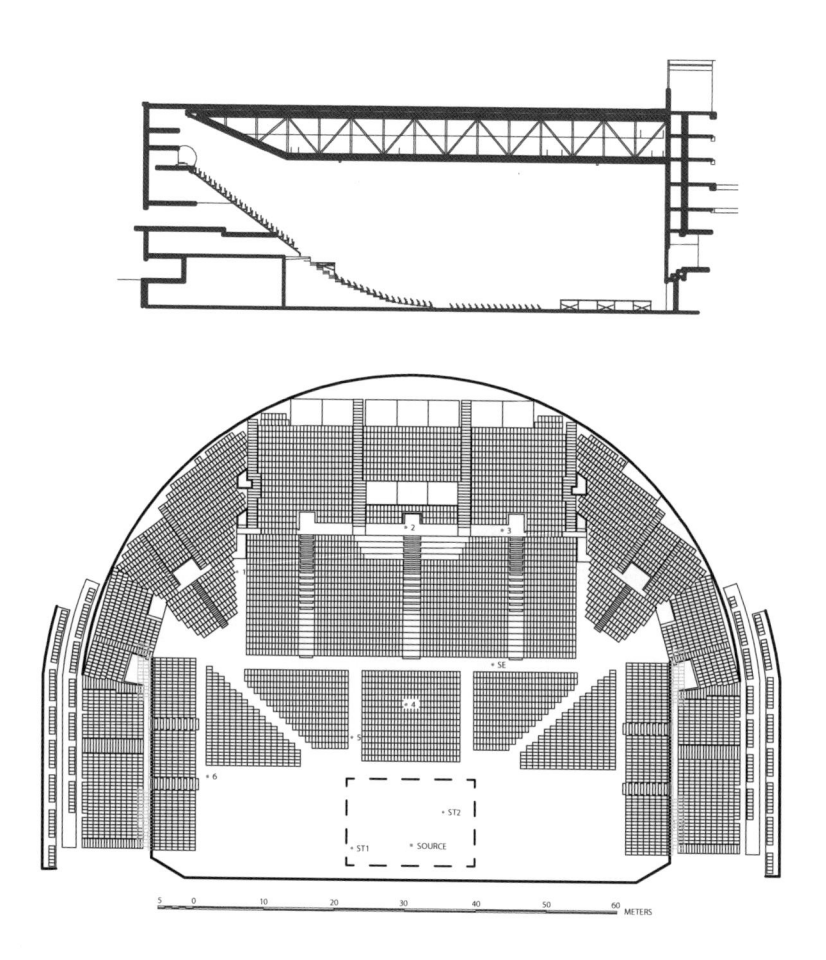

Oslo Spektrum Arena

Geometrical data	
Volume	150,000 m^3
$L \times W \times H_{(max)}$	71 × 94.5 × 26.5 m

Acoustical data	
Audience area	
$T_{30, 125-2k}$	1.61
EDT_{125-2k}	1.66
$C_{80, 125-2k}$	0.87
BR_{63} versus 0.5–1k	1.64
BR_{125} versus 0.5–1k	1.23

Materials Used

Audience Area

Floor: Concrete.
Ceiling: Suspended 5-cm thick mineral wool ceiling with a large cavity before a thin nonperforated steel trapezoid ceiling.
Walls: 10-cm thick mineral wool slabs with 10-cm cavity behind.

State When Measured

No additional chairs.

T30 in audience area

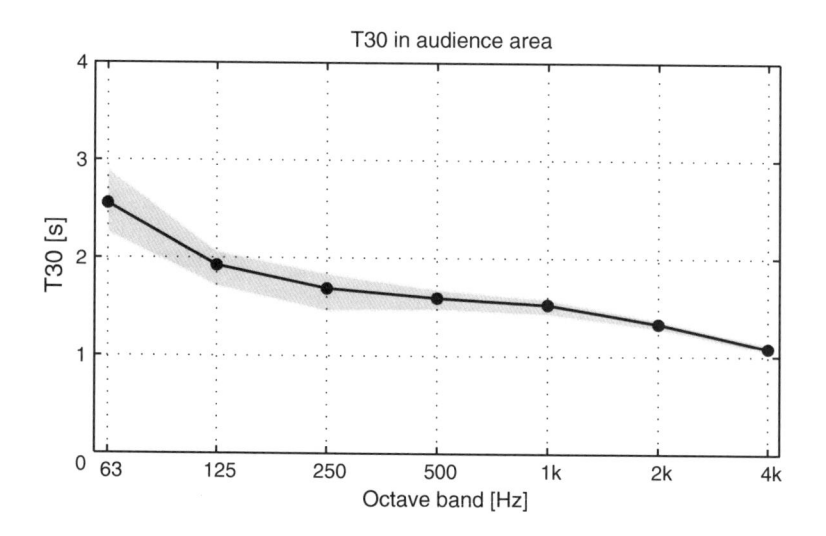

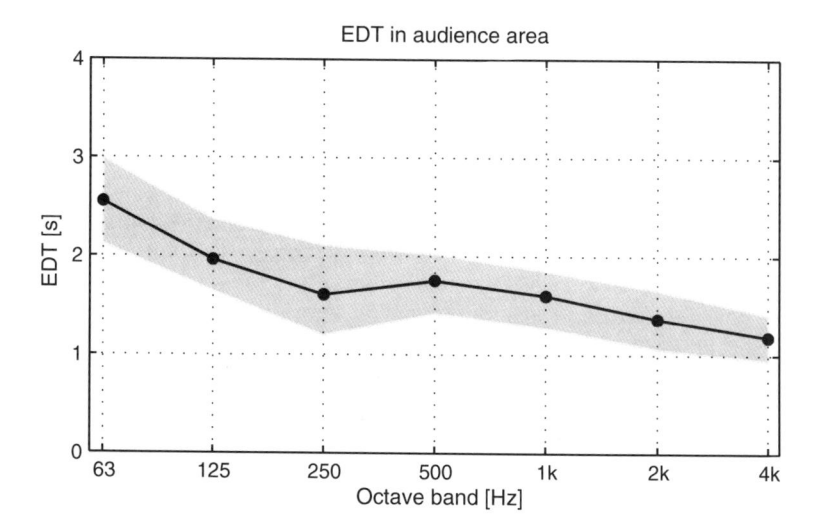

EDT in audience area

Palau Sant Jordi

Barcelona
Number of concerts per year: N/A
Founded: 1990
Capacity: 17,960
Architect: Arata Isozaki
Acoustician: N/A
Owner: Ajuntament de Barcelona (Barcelona City Hall)

Of all the buildings of the Olympic site, Arata Isozaki's Palau Sant Jordi is the most surprising, an undeniable symbol of Barcelona. Its immense dome was built at ground level with the most advanced techniques and then raised by a hydraulic mechanism in 10 days until its height was fixed at 45 m. The Palau, conceived as a versatile and multifunctional enclosure, is equipped with state-of-the-art technology. It is an intelligent building where everything is self-adjusting: temperature, light, air, sound, and image screens. It also has numerous other sections: services for athletes, press, bars, VIPs, and the offices of the company that administers the venue.

The Main Hall is totally versatile, with a lower level of retractable seats. Up to the moment, they have been transformed into a giant swimming pool, an ice rink, a theater, or a banqueting hall.

All the pop-rock, the best singers, the best groups, the biggest and best productions, the ones that are in style and the legends. U2, Dire Straits, Madonna, AC DC, Metallica, Lady Gaga, Coldplay, Bruce Springsteen, Shakira, Fito and Fitipaldis, Estopa, and Alejandro Sanz have played there.

Shows for all Audiences Choirs performed and sung operas, the best melodic singing, the great musicals, the singer/songwriters, classical and popular dancing, flamenco, and even very particular ways of interpreting music. Montserrat Caballé, Luciano Pavarotti, Placido Domingo, Frank Sinatra, Julio Iglesias, Joan Manel Serrat, Joaquin Sabina, and the great choreographies of Igor Moiseyev, Nacho Duato, Joaquín Cortés.

Many nonmusical events are taking place there from Disney's classics to Sponge Bob Square Pants, from the Cirque du Soleil to large television productions. It has been possible to follow the motocross competitions every year with the best competitors in the world, and to bring into the city water sports such as jet-ski racing and even windsurfing. They've had mountain sports including snow skiing and motorcycle trials, as well as social, political, religious and company events.

Palau Sant Jordi

Palau Sant Jordi was built for the 1992 Olympic games. The lowest level of seats are removable.

The beautiful dome of zinc corrugated iron rises on the Olympic Esplanade just a couple of kilometers from the Barcelona city center.

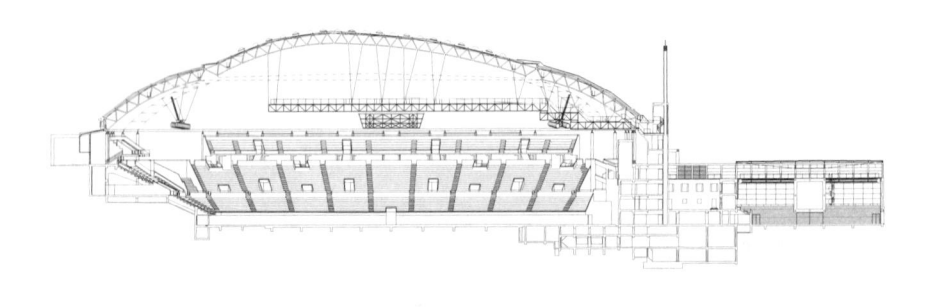

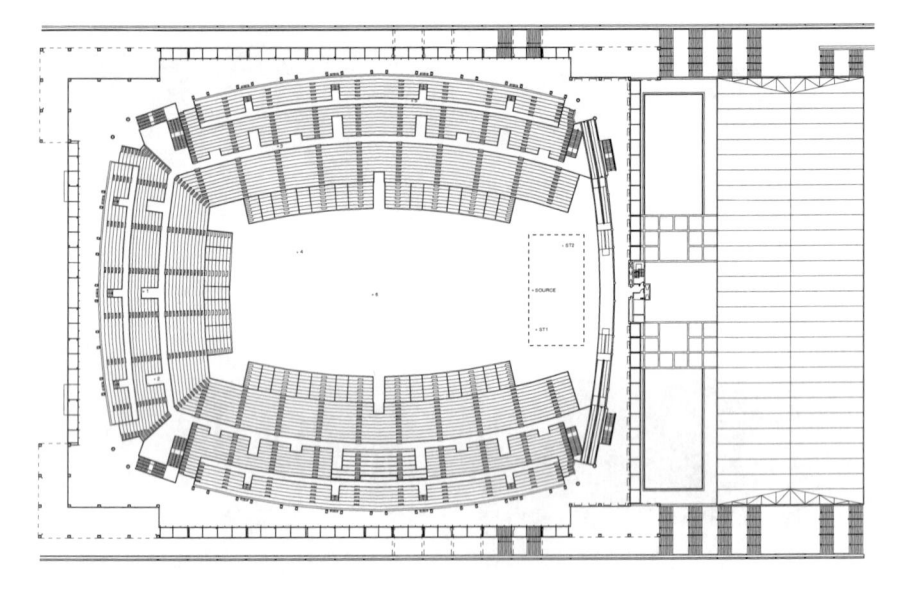

350 7 Gallery of Halls that Present Pop and Rock Music Concerts

Geometrical data	
Volume	approximately 400,000 m^3
$L \times W \times H_{(max)}$	$120 \times 102 \times 42$ m

Acoustical data	
Audience area	
$T_{30, 125-2k}$	4.98
EDT_{125-2k}	4.4
$C_{80, 125-2k}$	-4.62
$BR_{63 \text{ versus } 0.5-1k}$	0.76
$BR_{125 \text{ versus } 0.5-1k}$	1.03

Materials Used

Audience Area

Floor: Concrete.
Ceiling: The circumference is thin perforated plate with a large cavity and mineral wool behind. The large area in the center of the ceiling is covered with methacrylate skylights.
Walls: Concrete. End walls of sandstone typical for the region. Direct coupling to the hard surfaced hallways behind the seats.
Seats are not upholstered.

State of Hall When Measured

Empty; no additional seats mounted.

The reverberation time at mid and high are bound to change very significantly indeed in venues where unupholstered seats are covered by several thousand audience members.

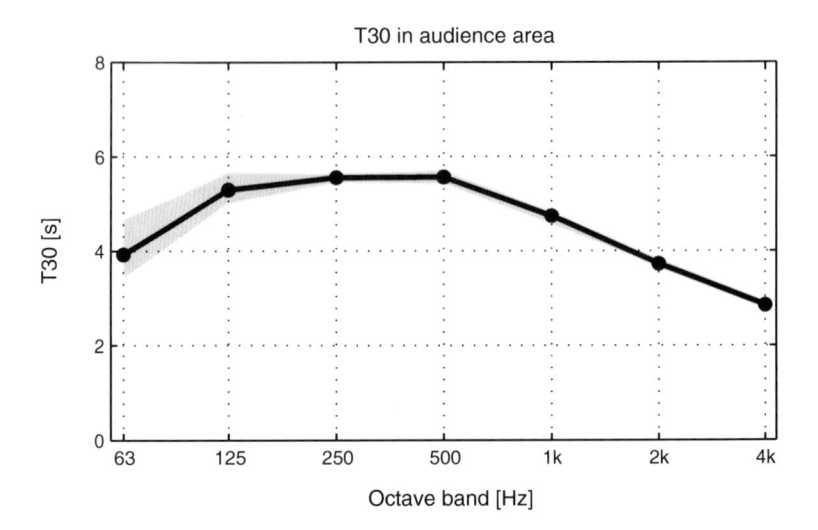

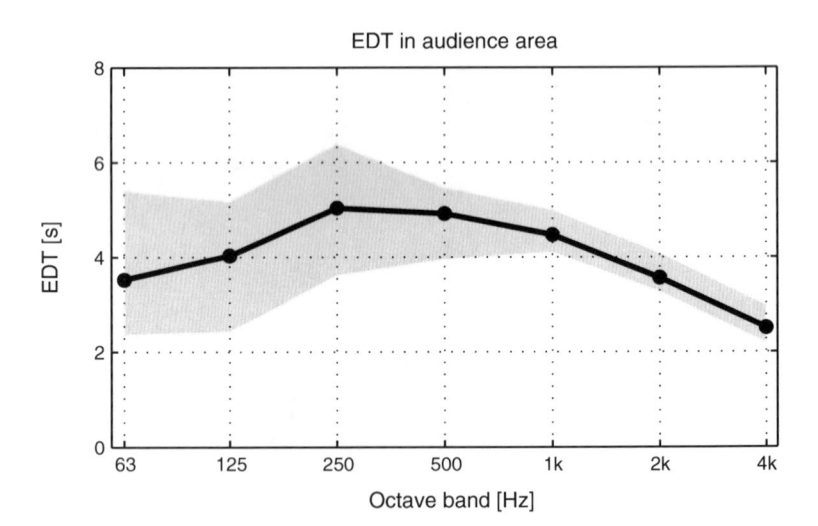

352   7 Gallery of Halls that Present Pop and Rock Music Concerts

T30 in audience area

EDT in audience area

Paradiso

Amsterdam
Number of concerts per year: 120. In all halls in the venue: 320
Founded: 1968
Capacity: 1,500
Architect: N/A
Acoustician: N/A

It is housed in a converted former church building that dates from the nineteenth century and that was used until 1965 as the meeting hall for a liberal Dutch religious group known as the *Vrije Gemeente* (Free Congregation). It is located on *de Weteringschans*, bordering one of the nightlife and tourism centers of the city. The main concert hall in the former church interior has high ceilings and two balcony rings overlooking the stage area, with three large illuminated church windows above the stage. In addition to the main concert hall, there are two smaller cafe stages, on an upper floor and in the basement.

Paradiso was squatted by hippies in 1967 who wanted to convert the church to an entertainment and leisure club. The police ended the festivities the same year. In 1968 the city opened Paradiso as a publicly subsidized youth entertainment center. Along with the nearby Melkweg (Milky Way), it soon became synonymous with the hippie counterculture and the rock music of that era. It was one of the first locations in which the use and sale of soft drugs was tolerated. From the mid-1970s, Paradiso became increasingly associated with punk and new wave music, although it continued to program a wide variety of artists. Starting in the late 1980s, raves and themed dance parties became frequent. In recent years, the venue has settled into an eclectic range of programming, which, besides rock, can include lectures, plays, classical music, and crossover artists. Long associated with clouds of tobacco and hashish smoke, Paradiso banned smoking in 2008 in accordance with a nationwide ban on smoking in public venues.

Artists who have recorded or filmed concerts at the Paradiso include the rolling stones, Joy Division, Willie Nelson, Arcade Fire, Nightwish, Bad Brains, Kayak, Loudness, Nirvana, John Cale, The Cure, Soft Machine, Emilíana Torrini, Jalebee Cartel, Link Wray, Omar and the Howlers, Nick Cave and the Bad Seeds, Beth Hart, Dayna Kurtz, Dave Matthews, Smoosh, Suzanne Vega, Amy Winehouse, Fiction Plane, Epica, Editors, Motorpsycho, Pain of Salvation, Riverside, and Live. Glen Matlock played his last gig with the Sex Pistols at the Paradiso.

On May 26–27, 1995, the Rolling Stones played two semiacoustic concerts at the Paradiso. Scalped tickets reportedly sold for many thousands of dollars. Recorded tracks from these concerts were released on the Stones' *Stripped* album later that year. Keith Richards said that the Paradiso concerts were the best live shows the Stones ever did. In the 1990s, the future of Paradiso became something of a political issue in Amsterdam, because there was some political resistance to the continuation of the subsidies that allowed the venue to operate in its central city location [2]. More recently, supporters have successfully argued that the Paradiso subsidy is reasonable in comparison with subsidies given to other performance venues.

Some of the only surfaces with absorption material are underneath balconies and many surfaces are diffusive. The sound engineer's position is in the rear corner but not under the balcony.

Paradiso

The beautiful old church room.

There is in fact two levels of balconies.

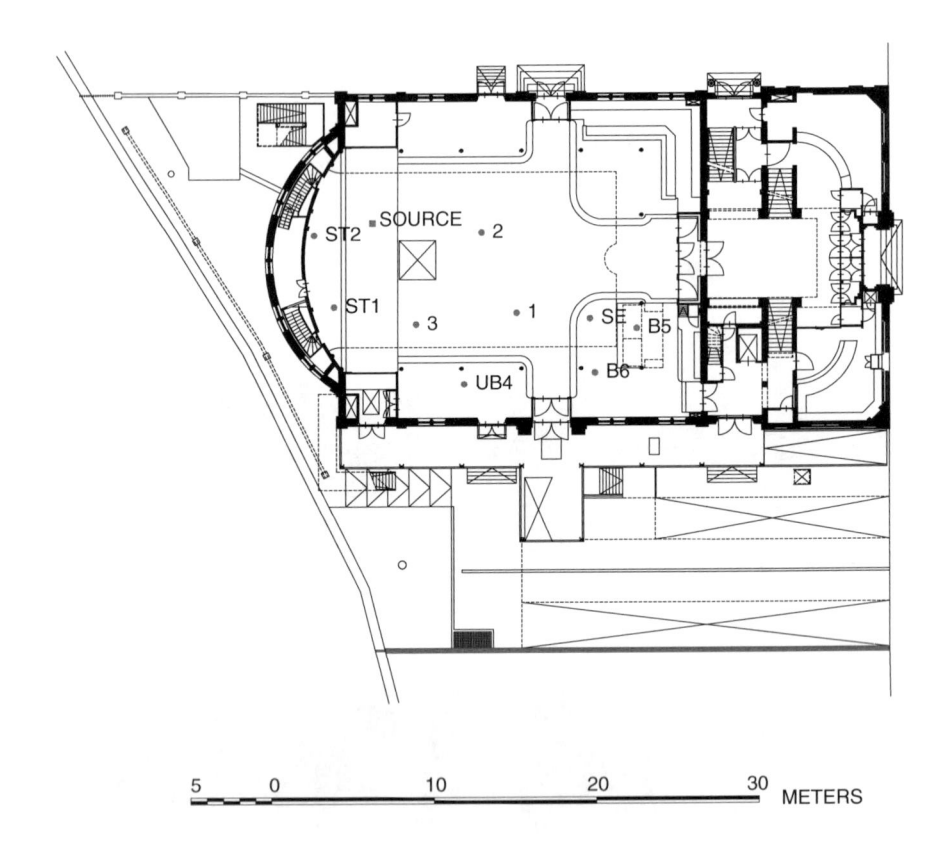

7 Gallery of Halls that Present Pop and Rock Music Concerts

Geometrical data	
Volume	6,000 m³
Height, audience area	14.2 m
$L \times W \times H$	$21.5 \times 19 \times 12$–14.2 m

Acoustical data	
Audience area	
$T_{30, 125-2k}$	1.74
EDT_{125-2k}	1.66
$C_{80, 125-2k}$	−2.34
$BR_{63 \text{ versus } 0.5-1k}$	0.57
$BR_{125 \text{ versus } 0.5-1k}$	0.74
Stage area	
EDT_{125-2k}	1.59
$D_{50, 125-2k}$	0.59
$BR_{63 \text{ versus } 0.5-1k}$	0.38
$BR_{125 \text{ versus } 0.5-1k}$	1.27

Materials Used

Audience Area

Floor: Concrete.
Ceiling: Wood on joists. Thick perforated plates with cavity behind lower balconies.
Walls: Masonry.

Stage Area

Floor: Wood on cavity.
Ceiling: Wood on joists.
Walls: Masonry, curtains.

State of Hall When Measured

Empty.

T30 in audience area

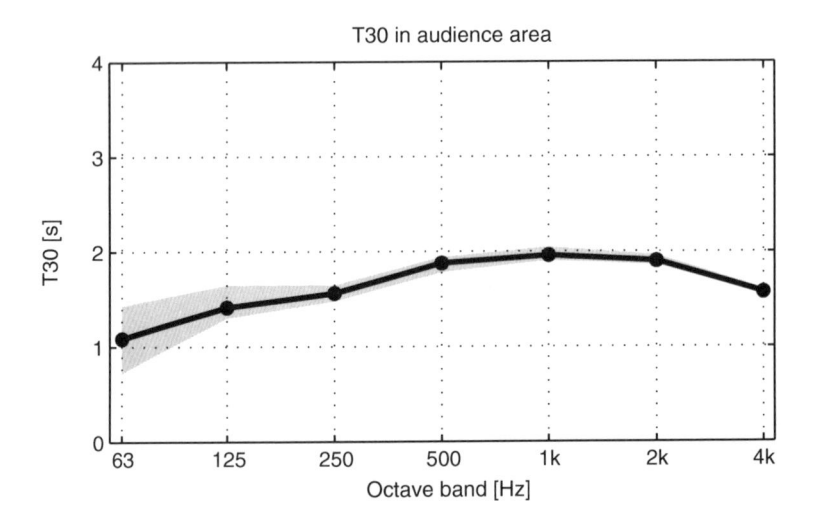

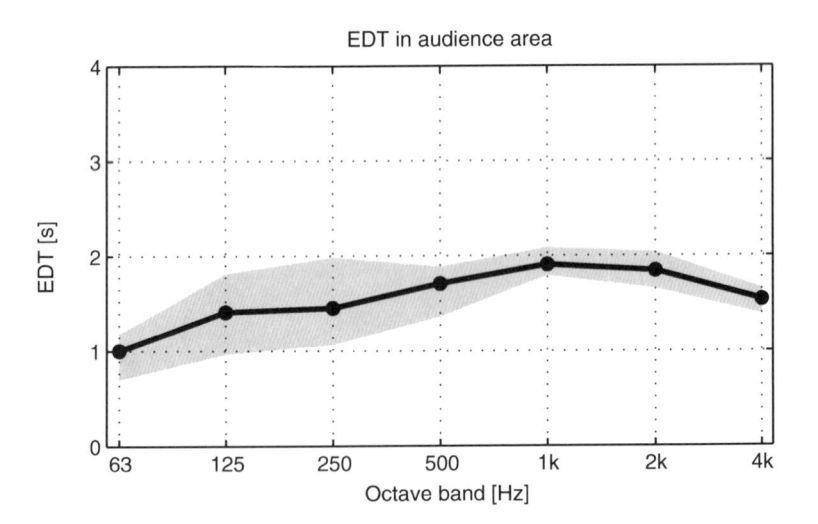

EDT in audience area

EDT on stage

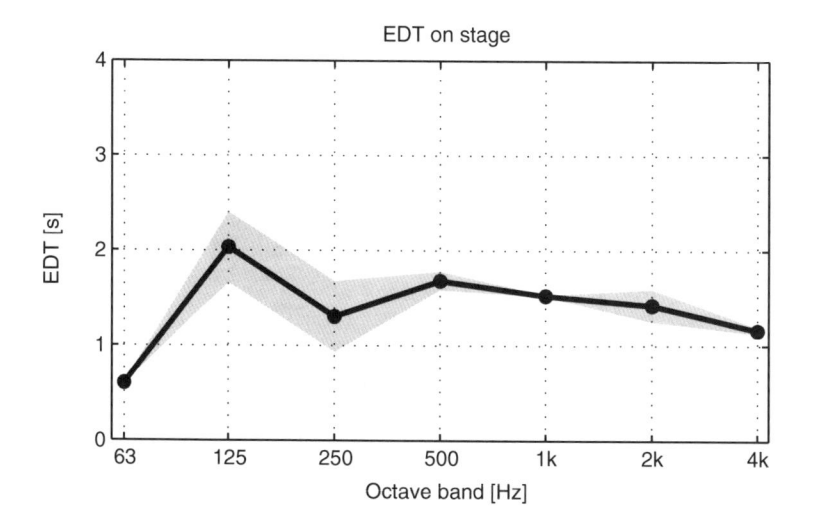

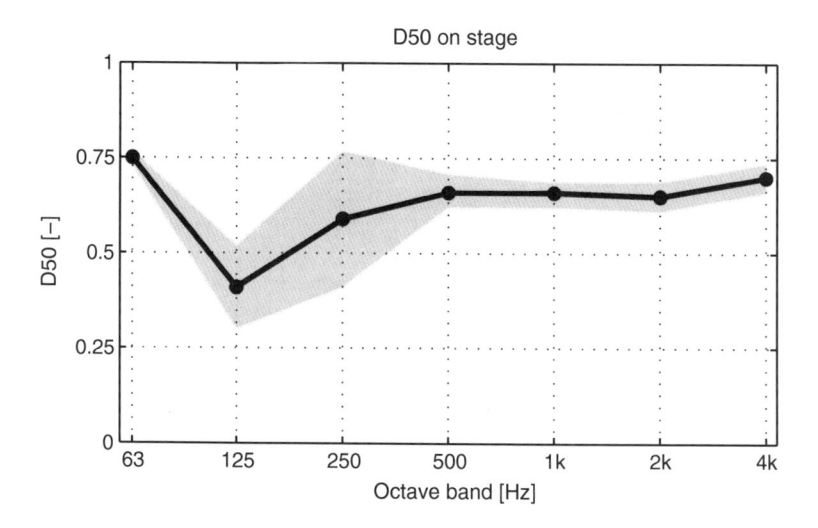

D50 on stage

Porsche Arena

Stuttgart
Number of concerts per year: 60
Founded: 2006
Capacity: 6,000
Architect: ASP Arkitekten, Stuttgart
Acoustician: NA

The seating capacity of the arena varies, from 5,100 to 8,000 people and it was opened in 2006, after 14 months of construction. To fund the construction, costs had already been planned as preconstruction sales of the name rights. Dr. Ing h.c. F. Porsche AG bought the name rights, for a €10 M, for a term of 20 years. The arena is part of a sport complex that includes the adjacent Mercedes-Benz Arena and Schleyerhalle. It is the venue for the Porsche Tennis Grand Prix, a WTA Tour event, and also hosted some matches at the 2007 World Men's Handball Championship.

NeckarPark Stuttgart is one of the biggest and most attractive event sites in Europe. Five state-of-the-art event locations for top international sports, cultural, business, and political events line the Mercedesstraße in the district of Bad Cannstatt: the Gottlieb-Daimler-Stadium, the Carl Benz Center, the Mercedes Benz Museum, and the Hanns-Martin-Schleyer-Halle and Porsche-Arena hall duo. Extensively modernized and enlarged in 2006, the 15,500 capacity Hanns-Martin-Schleyer-Halle is the largest indoor arena in south Germany. When officially opened in 1983 it was Europe's first multifunctional hall and together with the Porsche-Arena, which opened in May 2006, it forms a unique hall duo in the whole of Europe.

A light-flooded and airy lobby unites both halls. The elongated Porsche-Arena is elegantly connected to the glass construction through which people stream into both halls. Flexibility is the key and this is also mirrored in the diversity of the events. From superstars on the national and international music scene to sports events and big show productions; the program of events is as star-studded as it is emotional. More than 14 million visitors to the Hanns-Martin-Schleyer-Halle are clear proof of its attractiveness.

The diversity of events can be enjoyed as a double pack in the Hanns-Martin-Schleyer-Halle and the Porsche-Arena whereby the prerequisite is a perfect and professional organizational structure. Working behind the scenes, it ensures major performances go off smoothly. This applies to a special degree to company presentations, congresses, annual general meetings, and party conferences. An outstanding example is the Porsche Annual General Meeting, which, combined with a big presentation in the Schleyer-Halle, celebrated its premiere in the Porsche-Arena. The hall duo functions in a variety of ways. Spotlight on for a concert in the Schleyer-Halle and at the same time a first-class handball or ice-hockey match in the Porsche-Arena.

Porsche Arena, a modern and well-designed smaller arena.

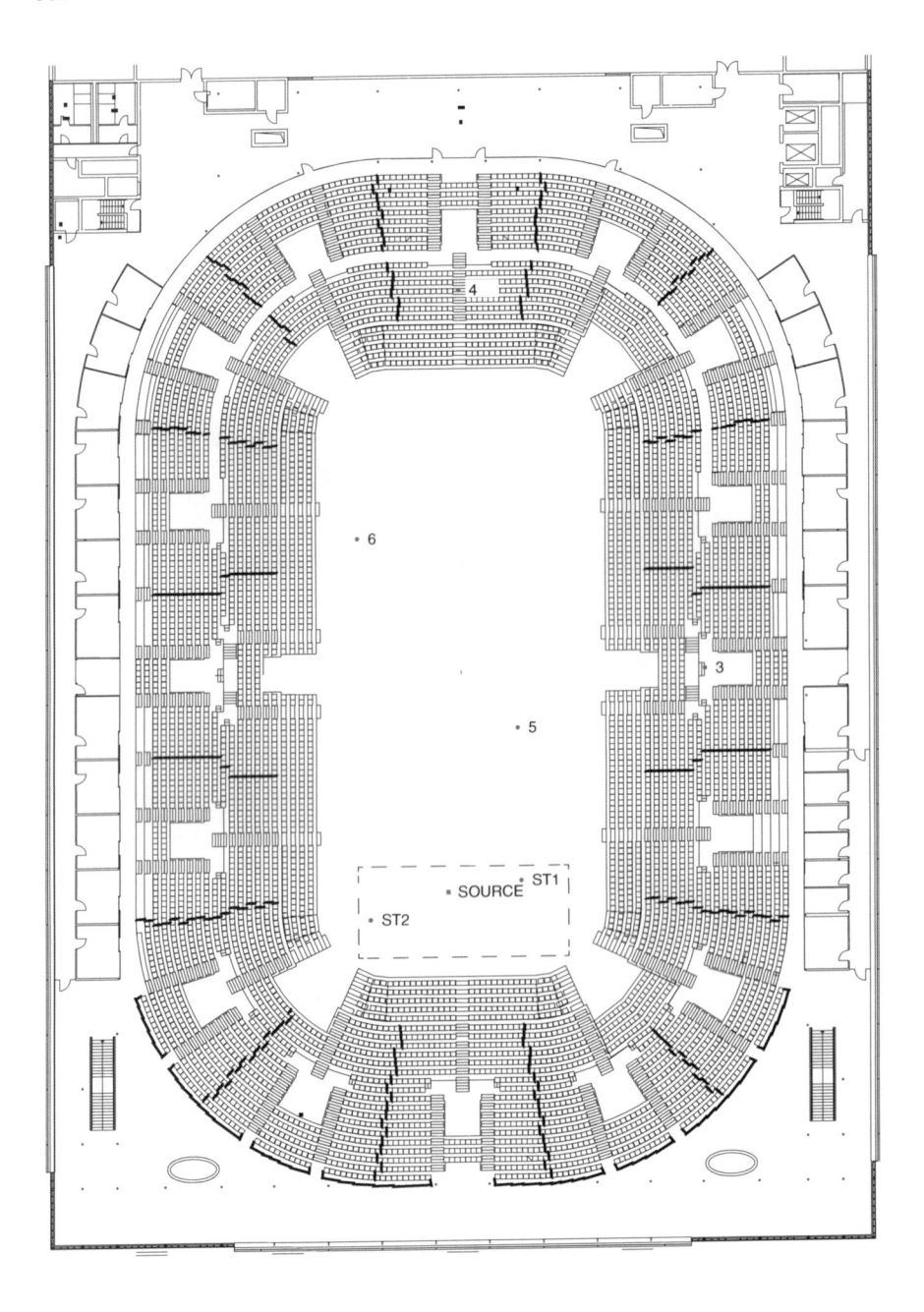

7 Gallery of Halls that Present Pop and Rock Music Concerts

Materials Used

Geometrical data	
Volume	Approximately 80,000 m³
$L \times W \times H$	m floor level max. 63 × 33 m, whole arena: 63 × 94 m, height up to 20.5 m, 12.5 m clear height

Acoustical data	
Audience area	
$T_{30, 125-2k}$	2.46
EDT_{125-2k}	2.33
$C_{80, 125-2k}$	−2.58
$BR_{63 \text{ versus } 0.5-1k}$	1.62
$BR_{125 \text{ versus } 0.5-1k}$	1.15

Materials Used

Audience Area

Floor: Wood.
Ceiling: Thin trapezoid metal with perforation.
Walls: Above the seats there is glazing around the hall.
Seats upholstered on seat not on back and perforated underneath seats (see photo).

State of Hall When Measured

Empty; no additional seats mounted.

Intelligent design of chairs if they need to be absorbent while not in use. Porsche Arena.

T30 in audience area

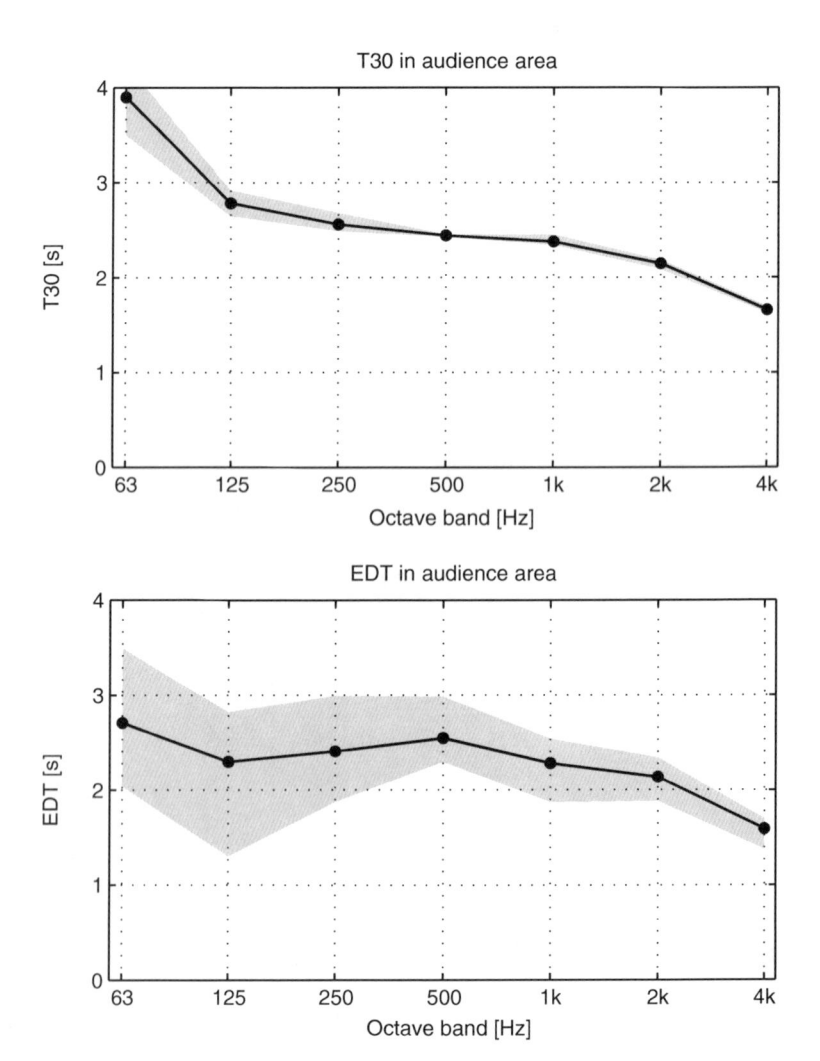

EDT in audience area

Rote Fabrik, Aktionshalle

Zürich
Number of concerts per year: 80
Founded: 1980
Capacity: 1,200
Architect: original: Carl-Arnold Séquin-Bronner (original)
Renovation 1994: Arbeitsgemeinschaft Rote Fabrik «ARFA»:

- Claude Vaucher, Architekt SIA/SWB (Büro Z, Zürich).
- Bob Gysin + Partner, Dübendorf.
- Architekturgenossenschaft Bauplan, Zürich (Renovation).

Acoustician: Originally none; Bruno Gandet during the renovation 1994.

The Rote Fabrik (Red Factory, so named because of its red bricks) came into use as a cultural center in 1980 as a result of widespread student protests in Zürich.

The buildings housing the Rote Fabrik were originally constructed as a silk weaving mill between 1892 and 1896 by one of the period's most important architects in the field of factory construction, Carl-Arnold Séquin-Bronner. With the decline of the Swiss textile industry, the location came to be used as a telecommunications factory owned by Standard Telephon and Telegraph, a subsidiary of International Telephone and Telegraph (ITT) starting in 1940. In 1974, the buildings were bought by the city of Zürich, the idea being at the time to tear the ageing factory down and widen the freeways leading along Lake Zürich.

A public vote in 1977 demanded that the Rote Fabrik be converted into a cultural center; in spite of the vote being passed by the public, the project got lost in bureaucracy. It was not until May 30th 1980 when a demonstration against a planned credit of 61 million Swiss francs for the Zürich Opera House turned into widespread demonstrations that the project was reopened. Squatters, left-wing activists, and thousands of spectators returning from a Bob Marley concert combined to form one of the largest (and also most violent) demonstrations the city had ever seen.

Under the impression of the public outrage following the demonstration, the city council put the buildings under historical protection and made them available to the newly founded IG Rote Fabrik, the association still running the Rote Fabrik today. Even though the location was never renovated under the guide of an acoustician (except for minor improvements made by the Rote Fabrik's own sound engineers), the Aktionshalle (Action Hall), the main hall of three on the premises, quickly became a favored concert spot for many of the most important bands in the 1980 and 1990s. Patty Smith, Nirvana, Mike Patton, the Melvins, and the Young Gods, to name only a few, have valued the Rote Fabrik not only for staying true to its roots, but at the same time providing a high-quality environment for playing gigs ranging from loud and bashing to quiet and fragile.

Rote Fabrik is situated on the lake side with jetset neighbours. Photo: Patrick Rimann.

The floor upstairs from the bar is not part of the concert room. Photo: Hans van Veen.

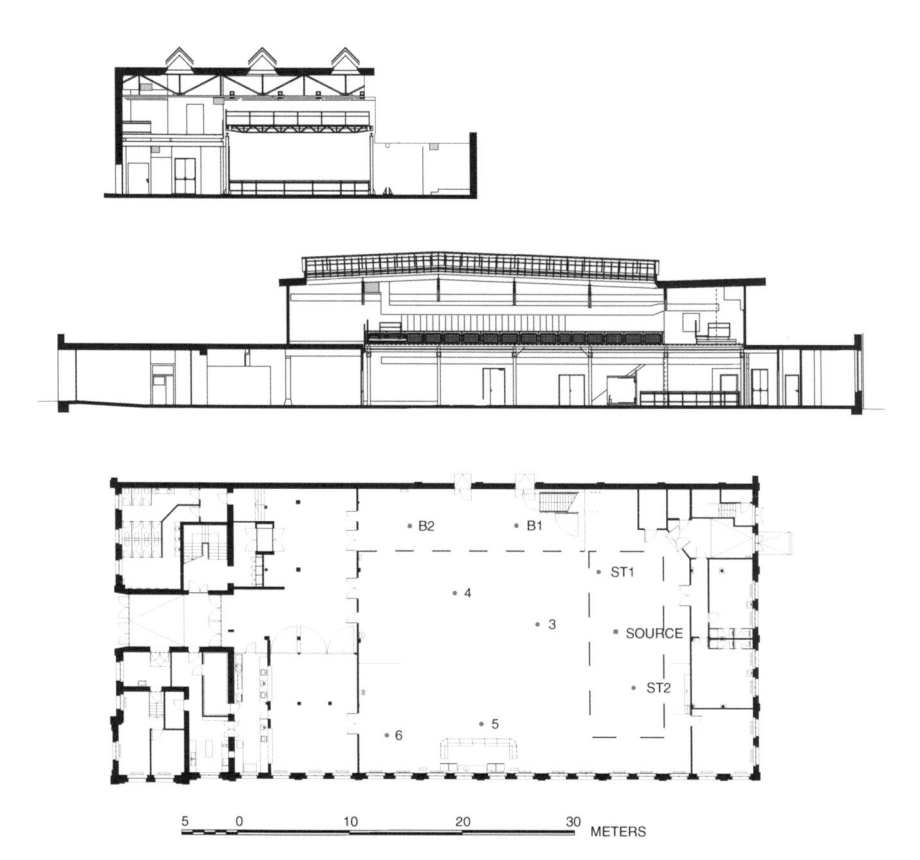

Rote Fabrik, Aktionshalle

Geometrical data	
Volume	5,000 m^3
$L \times W \times H$	30 × 25 × 8.5
Surface area of stage	80 m^2
Height of stage	1.20 m

Acoustical data	
Audience area	
$T_{30, 125-2k}$	1.08
EDT_{125-2k}	1.01
$C_{80, 125-2k}$	2.92
BR_{63} versus 0.5–1k	1.34
BR_{125} versus 0.5–1k	1.47
Stage area	
EDT_{125-2k}	0.53
$D_{50, 125-2k}$	0.86
BR_{63} versus 0.5–1k	1.64
BR_{125} versus 0.5–1k	1.37

Materials Used

Audience Area

Floor: Concrete.
Ceiling: Concrete with large areas of glazing. Suspended home-made reflectors.
Walls: Concrete. At each side of the front of the stage there are two big walls 20-cm thick with thin perforated plates on each side and mineral wool in between. Upper side wall on one side: perforated gypsum board. Back wall: painted brick.

Stage Area

Floor: Stage risers.
Ceiling: Concrete.
Walls: Curtains.

State of Hall When Measured

Empty; seating risers at the rear of the hall.

T30 in audience area

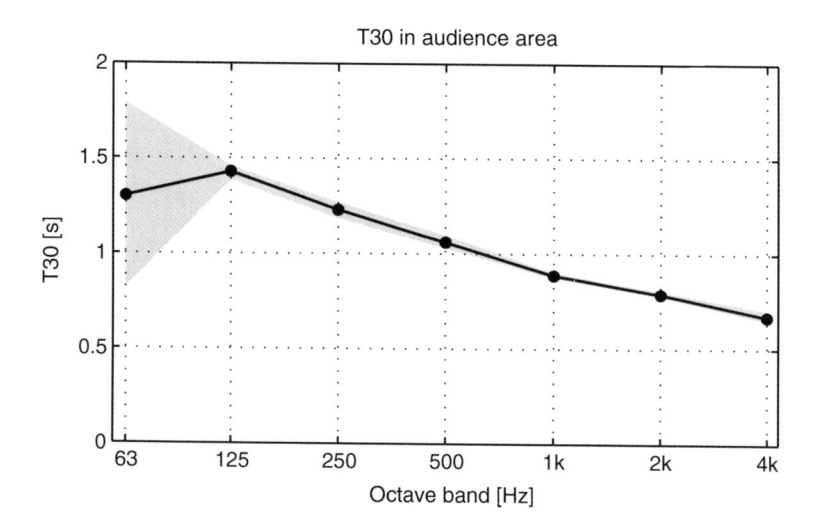

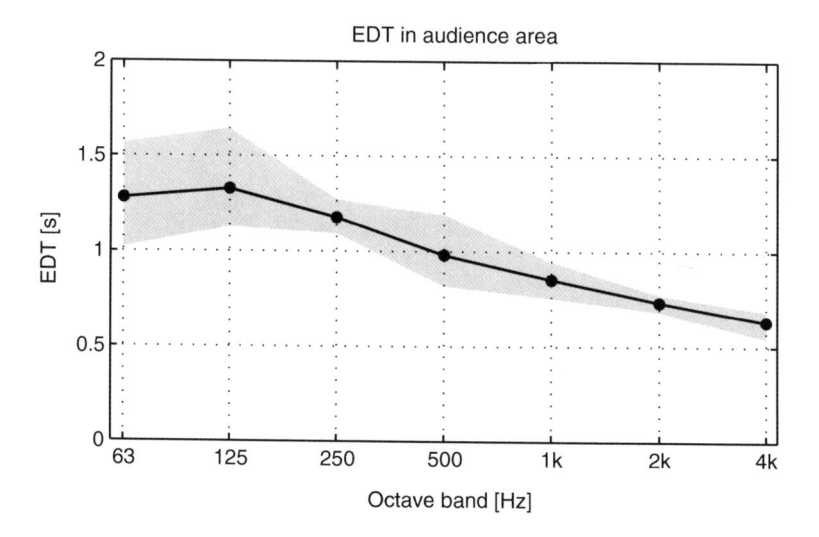

EDT in audience area

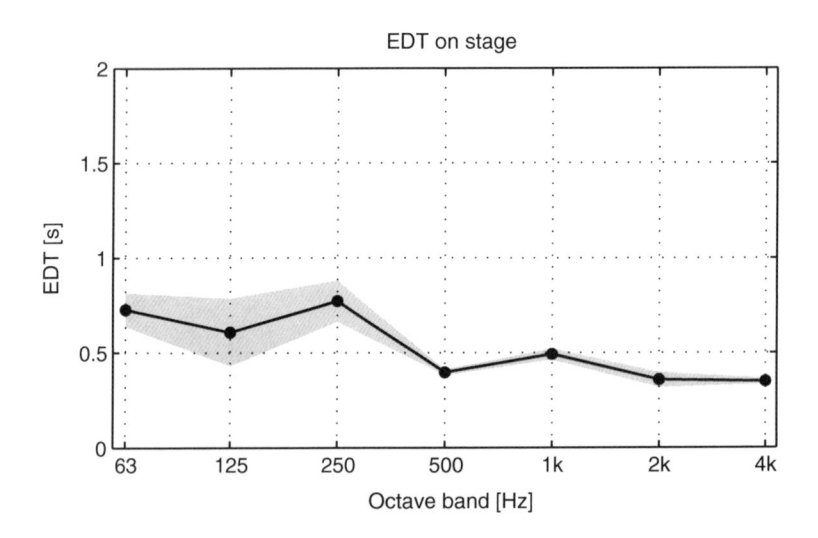

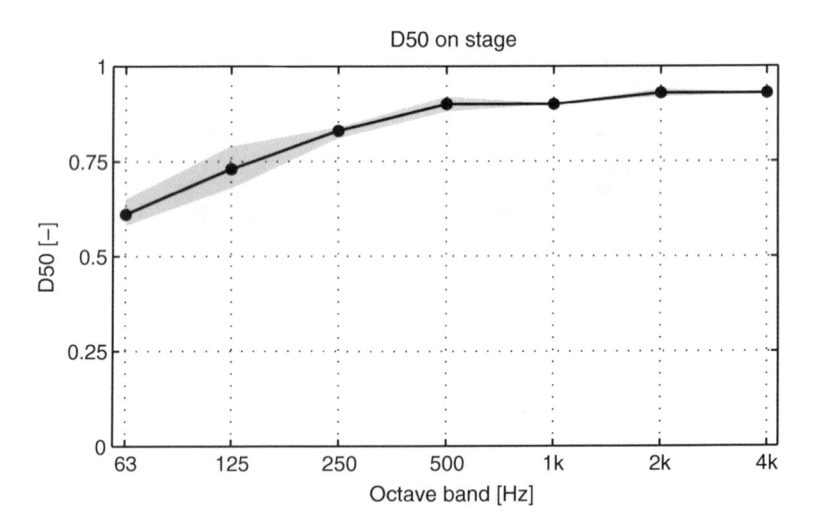

370 7 Gallery of Halls that Present Pop and Rock Music Concerts

Rote Fabrik Clubraum

Zürich
Number of events per year: 160
Founded: 1994
Capacity: 600
Architect: Original: Carl-Arnold Séquin-Bronner (original)
Renovation 1994: Arbeitsgemeinschaft Rote Fabrik «ARFA»:

- Claude Vaucher, Architekt SIA/SWB (Büro Z, Zürich).
- Bob Gysin + Partner, Dübendorf.
- Architekturgenossenschaft Bauplan, Zürich (Renovation).

Acoustician: Originally none; Bruno Gandet during the renovation 1994.

Clubraum is mostly used for smaller concerts, theater plays, and poetry slams, Photo: Hans van Veen.

372　　　　7　Gallery of Halls that Present Pop and Rock Music Concerts

And has as expensive an address as the bigger room next door. Photo: Jasmin Phasuk.

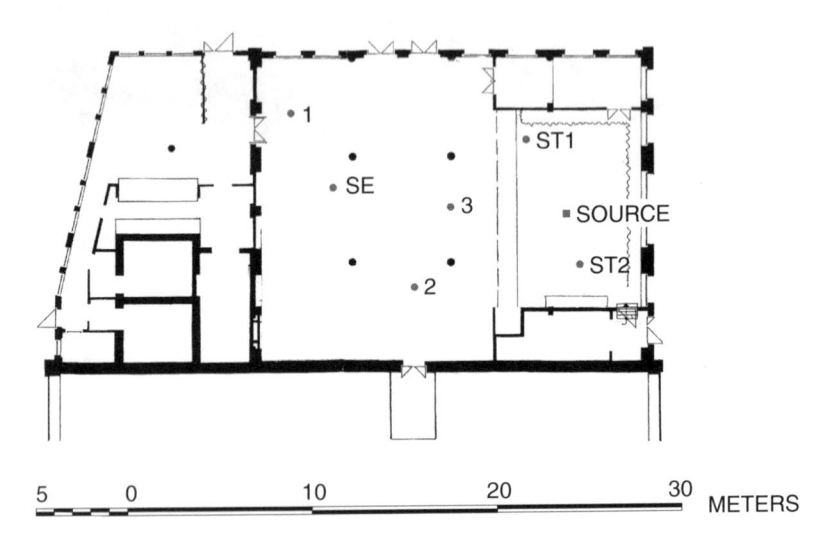

Geometrical data	
Volume	1,500 m³
$L \times W \times H$	$21 \times 16 \times 4.3$
Surface area of stage	70 m²
Height of stage	1 m

Acoustical data	
Audience area	
$T_{30,125-2k}$	0.96
EDT_{125-2k}	0.79
$C_{80,125-2k}$	5.70
BR_{63} versus 0.5–1k	1.42
BR_{125} versus 0.5–1k	1.21
Stage area	
EDT_{125-2k}	0.39
$D_{50,125-2k}$	0.88
BR_{63} versus 0.5–1k	3.51
BR_{125} versus 0.5–1k	2.76

Materials Used

Audience Area

Floor: Vinyl on concrete.
Ceiling: Wooden fiber on cavity.
Walls: Concrete, doors; back wall: painted wood fiber panels in some areas.

Stage Area

Floor: Stage risers.
Ceiling: Wooden fiber on cavity.
Walls: Concrete covered with curtains.

State of Hall When Measured

Empty.

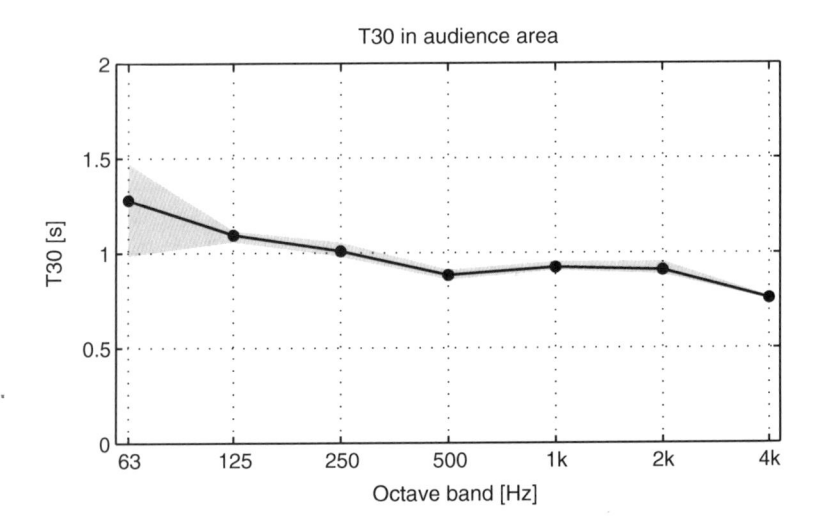

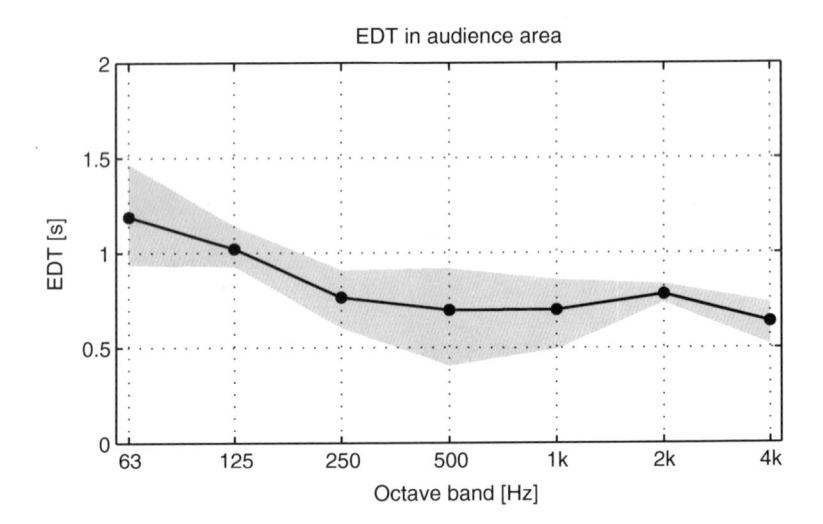

374 7 Gallery of Halls that Present Pop and Rock Music Concerts

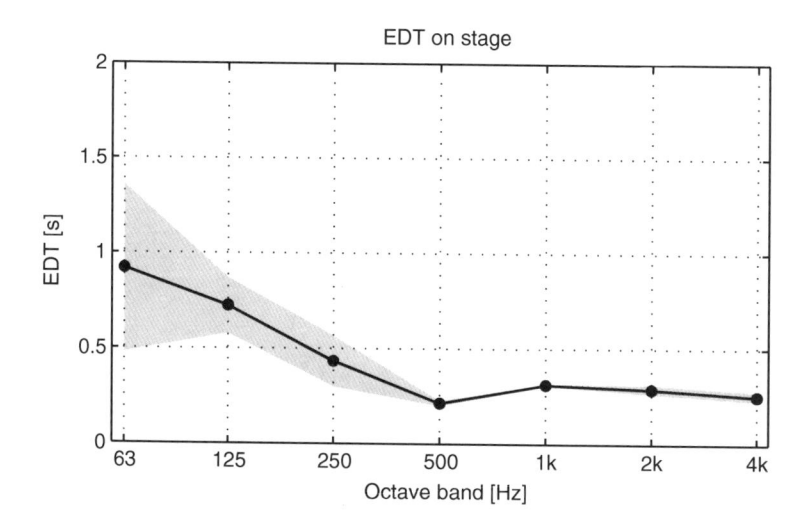

State of Hall When Measured 375

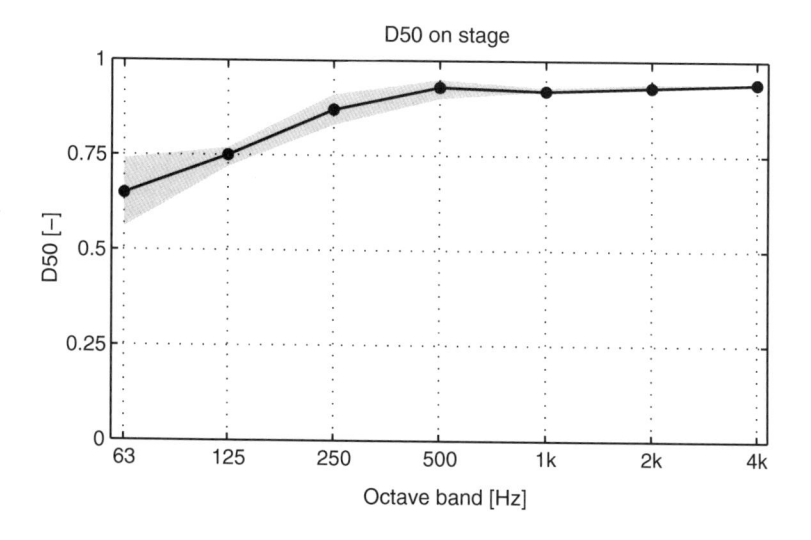

EDT on stage

D50 on stage

Rockefeller

Oslo
Founded: 1986
Capacity: 1,350
Number of concerts per year: 120
Architect: Movenstierne and Eide
Acoustician: N/A

Rockefeller combines the advantages of both the concert hall and the club. The capacity is 1,350 spectators, but the atmosphere is that of a friendly and intimate club. Thanks to the two balconies and balustrades everyone in the audience is assured a good view of the stage. Sound and lighting conditions are undoubtedly the best in Oslo. The other needs of the audience are catered by eight bars.

Rockefeller has a location matched by few concert halls. The building previously housed Totggata Bad, one of Oslo's oldest and most stylish public baths. The building is from 1925. The premises were completely renovated in 1982, and now house offices, restaurants, and pubs, with Rockefeller occupying the old swimming hall itself. Rockefeller's smaller venue, John DEE, is located one floor down. The location provides two distinct advantages: a famous address and excellent communications. Public transport to all parts of the city can be found within a few hundred meters.

Rockefeller has a high profile on the Norwegian music scene. This is due both to the quality of the more than 10,000 bands that have appeared on their stage, and to their willingness to develop new activities. In addition to concerts Rockefeller has its own Rock Cinema, a fully equipped 35-mm cinema where the guests can enjoy movies and drinks in an informal environment. Rockefeller is nonpolitical and nonracist, and has no fixed repertoire profile that prevents us from entering new projects or exploring new trends. In 1986 Rockefeller opened (capacity 1,000), founder Hans A. Lier. In 1990 the Rock Cinema opened and in 1991 additional bar areas opened (capacity now 1,200). In 1996 the upper gallery and roof terrace opened (capacity now 1,350) and in 1997 Rockefeller opened the smaller club, John DEE (capacity 400), one floor down. In 2006 Rockefeller took over Sentrum Scene (capacity 1,800) across the street from Rockefeller. All three stages are now run by the Rockefeller administration, including booking. "The Flat" opened in 2007 as a separate area by the roof terrace (capacity 100) and in 2011 Rockefeller celebrated 25 years as a concert hall on the 14th of March, with the same leadership: Hans A. Lier, Roar Gulbrandsen and Frits Løveng.

Some of the over 10,000 acts that have appeared at Rockefeller include A-ha, Beck, Blondie, Chaka Kahn, Chuck Berry, Crosby and Nash, Curtis Mayfield, David Byrne, Elvis Costello, Eminem, Faith No More, Grace Jones, Iggy Pop, Ike Turner, INXS, Jerry Lee Lewis, Jethro Tull, Kraftwerk, MC5, Morrissey, Motörhead, Ozzy Osbourne, Pantera, Patti Smith, PJ Harvey, Public Enemy, R.E.M, Radiohead, Rage Against The Machine, Ramones, Robert Plant, Run D.M.C., Santana, Snoop Dogg, Van Morrison, Willie Nelson, and ZZ Top.

Rockefeller fits 1,350 audience members on three levels.

The acoustic ceiling was suspended with a fairly large cavity by coincidence; they had to fit beneath some ventilation ducts.

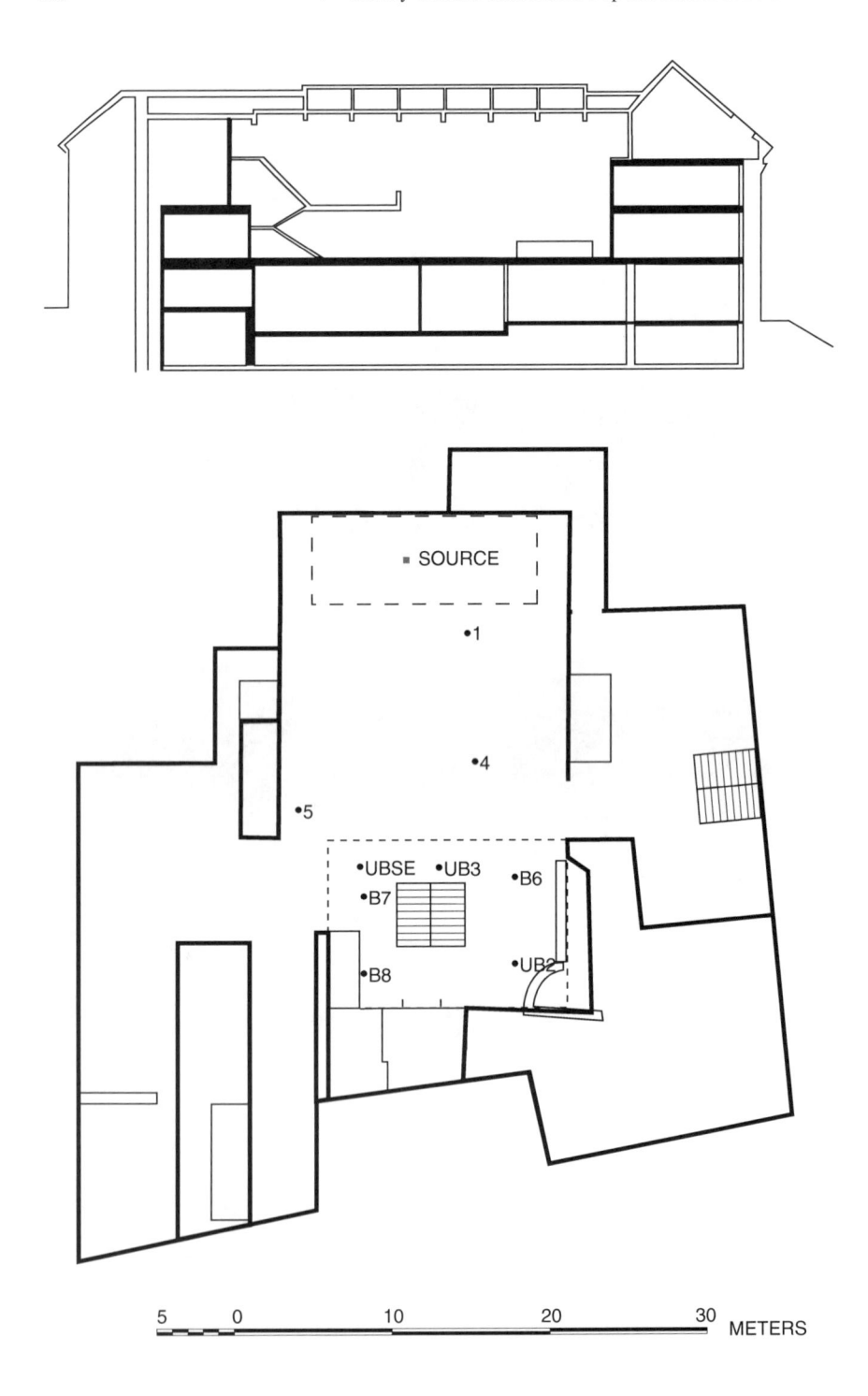

378   7 Gallery of Halls that Present Pop and Rock Music Concerts

5   0   10   20   30 METERS

Geometrical data	
Volume	5,300 m^3
$L \times W \times H$	$27.6 \times 19.6 \times 9.8$ m
Acoustical data	
Audience area	
$T_{30, 125-2k}$	0.95
EDT_{125-2k}	0.85
$C_{80, 125-2k}$	5.38
BR_{63} versus 0.5–1k	1.12
BR_{125} versus 0.5–1k	1.36

Materials Used

Audience Area

Floor: Concrete. Audience podiums of wood on cavity.
Ceiling: 2/3 suspended mineral wool ceiling with 1 m of cavity. 1/3 skylights. Underneath balconies: gypsum board on cavity.
Walls: Concrete except on balconies: glazing with mineral wool behind and curtains in front. Back wall: painted brick.

Stage Area

Floor: Stage risers with floor of 8 cm thick wooden slabs.
Ceiling: 2/3 suspended mineral wool ceiling with 1 m of cavity.
Walls: Curtains.

State of Hall When Measured

Empty.

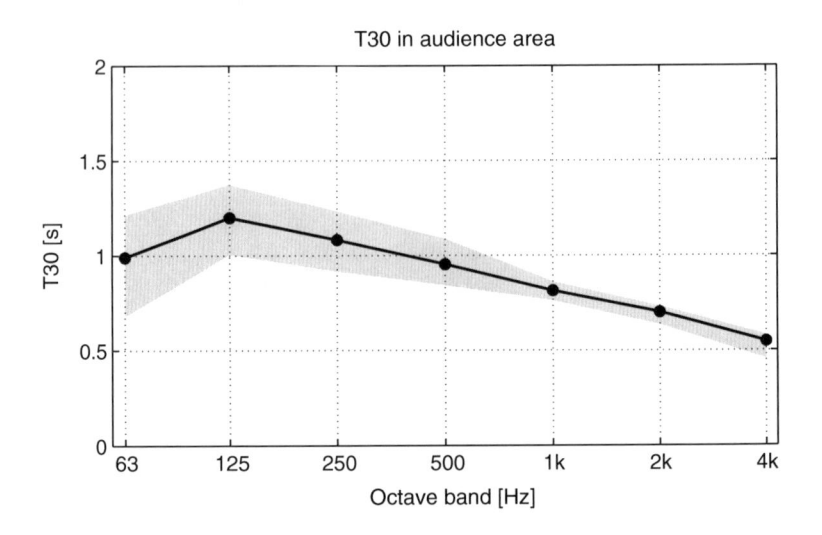

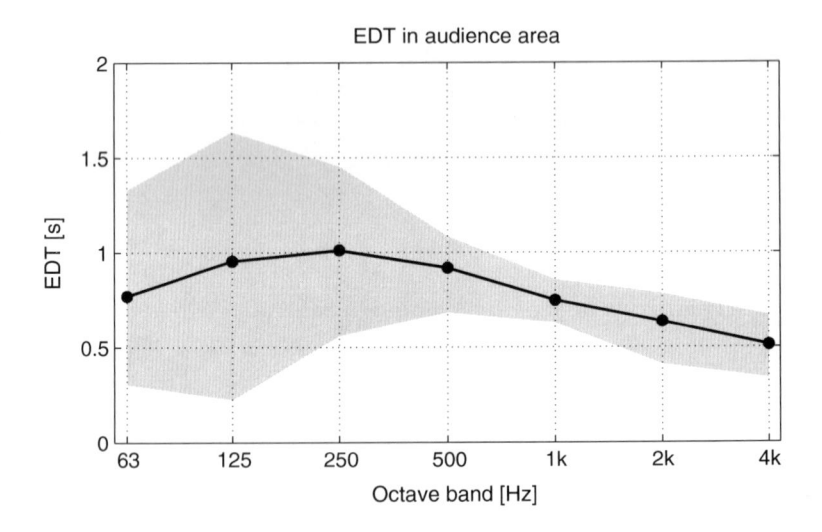

380   7 Gallery of Halls that Present Pop and Rock Music Concerts

Rockhal

Luxembourg
Founded: 2005
Capacity: 6,500
Number of concerts per year: 50–70. In all halls in the venue: 150
Architect: BENG Architecture
Acoustician: Dr. Albert Yaying Xu (XU-Acoustique)

Inaugurated in September 2005, the Rockhal is the most important concert venue in Luxembourg. The growing interest in contemporary popular music and the emergence of an increasingly professional local scene persuaded the Luxembourg government to provide the country a worthwhile structure. The Rockhal is under the patronage of and operates with the financial support of the Ministère de la Culture. It is located in Esch/Alzette on the historical and former industrial site Belval, which is now one of the country's most important urban development projects.

As its name implies, it was purposely built to host mainly rock and pop shows. In addition to the Main Hall it also accomodates the *Club*, a smaller concert venue with a capacity of 1,200 people. Also the Music and Resources center is an important part of the Rockhal. Its mission is to orient, inform, and assist amateur and professional musicians by providing them with a range of logistical and cultural tools and infrastructures. The Music and Resources center features a media library, six rehearsal rooms, and a recording studio and organizes conferences, workshops, and panels on a regular basis.

The Main Hall has standing room for 6,500 people and an all-seating capacity of 2,800. The flat base of the hall and the service grid covering the entire area ensure total flexibility of layout. It has a free height of 17 and is accessible for vehicles and machines up to 30 tons.

Due to its high technical standard and the ideal location in the heart of Europe, directly between Belgium, France, and Germany, the Rockhal also serves as a rehearsal venue for big arena productions. Recently productions such as those of Muse, Depeche Mode, Tokio Hotel, and The Chemical Brothers did tour rehearsals and launched their world tour with a show at the Rockhal. In 2011 it hosted among others productions Prince, Mark Knopfler and Bob Dylan, Lenny Kravitz, and Rammstein.

Rockhal—Centre de Music Amplifiée is situated on Boulevard du Rock and Roll maybe the hippest address of all venues in this book?

Rockhal is in many ways a rock temple. Seated concert.

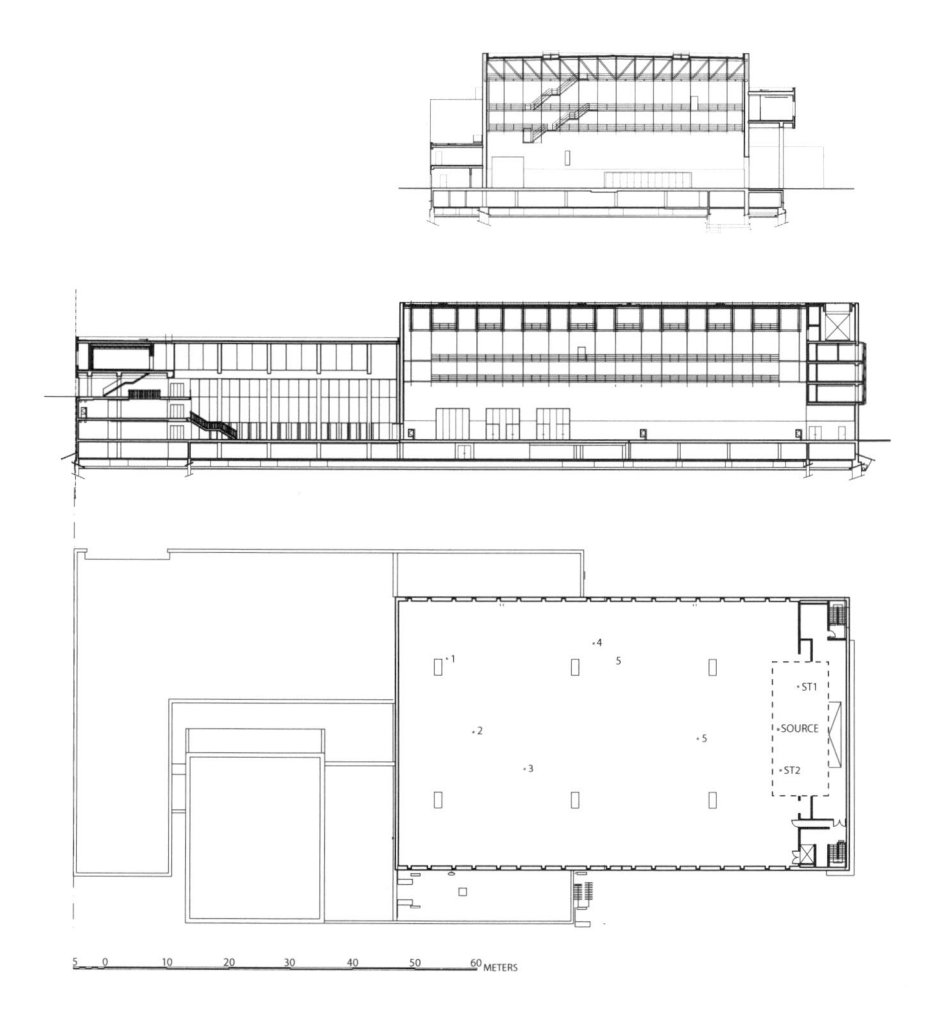

Rockhal

Geometrical data	
Volume (m^3)	58,000
$L \times W \times H$ (m)	$65 \times 42 \times 21$

Acoustical data	
Audience area	
$T_{30, 125-2k}$	1.47
EDT_{125-2k}	1.3
$C_{80, 125-2k}$	0.95
BR_{63} versus 0.5–1k	3.03
BR_{125} versus 0.5–1k	1.73

Materials Used

Audience Area

Floor: concrete.
Ceiling: Suspended 3-cm thick mineral wool with 45-cm cavity behind.
Walls: Wood fiber panels on a 6-cm cavity with mineral wool. Trapezoid metal roof.

State of Hall When Measured

Empty; a couple of vehicles in hall.

Rock temple.

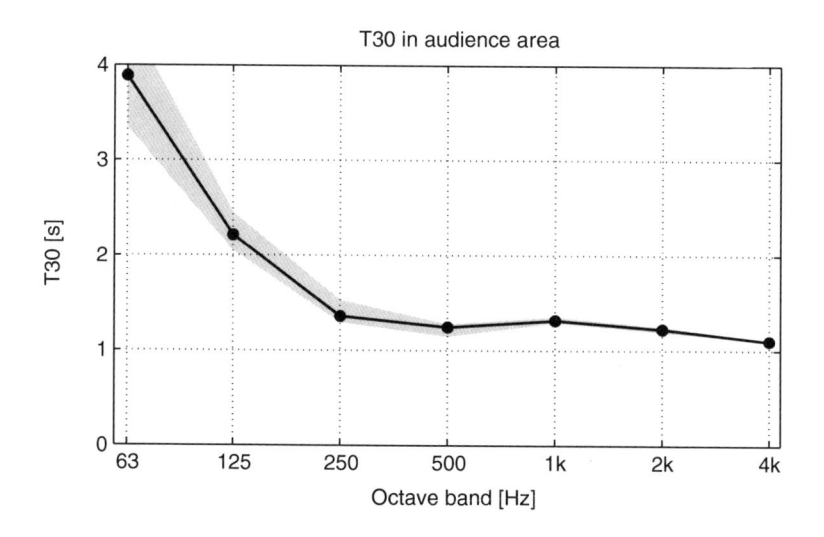

State of Hall When Measured

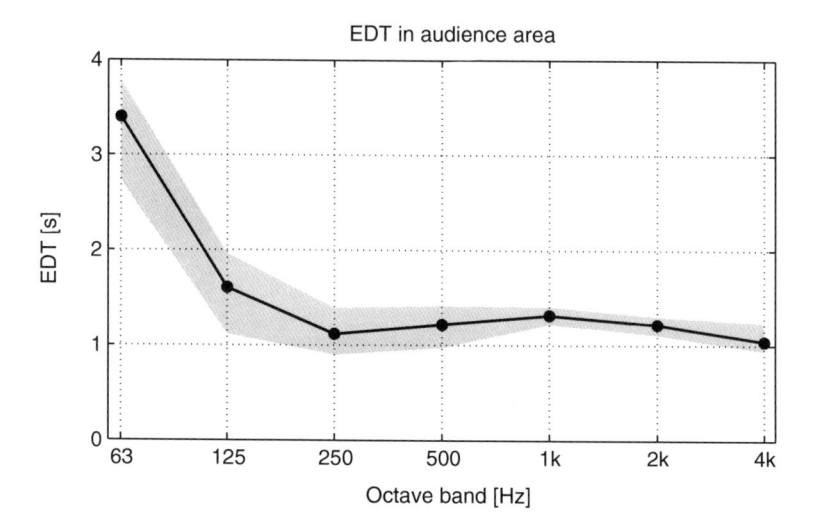

T30 in audience area

EDT in audience area

Razzmatazz 1

Barcelona
Number of concerts per year: NA
Founded: NA
Capacity: 1,500
Architect: N/A
Acoustician: N/A

Dwarfs have very good views from the 1.4-m tall opening to the concrete balcony corridors on both sides of the former factory room. So do sitting audiences.

Razzmatazz sala 1. Archetype rock venue.

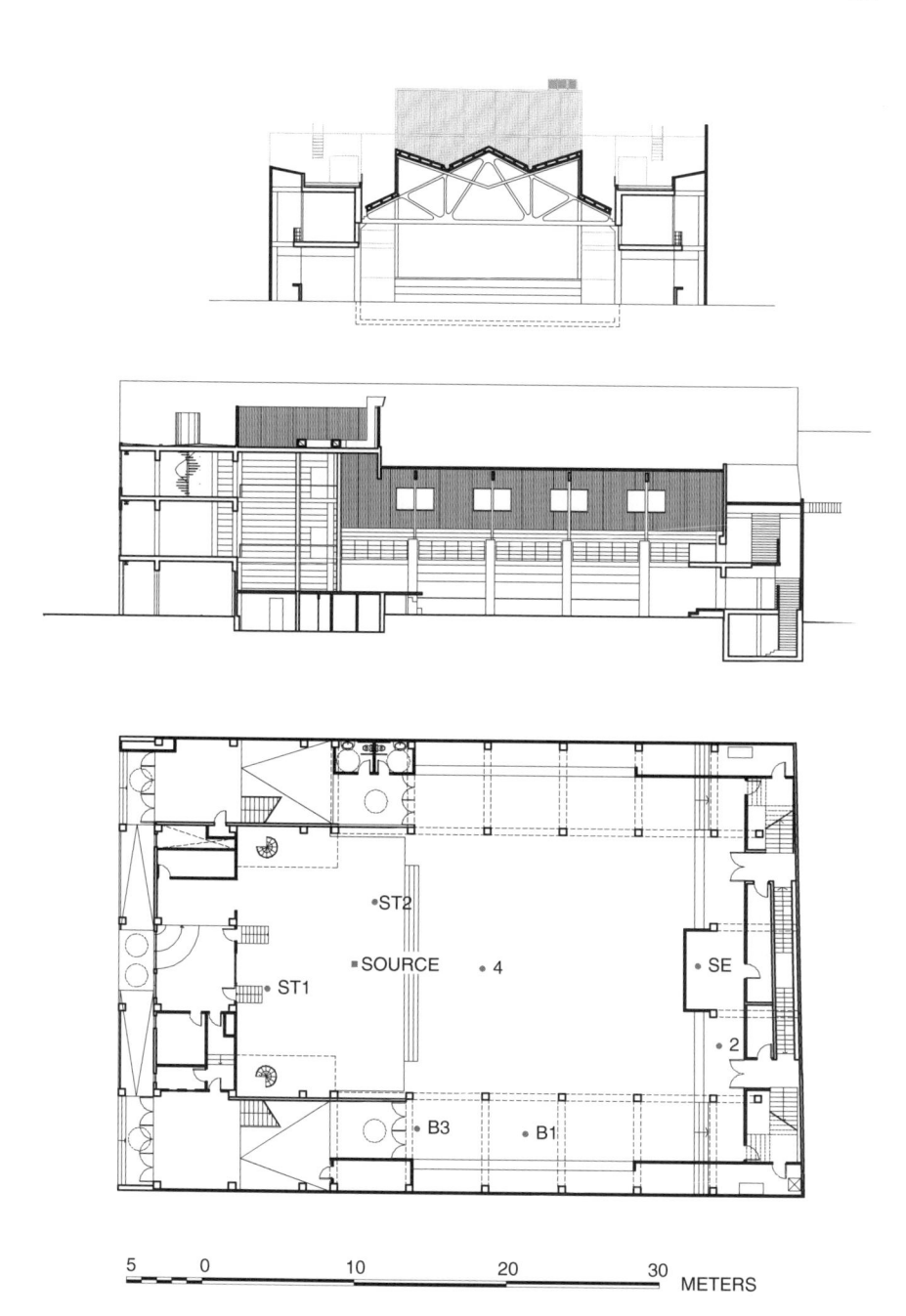

Razzmatazz 1

Geometrical data	
Volume	8,000 m^3
Height, audience area	9.8 m
$L \times B \times H$	$33 \times 26 \times 7\text{–}9.5$

Acoustical data	
Audience area	
$T_{30, 125-2k}$	1.51
EDT_{125-2k}	1.43
$C_{80, 125-2k}$	0.54
BR_{63} versus 0.5–1k	0.82
BR_{125} versus 0.5–1k	1.04
Stage area	
EDT_{125-2k}	0.84
$D_{50, 125-2k}$	0.71
BR_{63} versus 0.5–1k	1.23
BR_{125} versus 0.5–1k	1.12

Materials Used

Audience Area

Floor: Concrete.
Ceiling: Thin metal plate, nonperforated.
Walls: Concrete, curtains.

Stage Area

Floor: Wooden plates directly on concrete.
Ceiling: Thin metal plate, nonperforated.
Walls: Concrete, curtains.

State of Hall When Measured

Empty; no dwarfs.

T30 in audience area

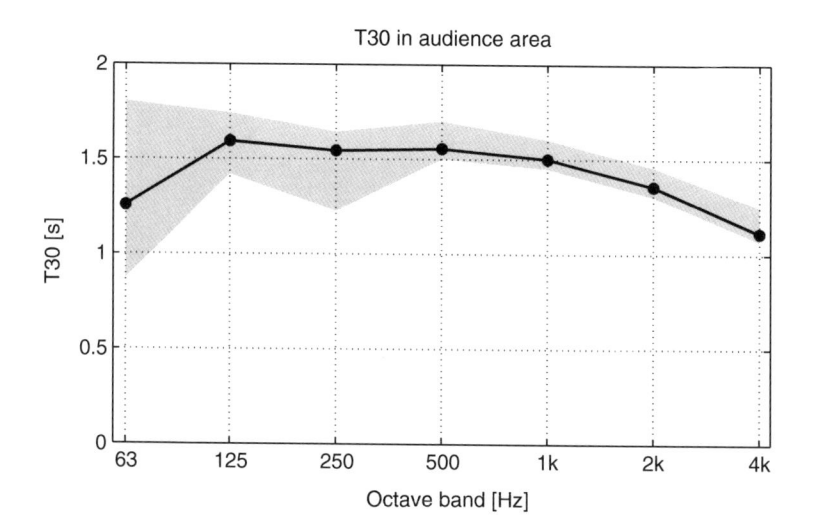

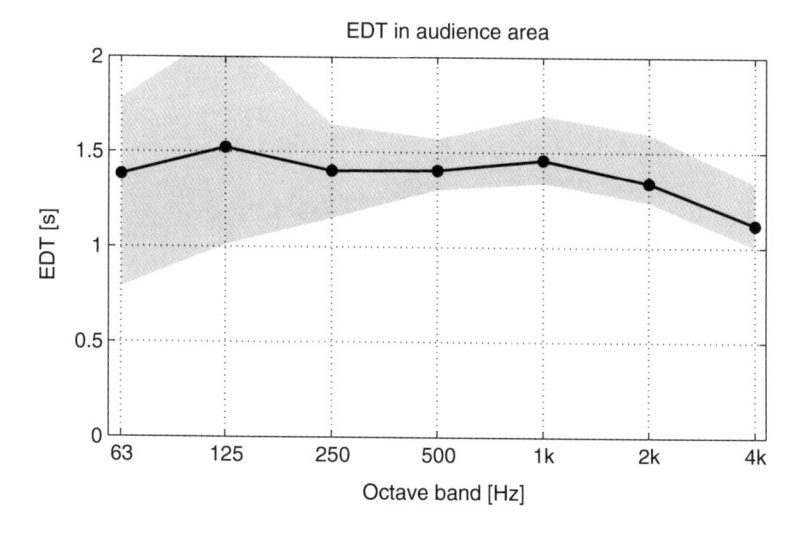

EDT in audience area

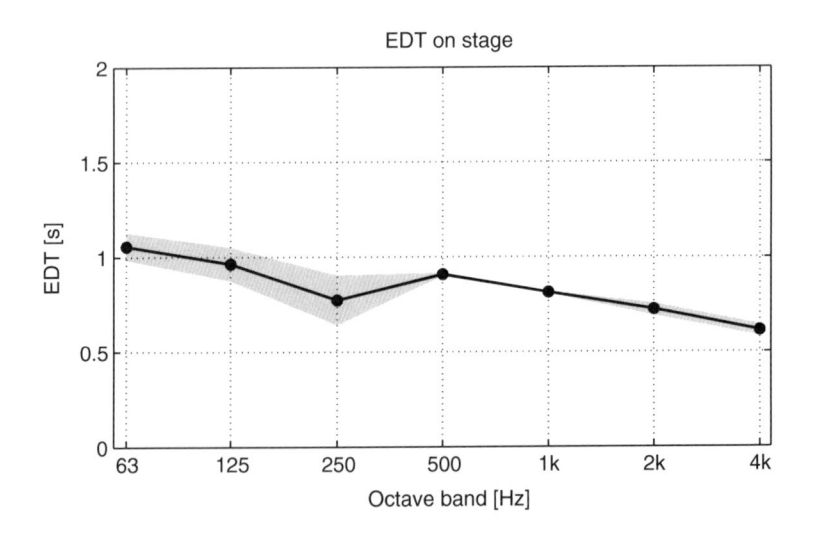

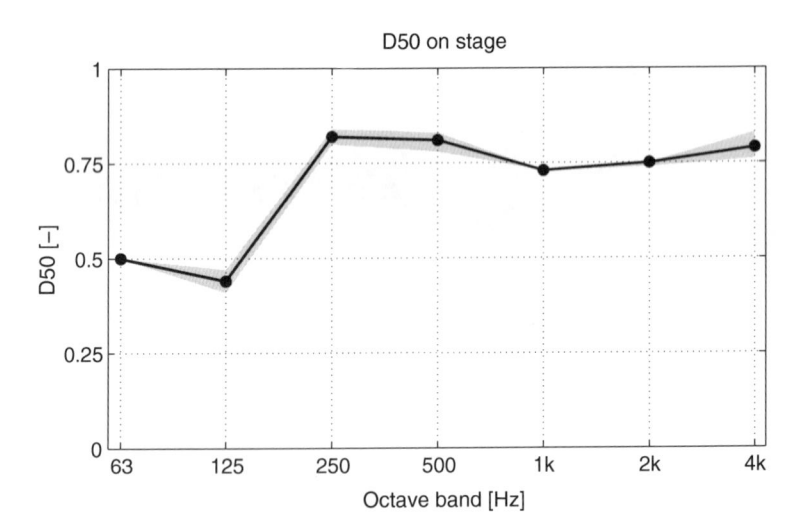

390   7 Gallery of Halls that Present Pop and Rock Music Concerts

Razzmatazz 2

Barcelona
Number of concerts per year: NA
Founded: NA
Capacity: 700
Architect: N/A
Acoustician: N/A

Razzmatazz 2. Typical underground rock club.

Razzmatazz sala 2.

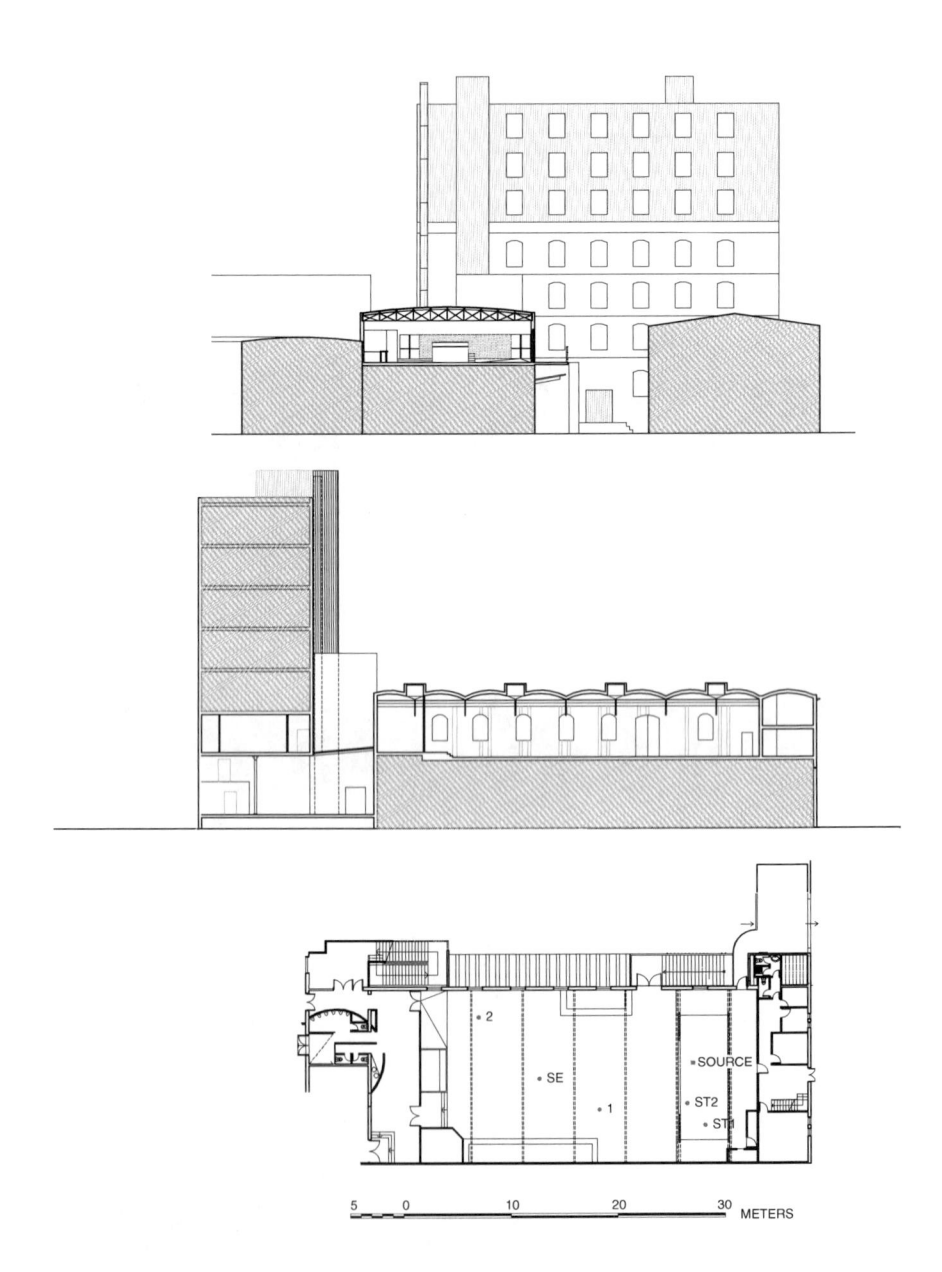

7 Gallery of Halls that Present Pop and Rock Music Concerts

Geometrical data	
Volume	2,600 m^3
$L \times W \times H$	$31 \times 16 \times 5.3$

Acoustical data	
Audience area	
$T_{30, 125-2k}$	1.68
EDT_{125-2k}	1.66
$C_{80, 125-2k}$	−0.52
BR_{63} versus 0.5–1k	1.03
BR_{125} versus 0.5–1k	1.01
Stage area	
EDT_{125-2k}	1.24
$D_{50, 125-2k}$	0.59
BR_{63} versus 0.5–1k	0.68
BR_{125} versus 0.5–1k	1.21

Materials Used

Audience Area

Floor: Concrete.
Ceiling: Painted plaster.
Walls: Concrete, a few windows; end wall: a bit of perforated bricks at DJ's location.

Stage Area

Floor: Wooden plates on risers.
Ceiling: As audience area.
Walls: As audience area, curtains.

State of Hall When Measured

Empty.

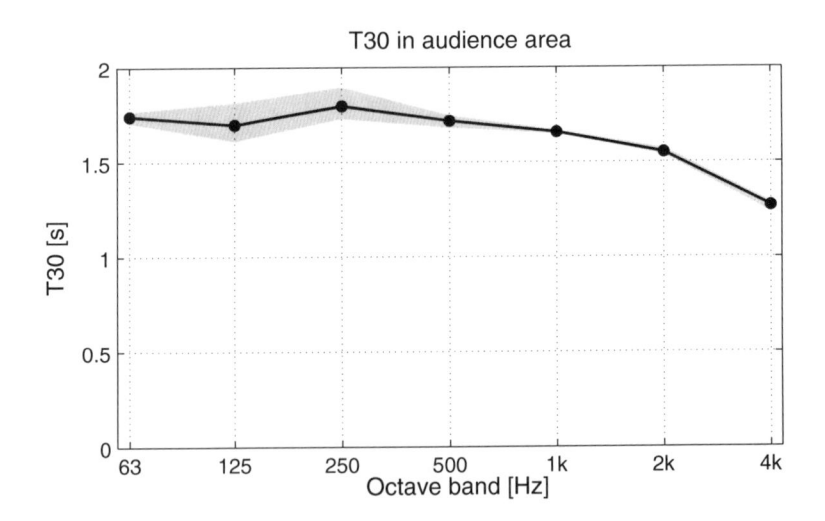

7 Gallery of Halls that Present Pop and Rock Music Concerts

T30 in audience area

EDT in audience area

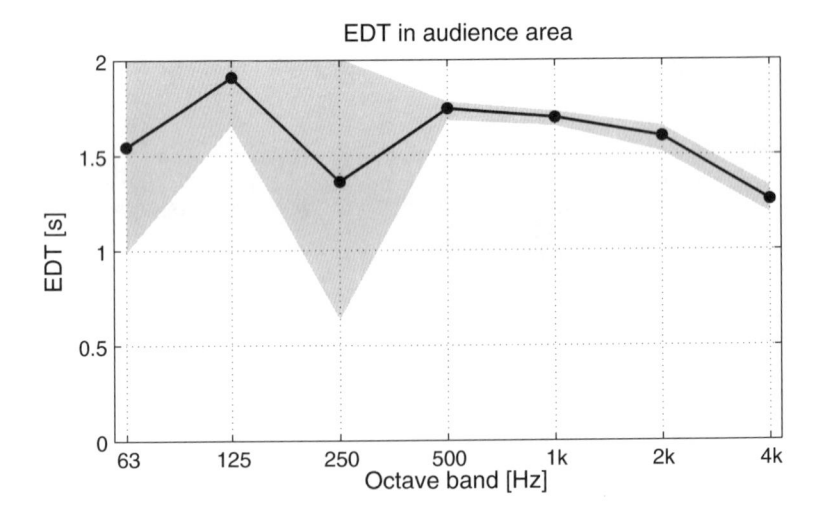

EDT on stage

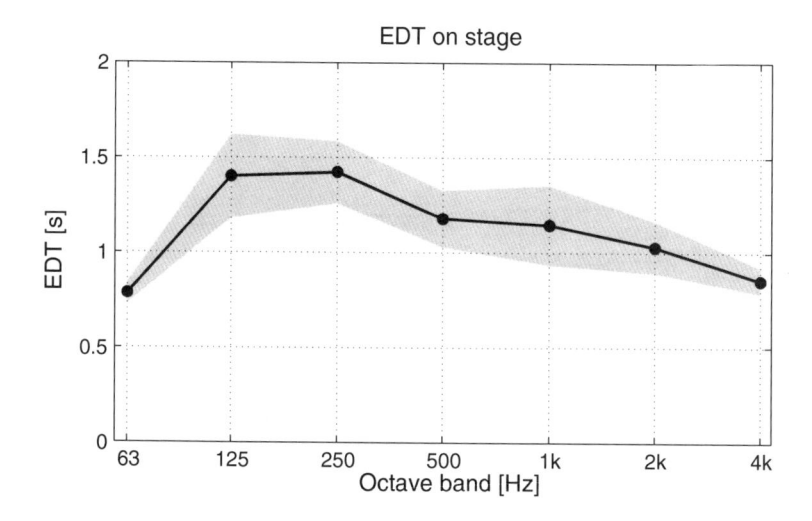

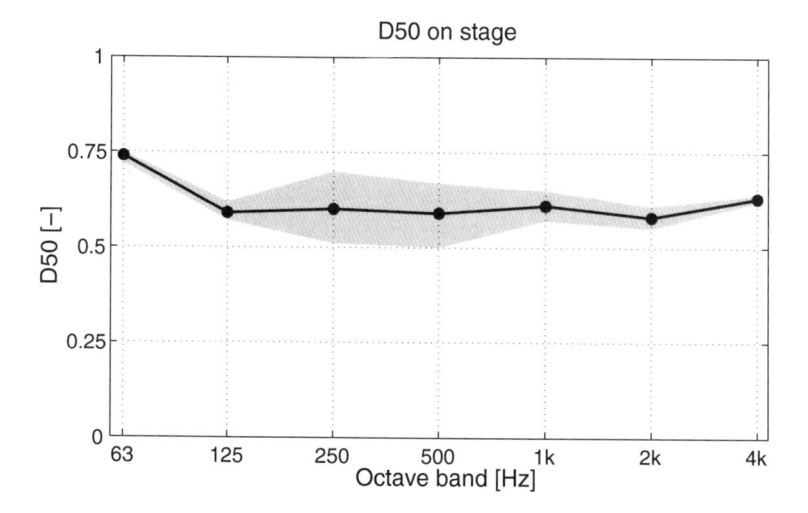

D50 on stage

Sala Barcelona'92/Sant Jordi Club

Barcelona
Number of concerts per year: 30
Founded: 1990
Capacity: 4,600
Architect: Arata Isozaki
Acoustician: N/A
Owner: Ajuntament de Barcelona (Barcelona City Hall)

The Sala Barcelona'92, annexed to the Palau Sant Jordi and well known as a space for organizing gala dinners and corporate events of different kinds, now has a new line of services designed for music events.

A completely open space that has let the imagination of creative people run riot and has fully satisfied event organizers.

An integral part of the Palau Sant Jordi, but at the same time totally independent, the recent installations of a stage and sound and lighting equipment have provided the venue with new technical services that facilitate corporate event programs.

With the name Sant Jordi Club, this medium-sized concert venue adds to the current number of musical venues available in Barcelona. With the aim of offering impeccable service, the Sant Jordi Club can count on upgraded and updated systems for concerts with a maximum capacity of 4,600 people.

The renovation of the Sant Jordi Club does not only feature structural improvements. With this new music venue concept, Palau Sant Jordi also commits itself to providing a wider and more select choice to the general public, as far as refurbishment is concerned, as well as access, car parks, and VIP services.

The number of concerts per year is 30, bands such as Arctic Monkeys, Jamie Cullum, Manowar, A-ha, Interpol, Alice Cooper, 30 s to Mars, Motorhead, Soulwax, and two Many DJ's are some of the shows that Sant Jordi Club held.

Curtains can be lowered to separate the long hall visually into smaller sections. The balcony is only on one side.

Sala Barcelona'92/Sant Jordi Club

The Venue is situated in conjunction with Palau Sant Jordi on The Olympic esplanade.

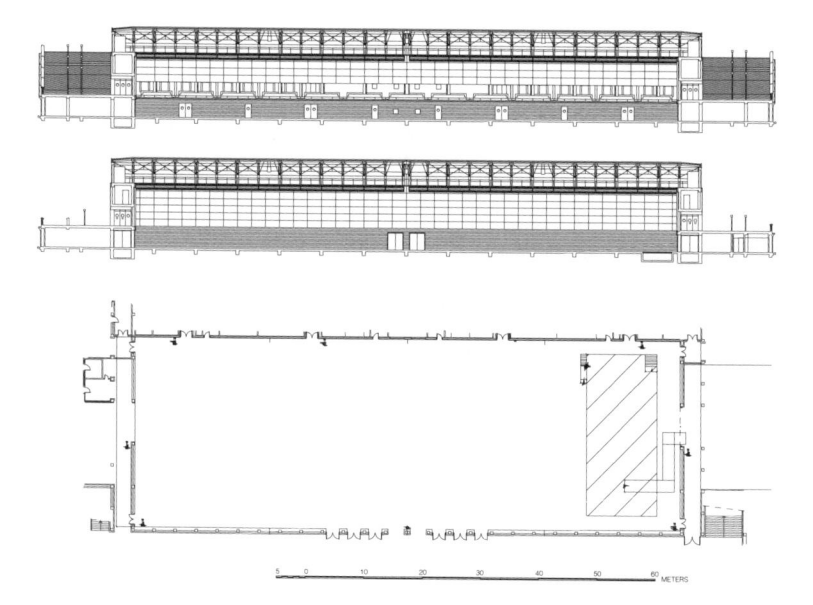

Geometrical data	
Volume	48,000 m^3
Height, audience area	16 m
Size	93 × 32 m

Acoustical data	
Audience area	
$T_{30, 125-2k}$	1.81
EDT_{125-2k}	1.67
$C_{80, 125-2k}$	1.65
BR_{63} versus 0.5–1k	0.81
BR_{125} versus 0.5–1k	1.07

Materials Used

Audience Area

Floor: Sports floor: wooden plates on joists and cavity.
Ceiling: Perforated thick plates with cavity behind.
Walls: Perforated thick plates with cavity behind. Lower walls are mainly hard thick plates (doors etc.).
Seats are not upholstered.

Stage Area

Curtains surround the temporary staging arrangement.

State of Hall When Measured

Empty; curtain lowered ¼ of the hall's length from the rear wall opposite the stage.

T30 in audience area

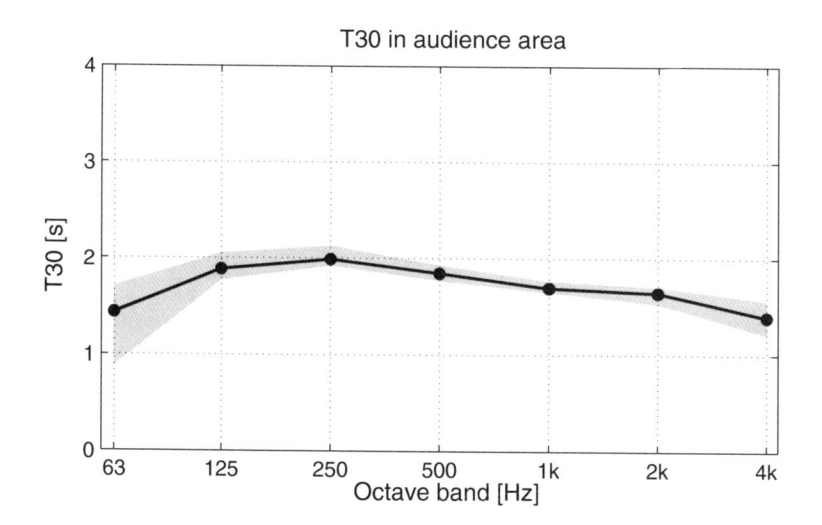

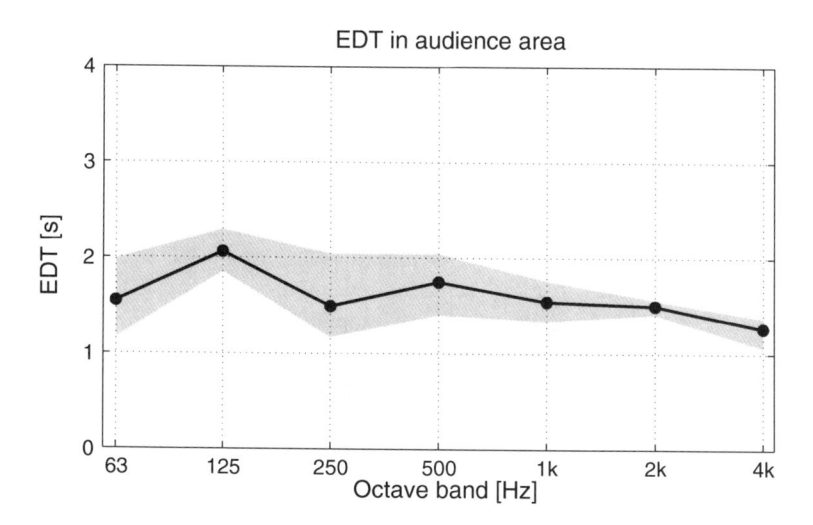

EDT in audience area

Scala

London
Number of concerts per year: 150+
Founded: 1999
Capacity: Approximately 650
Architect: N/A
Acoustician: N/A

The Scala was originally built as a cinema to the designs of H. Courtney Constantine, but construction was interrupted by the First World War and it spent some time being used to manufacture aircraft parts, and as a labor exchange for demobilized troops before opening in 1920 as the King's Cross Cinema. The cinema changed hands and names several times through its life and also changed focus, ranging from mainstream to art-house to adult film over 70 years, as well as spending a short time as a primatarium.

In the summer of 1972, the Scala (then known as the King's Cross Cinema) played host to the one and only UK concert by Iggy and The Stooges (who were in London recording the album *Raw Power*.) All photographs later featured in the *Raw Power* album sleeve (including the famous cover shot) were taken that night during the show.

In the early 1990s London's popular Scala Film Club showed the film, *A Clockwork Orange*, without permission from Stanley Kubrick or Warner Brothers. At Kubrick's insistence, Warners sued and won. As a result, Scala was almost bankrupt and closed in 1993, however, the club was reopened in 1999. The cinema had been refitted, with the lower seating area incorporating the new stage, DJ booth, and dancefloor, and the upper seating area incorporated a second room and a DJ booth.

Scala now plays host to many eclectic club nights, including Ultimate Power, Face Down, University of Dub, and Pure Temptation and has featured live music acts including The Libertines, Rhianna, Jessie J, The Killers, Scissor Sisters, Robert Plant Foo Fighters, Moby, HIM, Wheatus, Adam Ant, Sheryl Crow, Sara Bareilles, Gavin DeGraw, Ray LaMontagne, Super Furry Animals, The Cutaway, The Chemical Brothers, and Avril Lavigne, Sonic Youth, Plan B, Tiesto, Enslaved, Gorgoroth, and Lacuna Coil.

Scala means stairs explains Jane as I carry my equipment the 200 stairs from street to stage.

Many later-to-be world famous acts have played the Scala. Some of them return years later.

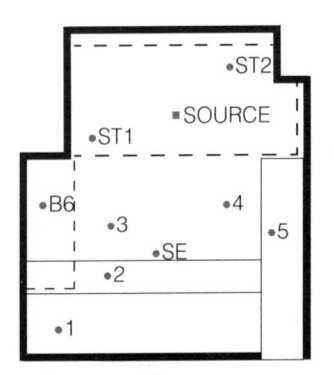

7 Gallery of Halls that Present Pop and Rock Music Concerts

Geometrical data	
Volume	2,000 m^3
$L \times W \times H$	$18 \times 16 \times 7.2$
Surface area of stage	75 m^2
Height of stage	0.5 m

Acoustical data	
Audience area	
$T_{30, 125-2k}$	1.48
EDT_{125-2k}	1.45
$C_{80, 125-2k}$	0.89
BR_{63} versus 0.5–1k	0.96
BR_{125} versus 0.5–1k	0.96
Stage area	
EDT_{125-2k}	1.09
$D_{50, 125-2k}$	0.71
BR_{63} versus 0.5–1k	0.75
BR_{125} versus 0.5–1k	0.79

Materials Used

Audience Area

Floor: Wood on joists.
Ceiling: Masonry.
Walls: Upper wall areas: concrete; lower wall surfaces: wood panels on 2-cm cavity.

Stage Area

Floor: Wood on cavity.
Ceiling: Masonry.
Walls: Curtains.

State of Hall When Measured

Empty.

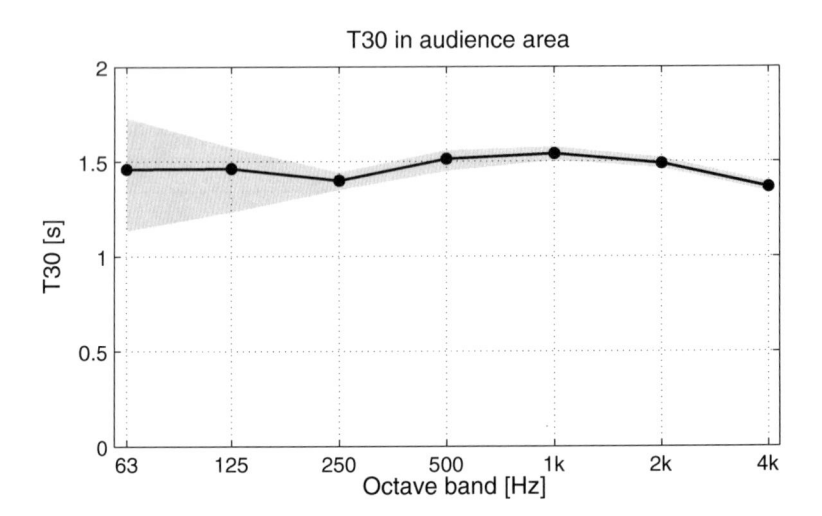

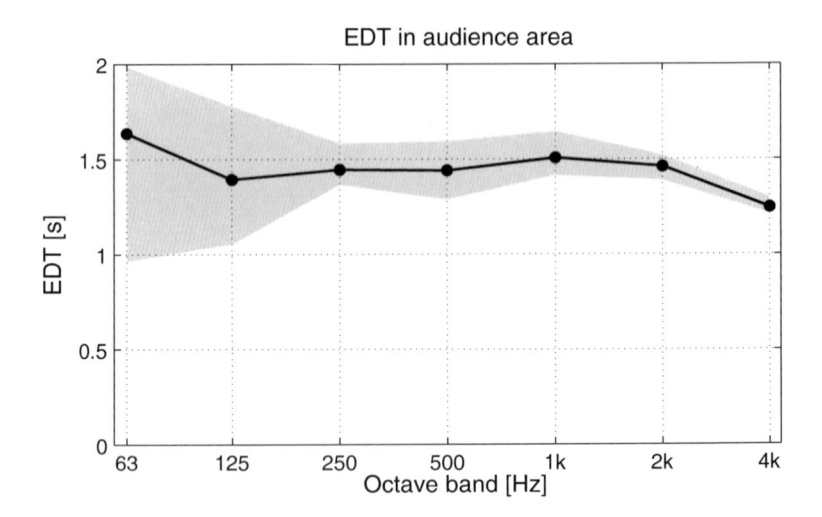

7 Gallery of Halls that Present Pop and Rock Music Concerts

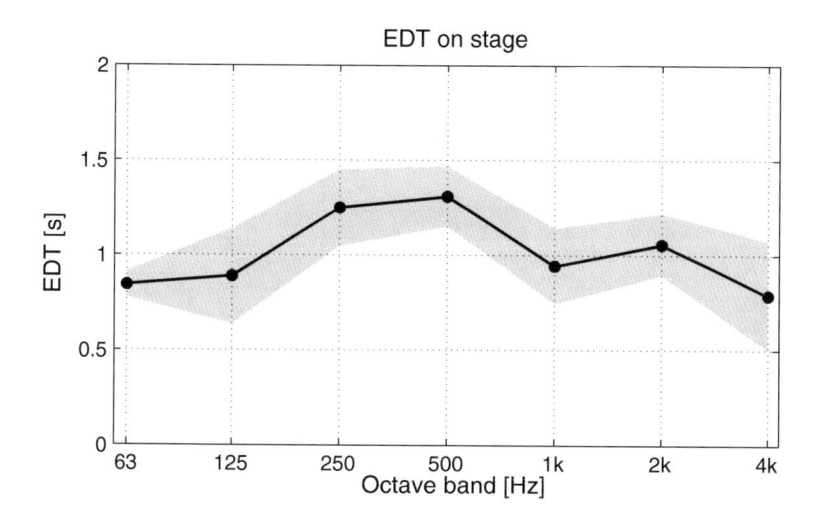

State of Hall When Measured

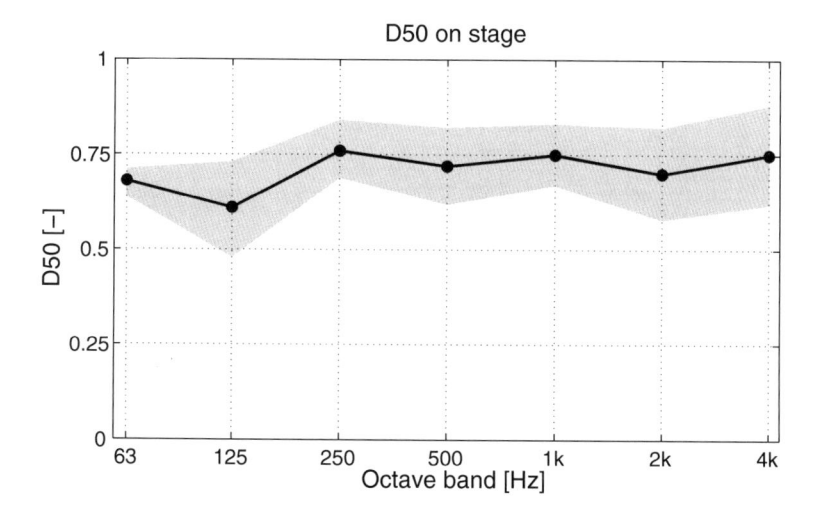

Tunnel

Milano
Number of concerts per year: 120
Founded: 1995
Capacity: 400
Architect: N/A
Acoustician: N/A

Formerly a warehouse of the Central Railway station built in 1936, during fascism, with its typical tunnel structure and barrel vault, the original industrial building has been brought to a new life in 1995, when it turned into Tunnel, an underground music sanctuary.

In its early years artists such as Skunk Anansie, The Cardigans, Uestlove (The Roots), Calexico, and Pan Sonic performed on its stage. The club also hosted some exciting secret shows (one of those featured a memorable live performance by Soulwax).

Tunnel was one of the first clubs in Milan that bet on alternative electronic sounds, inviting DJs like Gilles Peterson and Kid Loco. During the 1990s Tunnel rapidly became the place to be for the whole alternative music scene and a landmark for Milanese clubbers.

In 2009, after several management shifts, Tunnel went back to its roots, with a farsighted glance to the future: the new staff, headed by Diego Montinaro, has thoroughly transformed the venue: completely renewed and refreshed, boosted by an astonishingly powerful sound system, the club also features renewed interiors, and a new lighting set.

Since then the club has developed a solid reputation as an interesting and innovative underground club. Tunnel, with its 350-person capacity, is the small, intimate place where some of the best DJs and bands all over the world feel at home, bands like The Buzzcocks, The Black Angels, The Drums and Lanegan and Campbell, artists like Jon Spencer, Damo Suzuki (ex Can), and Lisa Germano with Phil Selway to name a few.

The club nights also have become really popular and the club has recently invited guest DJs including Moodymann, Etienne de Crecy, Audio Bullys, and Andrew Weatherhall. Far from being an empty, soulless case, Tunnel has many faces, always keeping the same underground flavor.

Tunnel 407

A tunnel is a nightmare for any acoustician because of focusing effects. Tunnel seems to do fine. Photo: Cecilia Giolli.

The porous light concrete and the absence of a back wall is believed to be the reason why Tunnel does not seem to have significant bass problems. Photo: Cecilia Giolli.

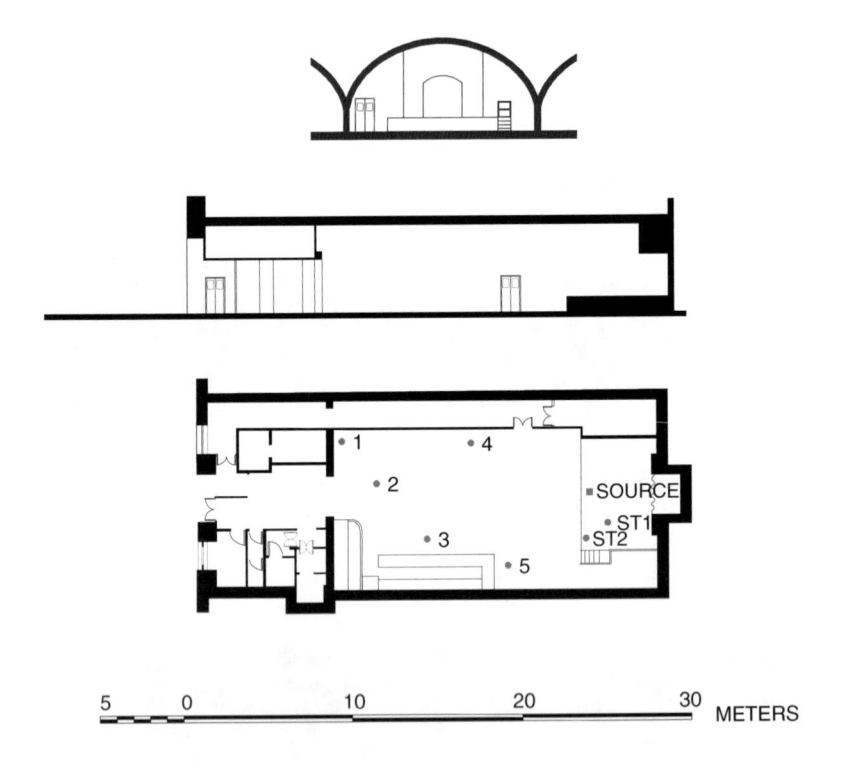

7 Gallery of Halls that Present Pop and Rock Music Concerts

METERS

Geometrical data	
Volume	700 m^3
$L \times W \times H$	$19 \times 9 \times 1.2$–5.4
Acoustical data	
Audience area	
$T_{30, 125-2k}$	0.94
EDT_{125-2k}	0.76
$C_{80, 125-2k}$	5.69
$BR_{63 \text{ versus } 0.5-1k}$	0.82
$BR_{125 \text{ versus } 0.5-1k}$	0.94
Stage area	
EDT_{125-2k}	0.4
$D_{50, 125-2k}$	0.89
$BR_{63 \text{ versus } 0.5-1k}$	2.36
$BR_{125 \text{ versus } 0.5-1k}$	0.95

Materials Used

Audience Area

Floor: Tiles on concrete.
Ceiling: Concrete.
Walls: Light concrete; light concrete covered with a painted porous insulation product.

Stage Area

Floor: Tiles on concrete.
Ceiling: 5-cm foam product on concrete.
Walls: 5-cm foam product on concrete.

State of Hall When Measured

Empty; a DJ setup on stage.

T30 in audience area

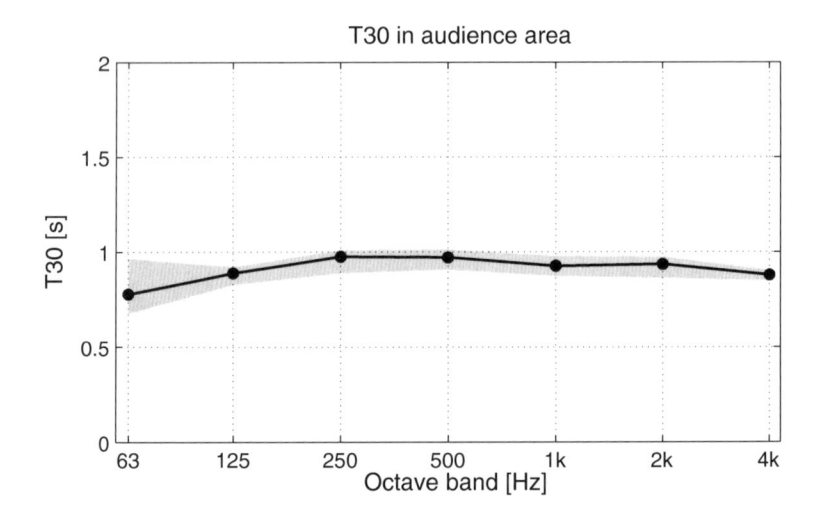

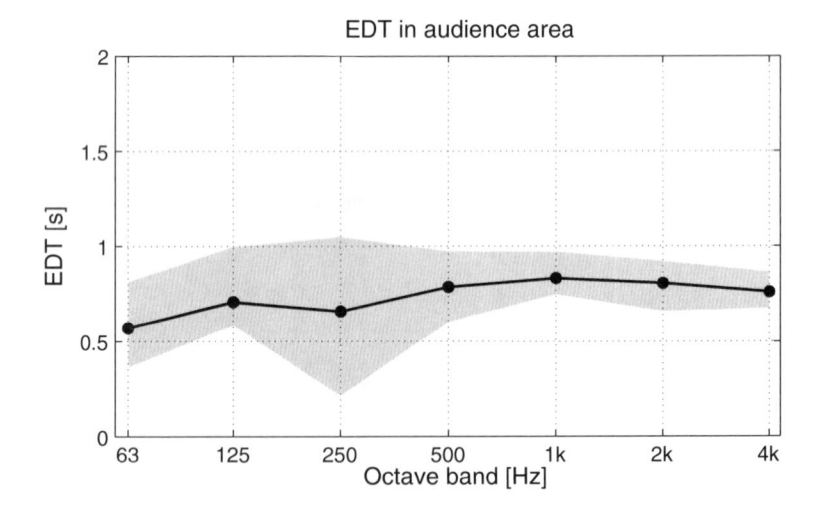

EDT in audience area

EDT on stage

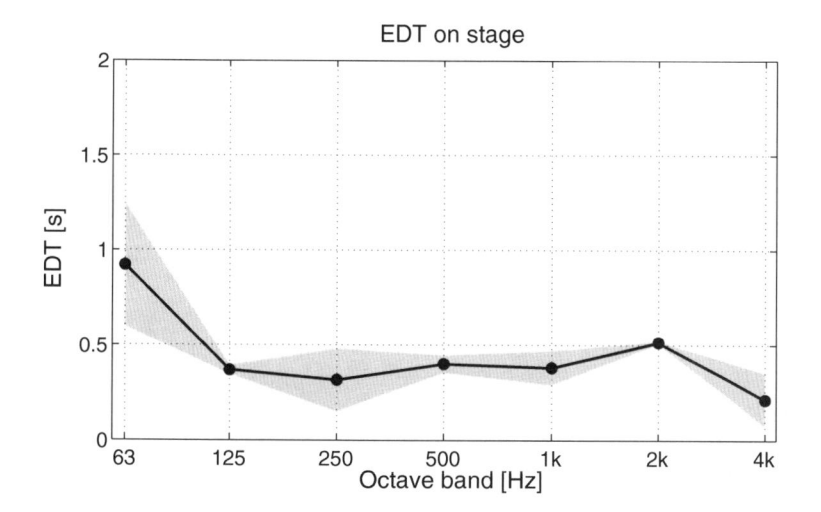

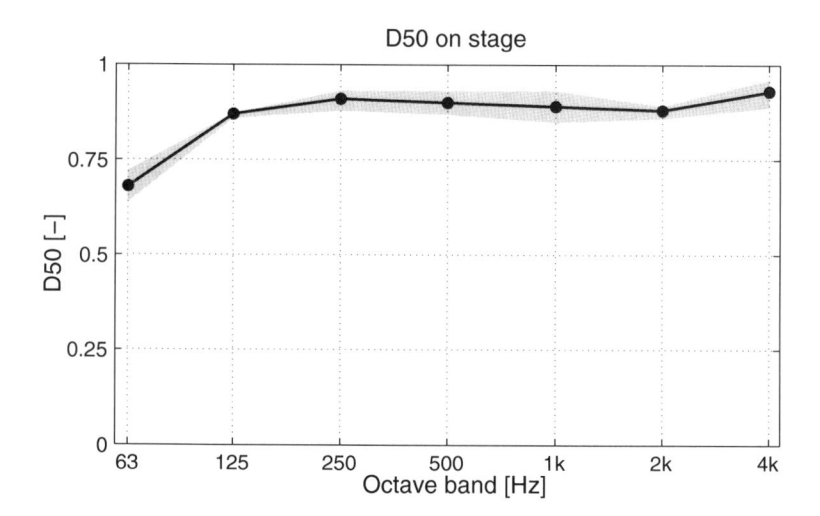

D50 on stage

Vega

Copenhagen
Number of concerts per year: 120. In both halls in the venue: 260
Founded: 1996, building: 1956
Capacity: 1,500
Architect: Vilhelm Lauritzen
Acoustician: Jordan/Jan Voetmann

VEGA is a regional venue, owned and operated by the foundation Koncertvirksomhedens Fond, which was created with the sole purpose of presenting concerts. Parts of the foundation's service is based on municipal and state grants.

In 2010 approximately 300,000 people visited VEGA and the venue has been named as the best concert arena in Europe by the international music magazine *Live*. With two separate concert halls VEGA presents a broad program of music, covering many different styles. The great hall, Store VEGA, has capacity for 1,500 guests, and the smaller hall, Lille VEGA, has space for 500 people.

Furthermore, the VEGA Lounge and Ideal Bar are at street level, and are both there to create the perfect setting for a night out in Copenhagen's music and nightlife. VEGA has a total of 12 bars serving the audience and musicians. VEGA's central location in Copenhagen and the close proximity to southern Sweden, make VEGA the obvious choice when indoor concerts by international artists are being planned.

Annually around 250 concerts and events are held at VEGA, reflecting VEGA's objective to find the right balance between new talent and more established names in rock, pop, soul, hip hop, electronic, and world music as well as presenting the audience with various club concepts. VEGA's own production of concerts and events each year is represented by about 150 concerts, club events, showcases, and special events.

VEGA has opened the doors to most of Scandinavia's leading artists as well as several international stars including Prince, David Bowie, Björk, Suede, Kylie Minogue, Norah Jones, Moby, Foo Fighters, Blur, Fatboy Slim, DJ Shadow Girl Talk, Erol Alkan, DJ Shadow, and others. VEGA's technical equipment, where particularly the sound system is of high quality, guarantee a perfect experience.

VEGA is known as one of the leading concert stages in Europe. This is not only due to optimal venue audio and lighting conditions, but also because the building is an exciting architectural gem, which provides the perfect setting for evocative arrangements. The original 1950s design gives the building a unique atmosphere, and VEGA's decor with dark wood paneling, mahogany floors, friezes, and the many original details of railings, balustrades, and lamps in typical Scandinavian style are the hallmarks of VEGA.

The building was originally named The People's House and was the stronghold of the Danish trade union movement. It was built in 1956 and designed by the famous Danish architect Vilhelm Lauritzen, who is also known for other period building works such as Broadcasting House and the award-winning old terminal at Copenhagen Airport. Since the unions pulled out, The People's House was deserted for many years and was beginning to decay. But after an extensive restoration in 1996 the building reopened as VEGA—House of Music. The property is one of the youngest listed buildings in Denmark.

- Most people know VEGA for concerts and VEGA's Night Club, but the atmospheric halls and labyrinthine hallways contain a wealth of other options. In addition to the two halls, Store and Lille VEGA, VEGA has many other rooms that each have a distinctive character and offer the opportunity to house small- and medium-sized companies. The beautiful premises can also provide a framework for, for example, press conferences, company parties, general meetings, and cultural events.

Headed by a board of seven members, VEGA has approximately 180 employees, most of whom are paid by the hour. In its daily operation and administration VEGA employs about 35 full-time employees.

Wooden panels in excess including the floor and a porous absorber with a 3-m cavity in the ceiling ensure control of the T30 including the crucial 125 Hz octave band. And it leaves a nice airy sound not too reverberant once the audience is in place.

Vega is without doubt the best established Danish rock venue of its size. The build-up of T30 below 100 Hz only becomes perceivable far back on the stage where one can feel a lack of early support due to the huge stage tower.

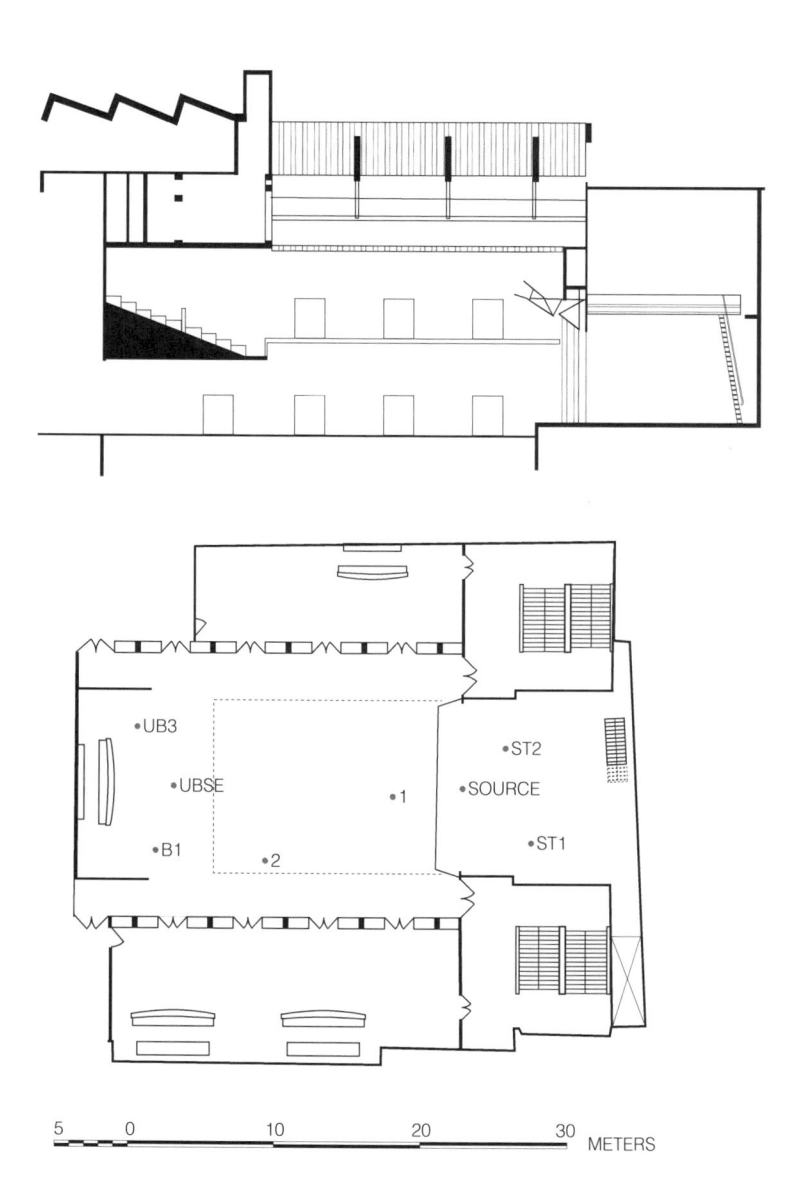

Geometrical data	
Volume	6,000 m^3
$L \times W \times H$	$38 \times 19 \times 10.5$

Acoustical data	
Audience area	
$T_{30, 125-2k}$	1.2
EDT_{125-2k}	0.97
$C_{80, 125-2k}$	4.97
BR_{63} versus 0.5–1k	1.34
BR_{125} versus 0.5–1k	0.98
Stage area	
EDT_{125-2k}	0.51
$D_{50, 125-2k}$	0.88
BR_{63} versus 0.5–1k	3.83
BR_{125} versus 0.5–1k	1.92

Materials Used

Audience Area

Floor: Concrete.
Ceiling: Concrete with large areas of glazing. Suspended home-made reflectors.
Walls: Concrete. At each side of the front of the stage there are two big walls 20-cm thick with thin perforated plates on each side and mineral wool in between. Upper side wall on one side: perforated gypsum board. Back wall: painted brick.

Stage Area

Floor: Stage risers.
Ceiling: Concrete.
Walls: Curtains.

State of Hall When Measured

Empty; seating risers at the rear of the hall.

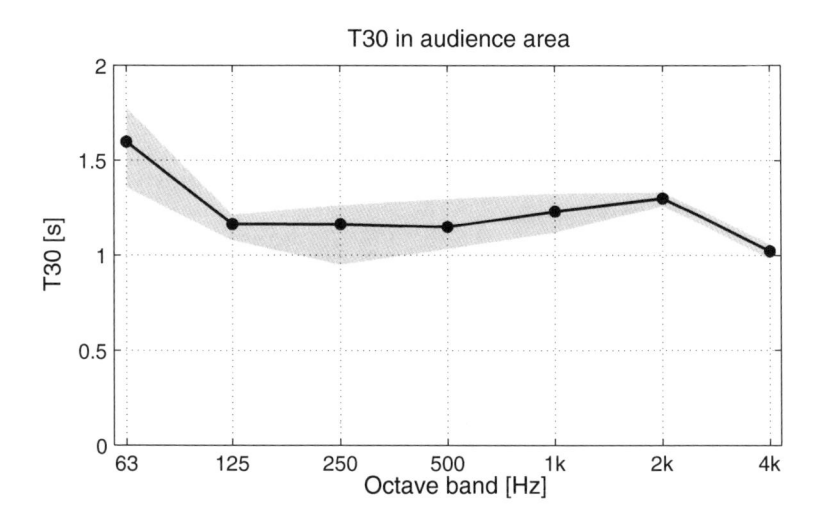

State of Hall When Measured

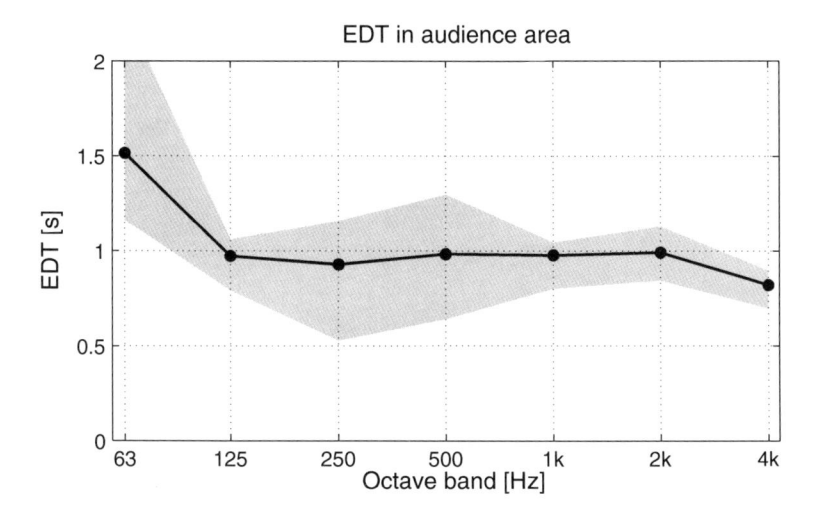

T30 in audience area

EDT in audience area

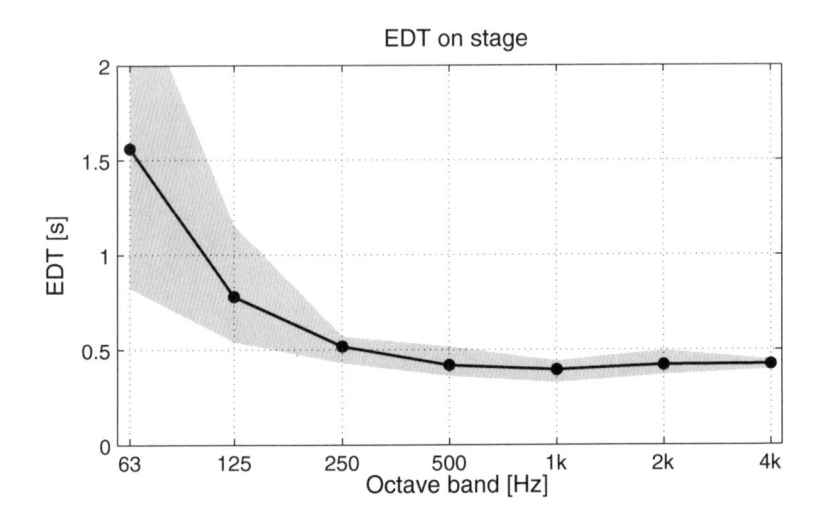

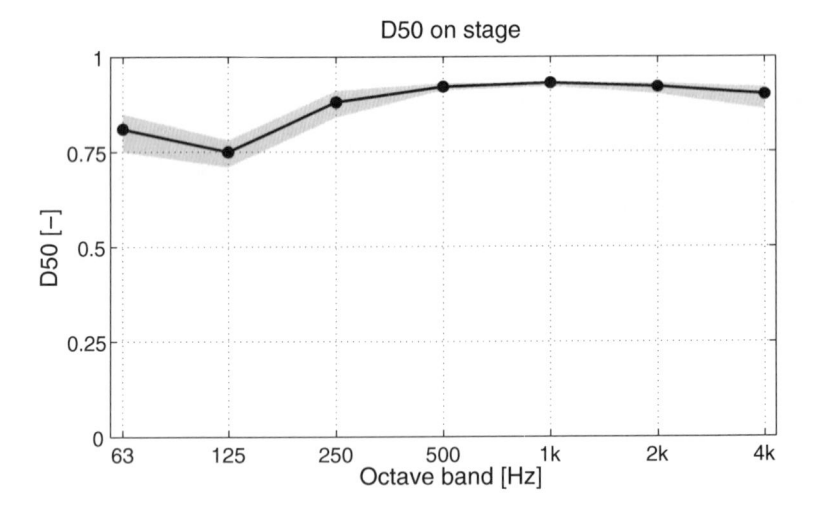

418   7 Gallery of Halls that Present Pop and Rock Music Concerts

Wembley Arena

London
Number of concerts per year: 50
Built: 1934; refurbished: 2006
Capacity: 12,000
Architect: Owen Williams
Acoustician: N/A

Wembley Arena, by many called the the UK's flagship live music venue, first opened its doors in 1934 as the Empire Pool and Sports Arena. At that time the building was considered a great innovation being one of the only buildings of its size with such a vast roof span and having no supporting columns anywhere within the auditorium, thus giving a full view to all guests. It opened as a multipurpose venue, hosting public swimming sessions as well as international swimming competitions, boxing, and a variety of sporting events. In 1948 it was the venue for the swimming events of the Olympic Games and went on to host numerous sporting and musical events. The famous Wembley lion was adopted by the home ice hockey team as their name (The Wembley Lions), as did the neighboring speedway team at the Stadium next door.

72 years and a £35 million refurbishment later, the venue reopened its doors in April 2006 with a twenty first-century look and has become synonymous with live music, welcoming scores of sell-out shows and one-off UK tour dates from major recording artists each year. Wembley Arena has played host to some of the greatest music acts of all time, but is also well known as a sports and entertainment venue. Playing host to ice skating shows such as *Disney on Ice*, *Holiday on Ice*, and *Dancing on Ice*, family events such as CBeebies and Thomas the Tank Engine and even becoming a luxury equestrian center for the pampered horses of the Spanish Riding School and Masters Snooker it retains its historic links with Wembley following its move from the Conference Centre in 2007.

The Arena continues to play host to the greatest global recording artists, including Madonna, Bruce Springsteen, and the Rolling Stones as well as popular sports personalities and most enchanting children's shows and is looking forward to celebrating many more great years of entertainment.

The Arenas first solo rock band concert was the Monkees in 1967 and also played host to the only time the Beatles and the Rolling Stones appeared on the same bill, at the NME Poll winners Party in 1964. When the venue was known as "Empire Pool," it hosted the annual *New Musical Express* Poll Winners' concert during the early 1960s. Audiences of 10,000 viewed acts including the Beatles (who performed there three times), Cliff Richard and The Shadows, Joe Brown and the Bruvvers, The Rolling Stones, The Who, Dave Dee, Dozy, Beaky, Mick and Tich, and many others, hosted by Jimmy Savile and Pete Murray. The individual performances were then finished by a famous personality joining the respective performer on stage and presenting them their award. The Beatles were presented one of their awards by actor Roger Moore and Joe Brown was joined on stage by Roy Orbison, to present him his. These were filmed and recorded and later broadcast on television.

A notable attendance record was set in the early 1970s, by David Cassidy, in his first tour of Great Britain in 1973, when he sold out six performances in one weekend. In 1978, Electric Light Orchestra sold out eight straight concerts (a record at the time) during their *Out of the Blue* Tour. The first of these shows was recorded, televised, and later released as a CD/DVD. ABBA also played six sold-out concerts, in one week in 1979 and one song from these concerts, "The Way Old Friends Do," is on their album, *Super Trouper*.

In April 1994, Barbra Streisand began her *Barbra—The Concert* Tour, with four performances at the arena. They marked her first performances in the United Kingdom after 28 years and were the only shows outside of the United States. The opening song on the first night, "As If We Never Said Goodbye," was recorded and transmitted on BBC TV's *Top of the Pops*. Christina Aguilera performed three shows at the arena during her *Stripped* world tour on 2–3 and 5 November 2003. They were filmed and later released as a DVD, titled *Stripped Live in the U.K*. Beyoncé performed, on two consecutive nights, at the arena during her *Dangerously in Love Tour* on November 10–11, 2003. Her show on the 10th was filmed and later released as a DVD, titled *Beyoncé: Live at Wembley*.

Pop band Busted sold out the arena a record 11 times in one year, in 2004 P!nk performed two shows at the arena during her *I'm Not Dead Tour* in October and December, 2006. They were filmed and later released as a DVD, titled *Pink: Live from Wembley Arena*. Pearl Jam holds the attendance record for one show, with 12,470 fans at their 2007 gig [1].

Cliff Richard holds the record for the most headline shows by one artist, having played his 61st concert at the arena in October 2009. Tina Turner is the female artist with the most shows at Wembley, with 26 and with 7 at Wembley Stadium.

Wembley Arena

Legendary. Curtains in the ceiling lower the RT only at high frequencies. Same constellation at measurement.

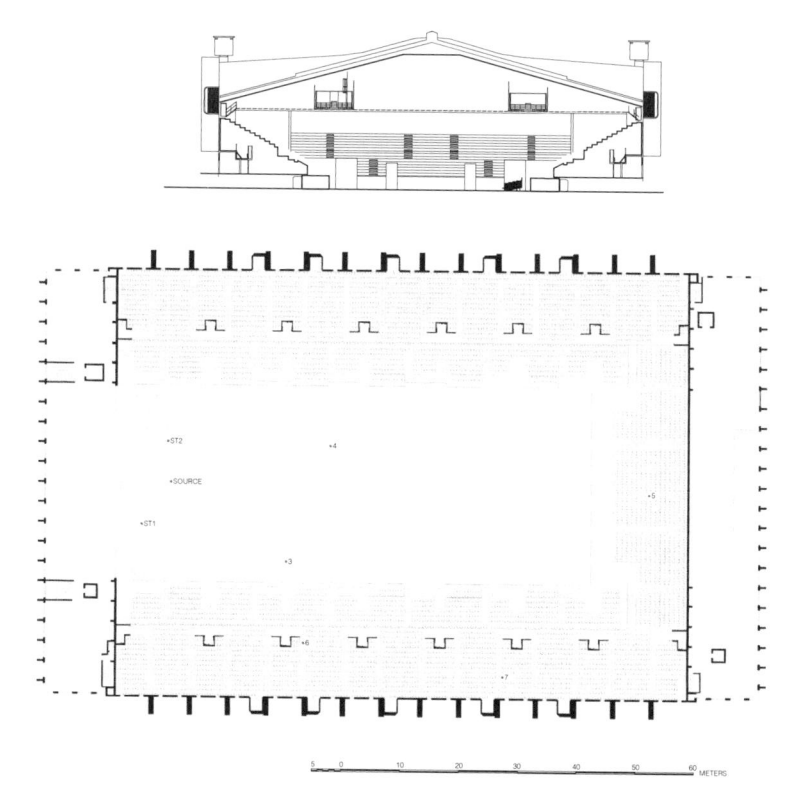

Geometrical data	
Volume	approximately 150,000 m^3
$L \times W \times H$	$96 \times 70 \times 24$ m

Acoustical data	
Audience area	
$T_{30, 125-2k}$	2.56
EDT_{125-2k}	2.41
$C_{80, 125-2k}$	−3.47
BR_{63} versus 0.5–1k	2.1
BR_{125} versus 0.5–1k	1.64

Materials Used

Audience Area

Floor: Concrete.
Ceiling: Perforated plate probably with some cavity behind.
Walls: Beneath the lower balcony: concrete and doors. At both end walls and on walls over the upper balconies: perforated thin plate with 5–10-cm porous absorption behind.
Seats are not upholstered.

State of Hall When Measured

Additional seats mounted on the entity of the floor. Curtains in ceiling as in photo but also on the sides 2/3 down from stage.

State of Hall When Measured 423

Wembley Arena. "…venues, it seems, do get better with age." (The Times).

"The arena looks great—it looks fantastic—and comfy seats too!!" (Music Week).

T30 in audience area

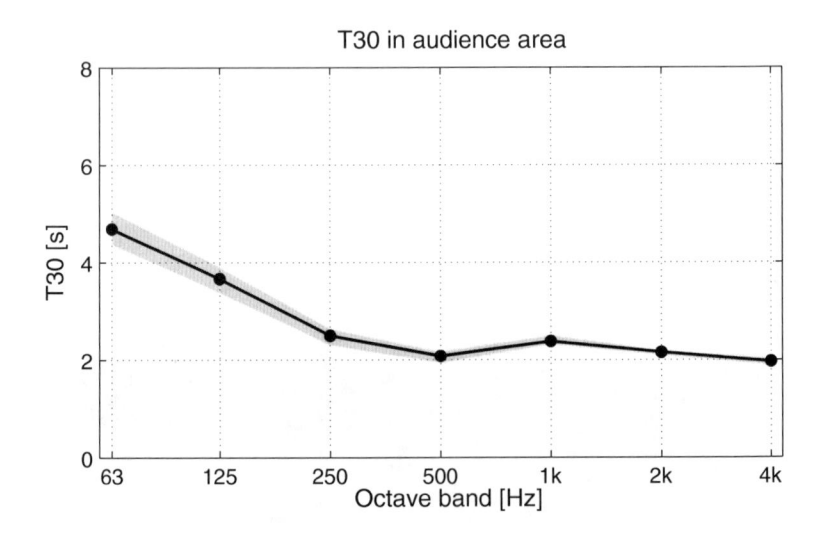

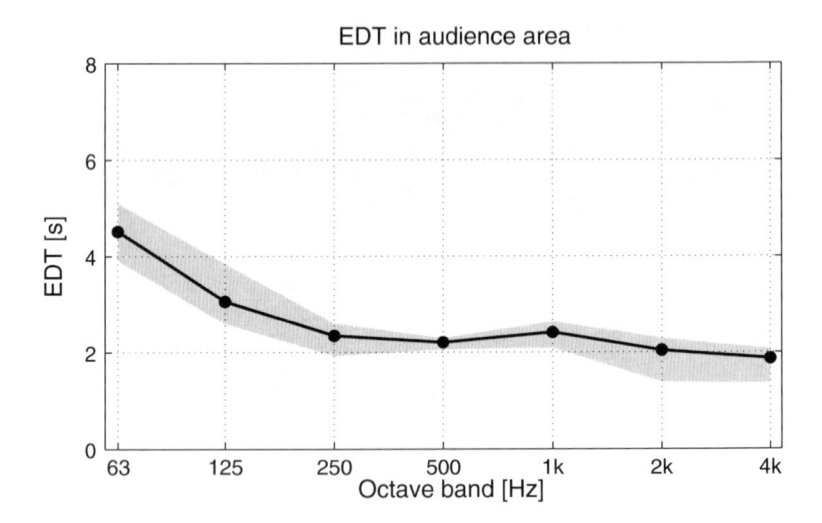

EDT in audience area

Werk

München
Number of concerts per year: N/A
Capacity: 1,500
Founded: N/A
Architect: N/A
Acoustician: N/A

Due to noise issues with neighbors the future of the club is uncertain.

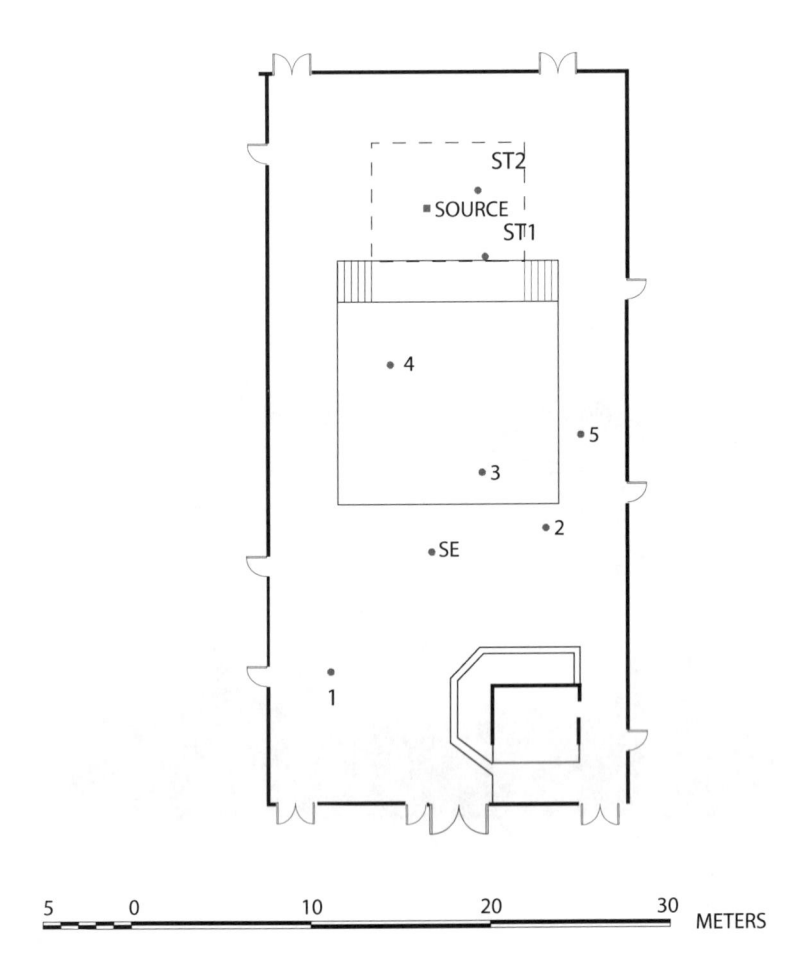

7 Gallery of Halls that Present Pop and Rock Music Concerts

Geometrical data	
Volume	4,000 m³
$L \times W \times H$	$40 \times 20 \times 5$ m
Acoustical data	
Audience area	
$T_{30, 125-2k}$	1.39
EDT_{125-2k}	1.46
$C_{80, 125-2k}$	−0.15
BR_{63} versus 0.5–1k	0.96
BR_{125} versus 0.5–1k	1.04
Stage area	
EDT_{125-2k}	0.8
$D_{50, 125-2k}$	0.7
BR_{63} versus 0.5–1k	0.77
BR_{125} versus 0.5–1k	1.07

Materials Used

Audience Area

Floor: Concrete.
Ceiling: Concrete with large areas of glazing. Suspended home-made reflectors.
Walls: Concrete. At each side of the front of the stage there are two big walls 20-cm thick with thin perforated plates on each side and mineral wool in between. Upper side wall on one side: perforated gypsum board. Back wall: painted brick.

Stage Area

Floor: Stage risers.
Ceiling: Concrete.
Walls: Curtains.

State of Hall When Measured

Empty; seating risers at the rear of the hall.

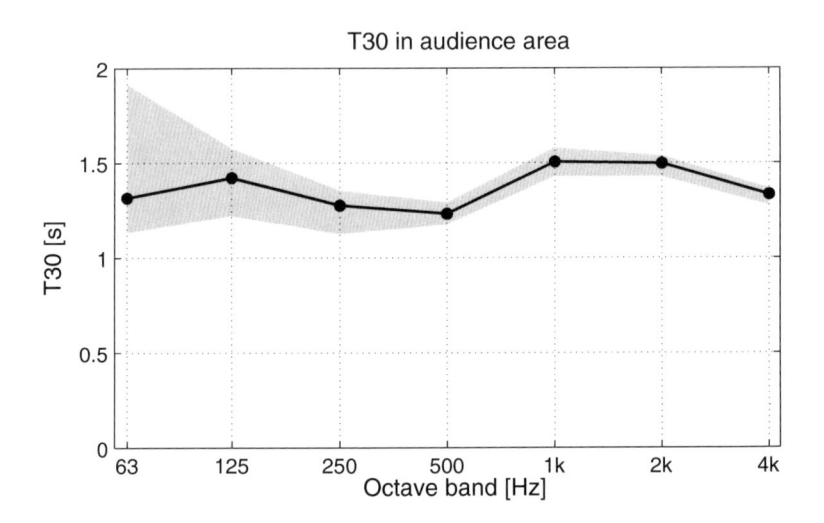

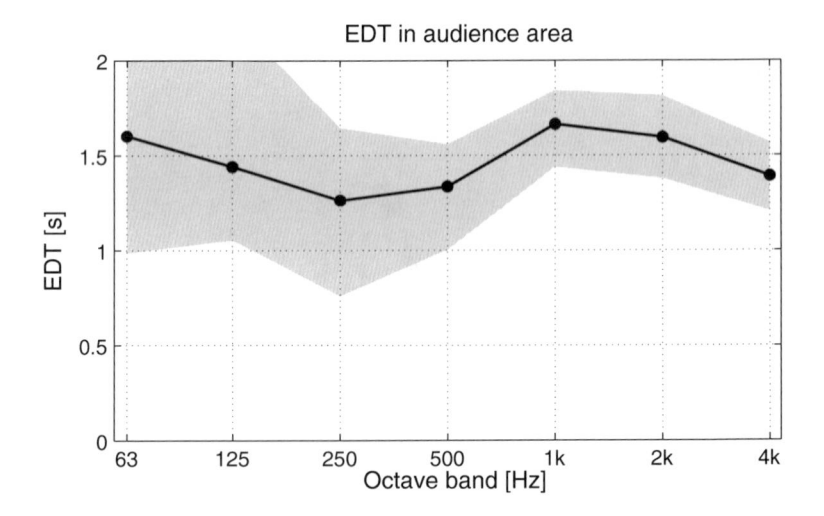

7 Gallery of Halls that Present Pop and Rock Music Concerts

EDT on stage

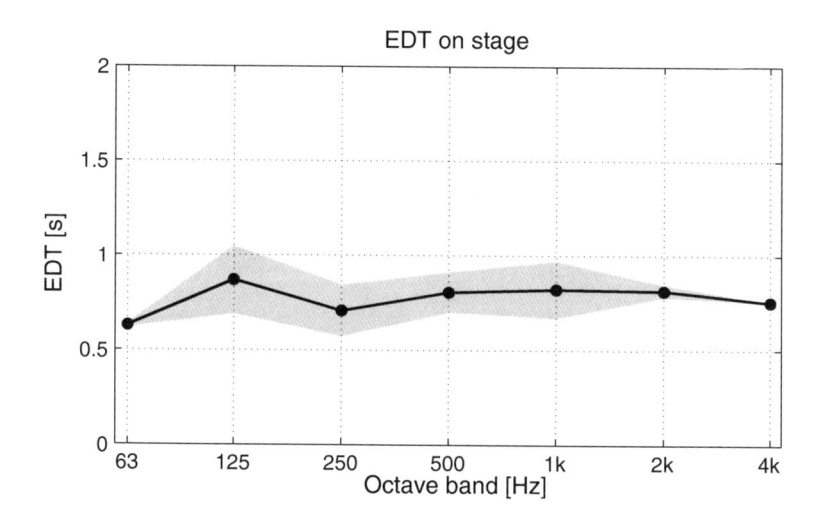

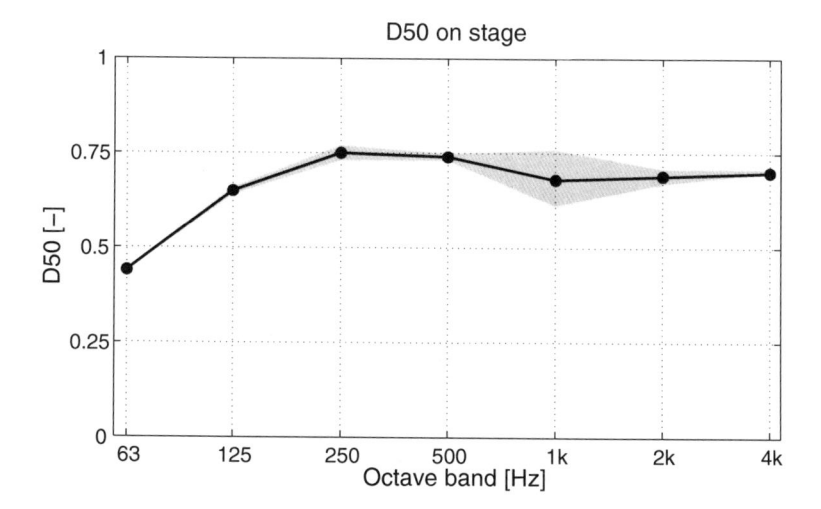

D50 on stage

Zeche

Bochum
Number of concerts per year: 150
Founded: 1981
Capacity: 1,100
Architect: N/A
Acoustician: N/A

Die *Zeche Bochum* ist ein Veranstaltungsmultiplex mit Diskothek in Bochum. Sie wurde im November 1981 eröffnet und war mit Vorreiter für Konzepte der Sozio-Kulturellen Zentren. Ihre Räumlichkeiten mit Veranstaltung-, Verzehr- und Tanzbereichen befinden sich in der ehemaligen Schlosserei der Zeche Prinz Regent in Bochum-Weitmar.

Anfänglich wurden Veranstaltungen in der Zeche Bochum teilweise bei kultureller Anerkennung öffentlich mitgefördert, jedoch musste sie sich kommerziell selbst von Beginn an von allein tragen und war ein Vorreiter für weitere Musikklubs in Deutschland. Die Musikrichtung war zunächst dem Underground verpflichtet, später verlegte man sich auf Pop und Mainstream. Die Zeche Bochum war von Anfang an regelmäßiger Veranstaltungsort für Konzerte. Zum Interieur zählt neben der Veranstaltungshalle mit Empore, eine Kneipe, ein Restaurant und ein kleiner Veranstaltungsraum.

Zeche is one of Germanys most legendary clubs of this size.

Zeche 431

Balcony and staircase levels make it possible for everybody to find a spot with good visibility.

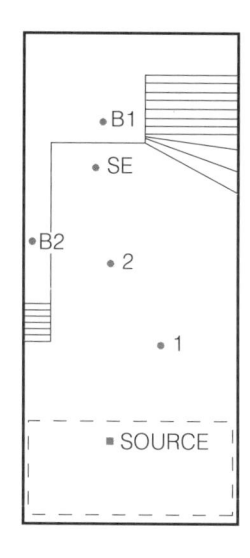

Geometrical data	
Volume	2,700 m^3
$L \times W \times H$	26.7 × 11.9 × 7.2–10.3 m

Acoustical data	
Audience area	
$T_{30, 125-2k}$	1.13
EDT_{125-2k}	1.17
$C_{80, 125-2k}$	0.45
BR_{63} versus 0.5–1k	1.13
BR_{125} versus 0.5–1k	1.23
Stage area	
EDT_{125-2k}	0.74
$D_{50, 125-2k}$	0.78
BR_{63} versus 0.5–1k	0.95
BR_{125} versus 0.5–1k	1.7

Materials Used

Audience Area

Floor: Wood direct on concrete.
Ceiling: Thin plate on cavity filled with mineral wool.
Walls: Concrete with areas of windows covered with wooden plate. Curtain on upper 2 m.

Stage Area

Floor: Modular stage risers.
Ceiling: Thin plate on cavity filled with mineral wool.
Walls: Curtains.

State of Hall When Measured

Empty; risers on stage.

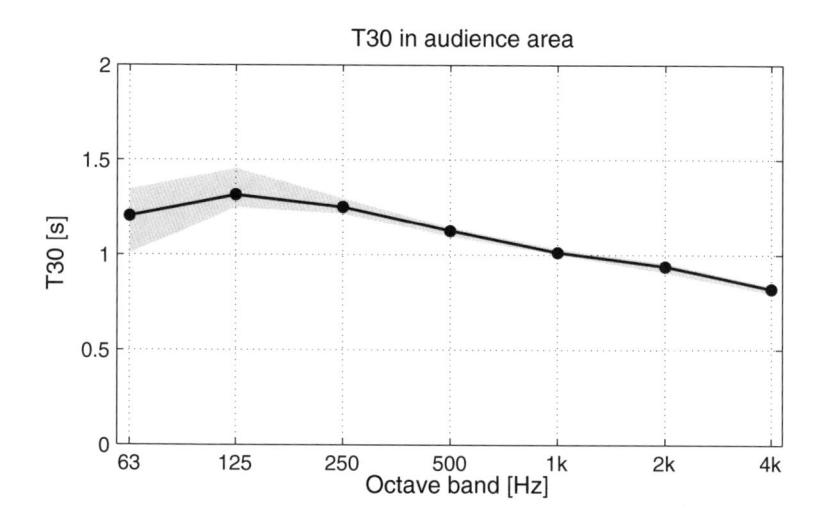

State of Hall When Measured

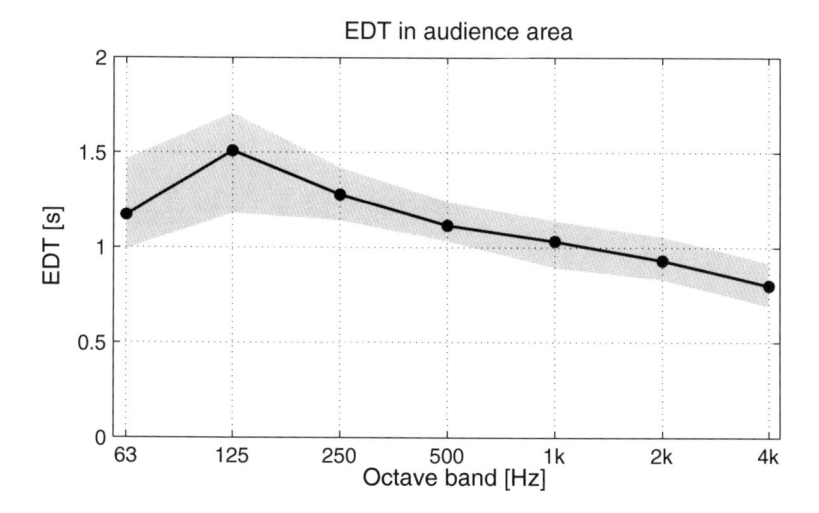

T30 in audience area

EDT in audience area

EDT on stage

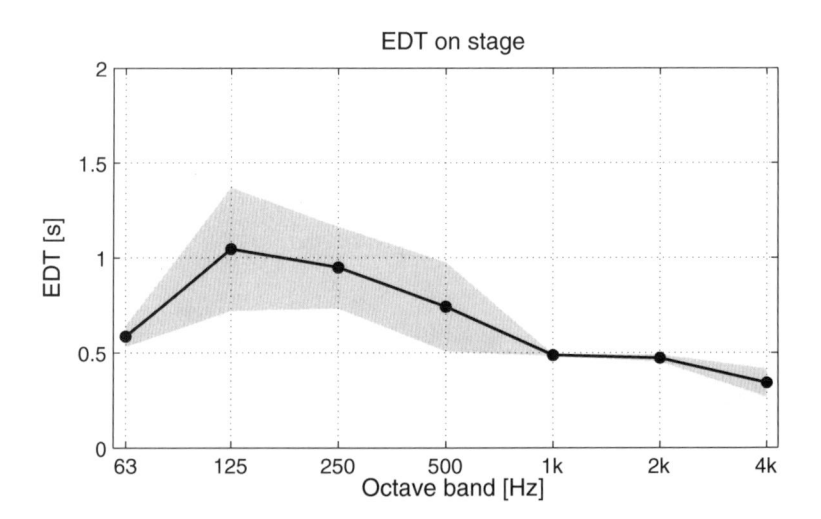

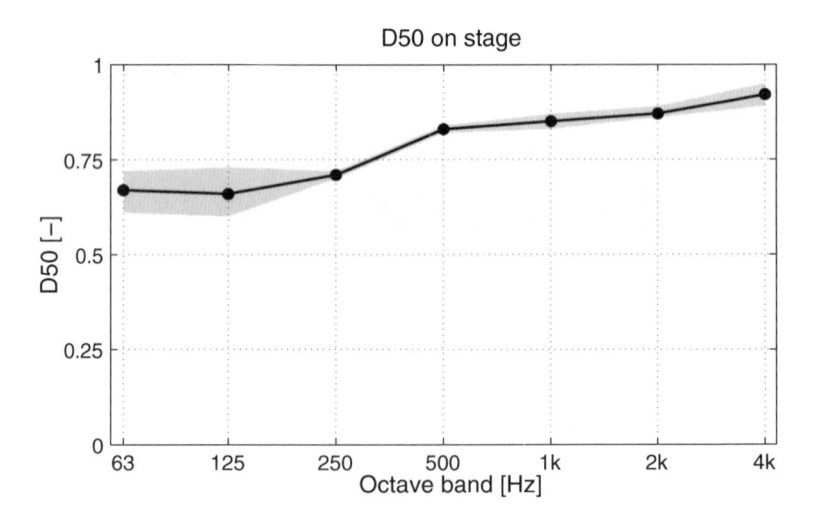

D50 on stage

Zeche Carl "Kaue"

Essen
Number of concerts per year: 60
Founded: 1970
Capacity: 600 (250 seated)
Architect: N/A
Acoustician: N/A

Situated in the north of Essen, Zeche Carl has a long and interesting history. Originally the site of a gas coal mine that opened in 1861, it boasts one of the oldest (and best preserved) Malakov Towers in the region. For a long time, Zeche Carl has been a "prototype" of the structural change witnessed by the entire Ruhr area. The mine closed in 1970, and the site was converted into a cultural center that same decade by an initiative made up of local citizens, youth groups, and the local Protestant congregation. Over the years, and with the backing of agencies including the city of Essen, the center grew to become one of the most significant sociocultural institutions in Germany. Activities were forced to a temporary halt when the organization behind the center went bankrupt, but a new association was found to run it and in the autumn of 2009 the Zeche Carl casino building reopened with a revised concept to continue its work as an intergenerational, cross-nationality forum for the citizens of Essen, and as a nationally influential cultural center for the entire urban area and the region as a whole. The events that take place at Zeche Carl include cabaret, concerts, parties, courses, workshops, readings, exhibitions, and much more, and social institutions and self-help groups also use it as their base.

Zeche Carl. Very typical T30 values across frequency.

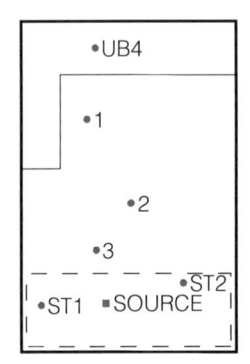

436   7   Gallery of Halls that Present Pop and Rock Music Concerts

Geometrical data	
Volume	Approximately 1,000 m^3
$L \times W \times H$	Approximately $18 \times 12 \times 5$

Acoustical data	
Audience area	
$T_{30, 125-2k}$	0.9
EDT_{125-2k}	0.72
$C_{80, 125-2k}$	5.55
$BR_{63 \text{ versus } 0.5-1k}$	1.97
$BR_{125 \text{ versus } 0.5-1k}$	1.54
Stage area	
EDT_{125-2k}	0.48
$D_{50, 125-2k}$	0.89
$BR_{63 \text{ versus } 0.5-1k}$	2.87
$BR_{125 \text{ versus } 0.5-1k}$	2.6

Materials Used

Audience Area

Floor: Tiles on concrete.
Ceiling: Concrete 20 % of which is covered with felt 1 m from ceiling.
Walls: Concrete.

Stage Area

Floor: Wood on joists.
Ceiling: Suspended mineral wool.
Walls: Backdrop at a 20-cm distance from rear wall.

State of Hall When Measured

Empty; a few chairs etc. on the floor. No measurement on balcony.

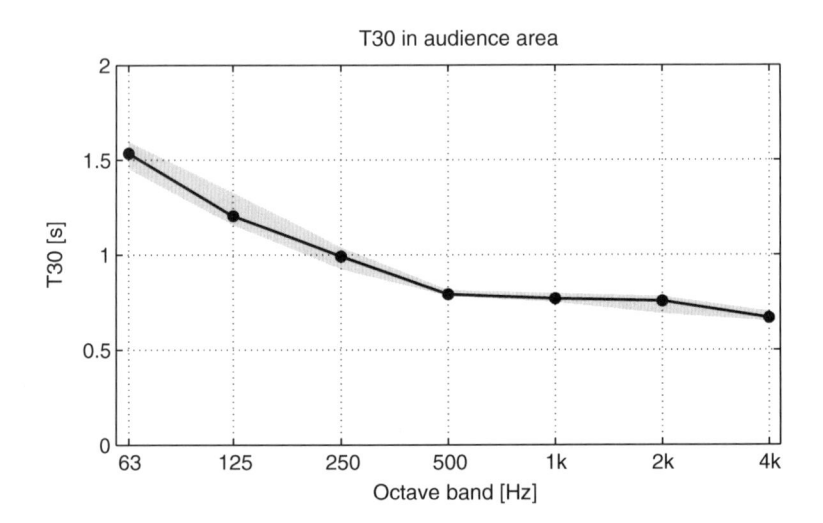

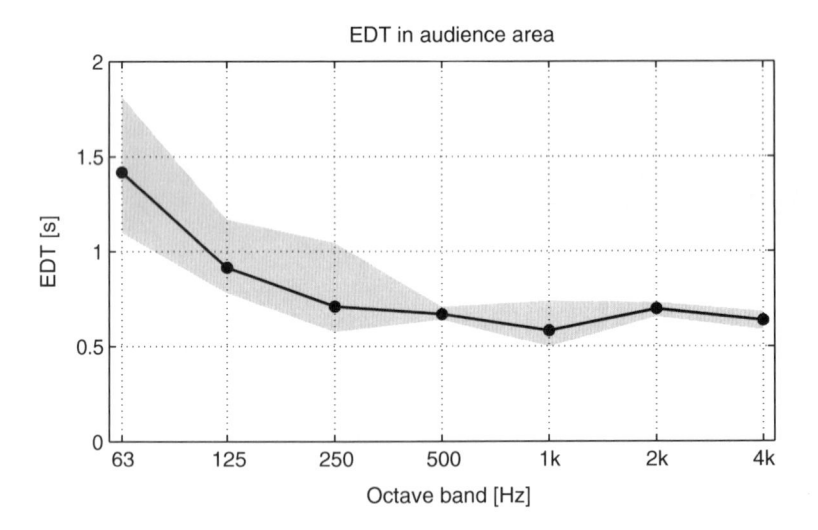

438   7 Gallery of Halls that Present Pop and Rock Music Concerts

T30 in audience area

EDT in audience area

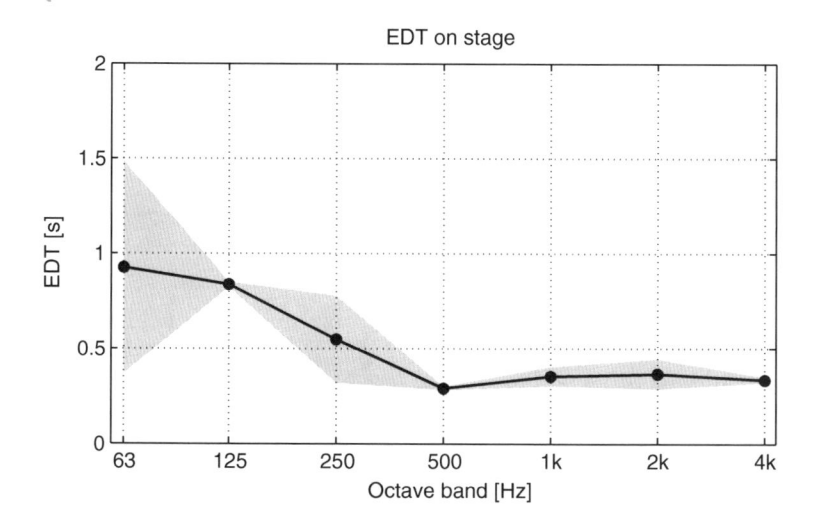

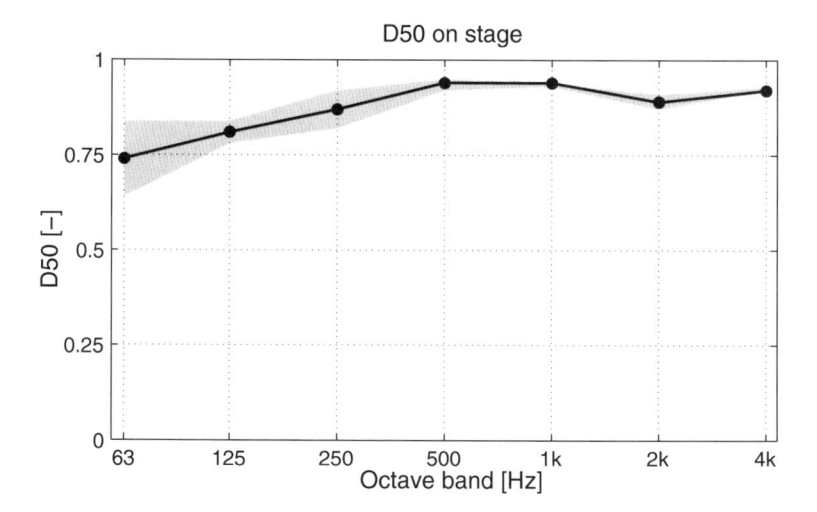

Zénith Paris—La Villette

Paris
Number of concerts per year: Around 110
Number of total events per year: Around 150
Founded: 1984
Capacity: 6,238
Architect: Philippe Chaix et Jean Paul Morel
Acoustician: N/A

In 1981, Jack Lang opened the doors of the ministry of culture to rock and also to pop music.

The politicians of the ministry together with professionals and artists, introduced the richness of today's music. The ZENITH burst forth from an idea, like all good ideas: conceiving a special room adapted for these types of music. It was brought into being with a theatrical professional, Daniel Colling (assisted by Daniel Keravec) and two architects, Philippe Chaix and Jean Paul Morel. They invented the concept of the ZENITH.

The public was not mistaken and immediately shaped the place for music. Recognized from the beginning as a temporary prototype, the ZENITH of Paris, situated on the Parc de la Villette, is still as lively 27 years later as when it first opened. Above all, it is because the concept of the ZENITH is simple and responds to true needs that it was an immediate success. Popular music does not necessarily rhyme with delicacy and the ephemeral. It must satisfy the public who wish, as is normal, ro see well, to hear well, to be well seated, and welcomed, and also to count on theatrical professionals: spectacles that are more and more sophisticated, but also very different from each other. There cannot be stationary rooms for fixed heavy scenic equipment. Rapport with the stage, the acoustic quality, and the movement of a great number of spectators must all be taken into account.

An intriguing thought in the trade has permitted the definition of criteria that guarantee functional usage closer to the demands of the artists and the public: these criteria were the object of a unique contract of its type. The Zenith of Paris has been the prototype for 17 other Zeniths in France.

Every other wall panel (*blue*) behind the audience is reflective; the others are absorptive. All have a thin steel plate on the back. There is nothing in the tent that reflects sound below some 150 Hz.

Zenith is a concept that has proven its success all over France. The size is excellent; there is a great proximity between the performer and even the farthest away audience members. Photo © Jean-Luc Bouchard.

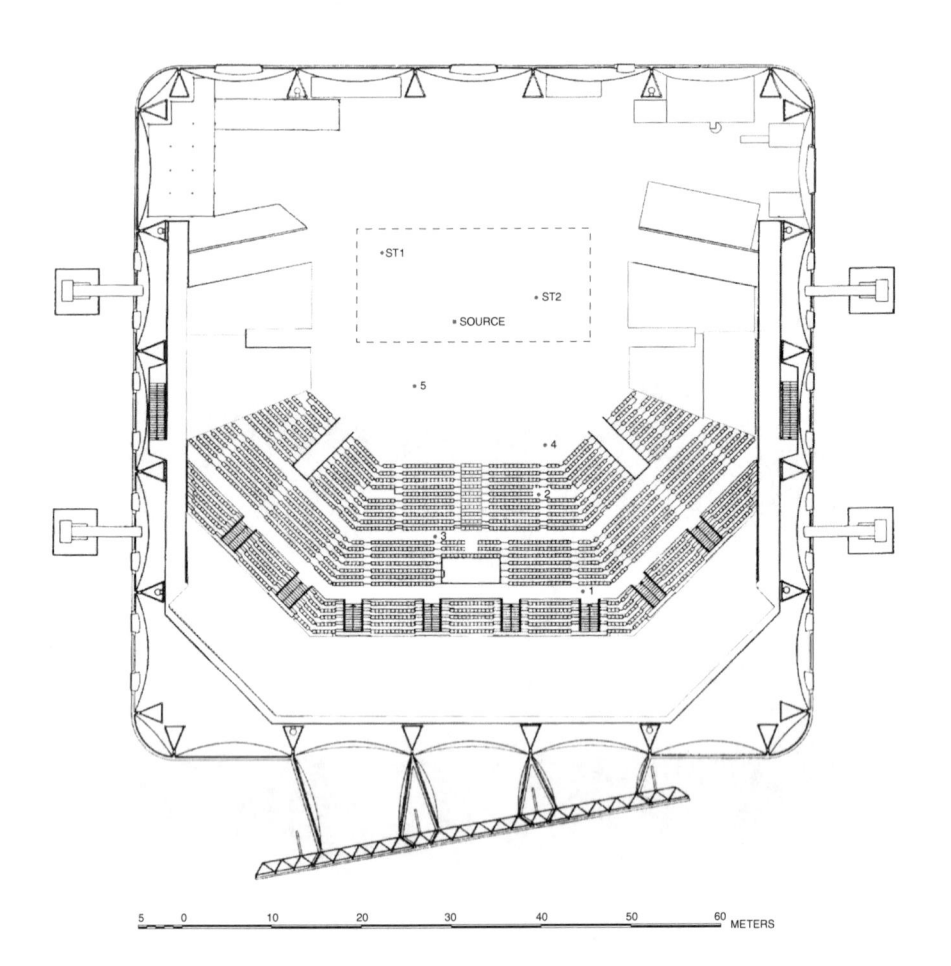

7 Gallery of Halls that Present Pop and Rock Music Concerts

Geometrical data	
Volume	Inside 90,000 m³
$L \times W \times H$	Inside $70 \times 70 \times 19$

Acoustical data	
Audience area	
$T_{30, 125-2k}$	1.8
EDT_{125-2k}	1.68
$C_{80, 125-2k}$	−0.19
BR_{63} versus 0.5–1k	0.82
BR_{125} versus 0.5–1k	1.07

Materials Used

Audience Area

Floor: Concrete.
Ceiling: Inner tent cover with cylinders ($d = $ ~15 cm) of foam stretched across 60 cm.
Walls: Inner walls are panels of perforated thin metal but every second one is blinded. There is a 7-cm cavity with mineral wool and a back plate of thin metal. Behind are the hallways before the outer tent.

Stage Area

Floor: Stage risers.
Ceiling: As audience area.
Walls: None. Backdrop behind stage.
Seats: Plastic not upholstered.

State of Hall When Measured

Empty as on upper photo; instruments on stage.

444   7 Gallery of Halls that Present Pop and Rock Music Concerts

As the Zenith de Paris was opened only as a temporary venue unconventional but quite efficient solutions are found in every corner. Here foam cylinders on wires in the ceiling.

Zenith de Paris is simply a tent. Photo © Jean-Luc Bouchard.

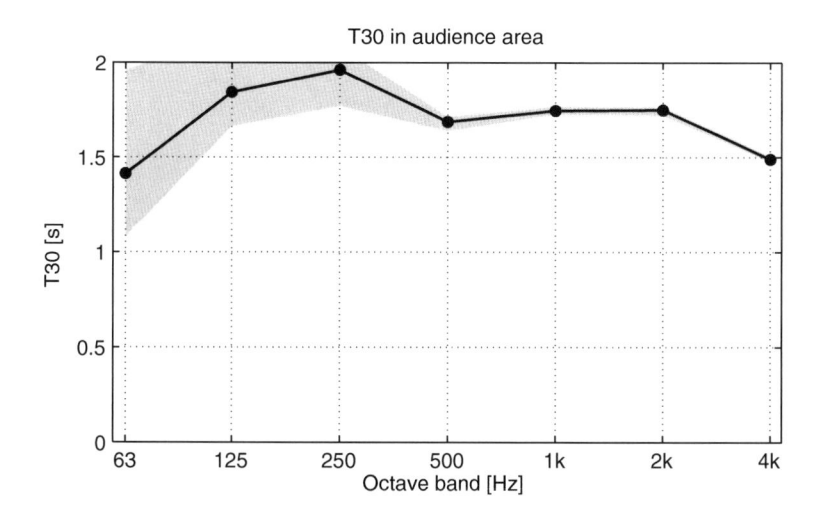

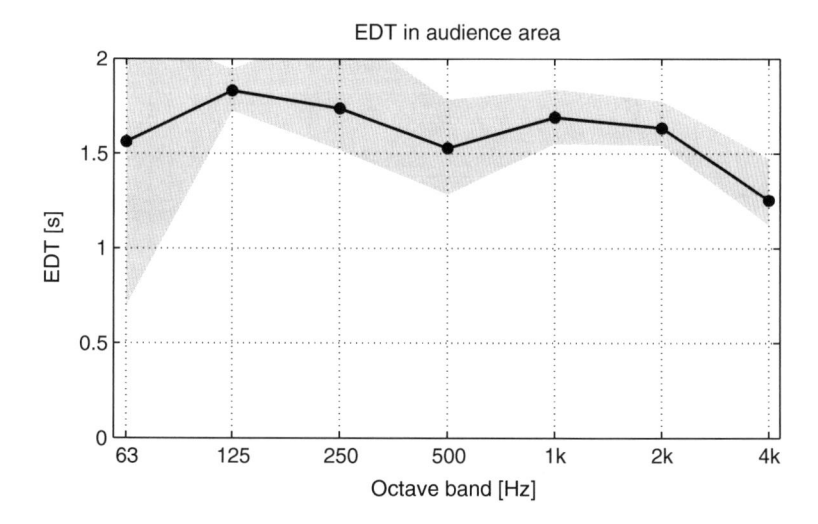

State of Hall When Measured

T30 in audience area

EDT in audience area

Zenith Strasbourg

Strasbourg
Number of concerts per year: 100. Total number of events: 120
Founded: 2008
Capacity: 12,000
Architect: Massimiliano Fuksas
Acoustician: Altia

A missing link in the large scale of event halls existed in the Strasbourg area, and the Zénith Strasbourg became a stopping-place shaped for large musical shows and international artists. The quality of the room, its scenic and acoustic devices, and its dimensions were thus perfectly adapted particularly to Anglo-Saxon artists, who made European Strasbourg a stop on their international tours.

Local and regional producers have also, from the beginning, held the idea of constructing such equipment, offering the public a choice of hospitable conditions, visibility, acoustical comfort and irreproachable security.

The Zénith can accommodate nearly 12,079 spectators. Above all designed to receive a variety of shows and live music, the equipment equally permits holding diverse demonstrations duch as conventions and large sporting events.

Graced with respect tor the environment, construction faithful to the Zenith concept (quality, modularity, functionality) favors durable, easily recyclable, energy-saving materials.

The Starsbourg Zenith is managed as a function of a public service delegation, by the SNC ZENITH DE STRASBOURG, an affiliate of VEGA.

Zenith Strasbourg was in 2011 the newest of all the French Zeniths.

To avoid sound focusing effects because of the wall's concave curvature special diffusive patterns have been applied on wall areas.

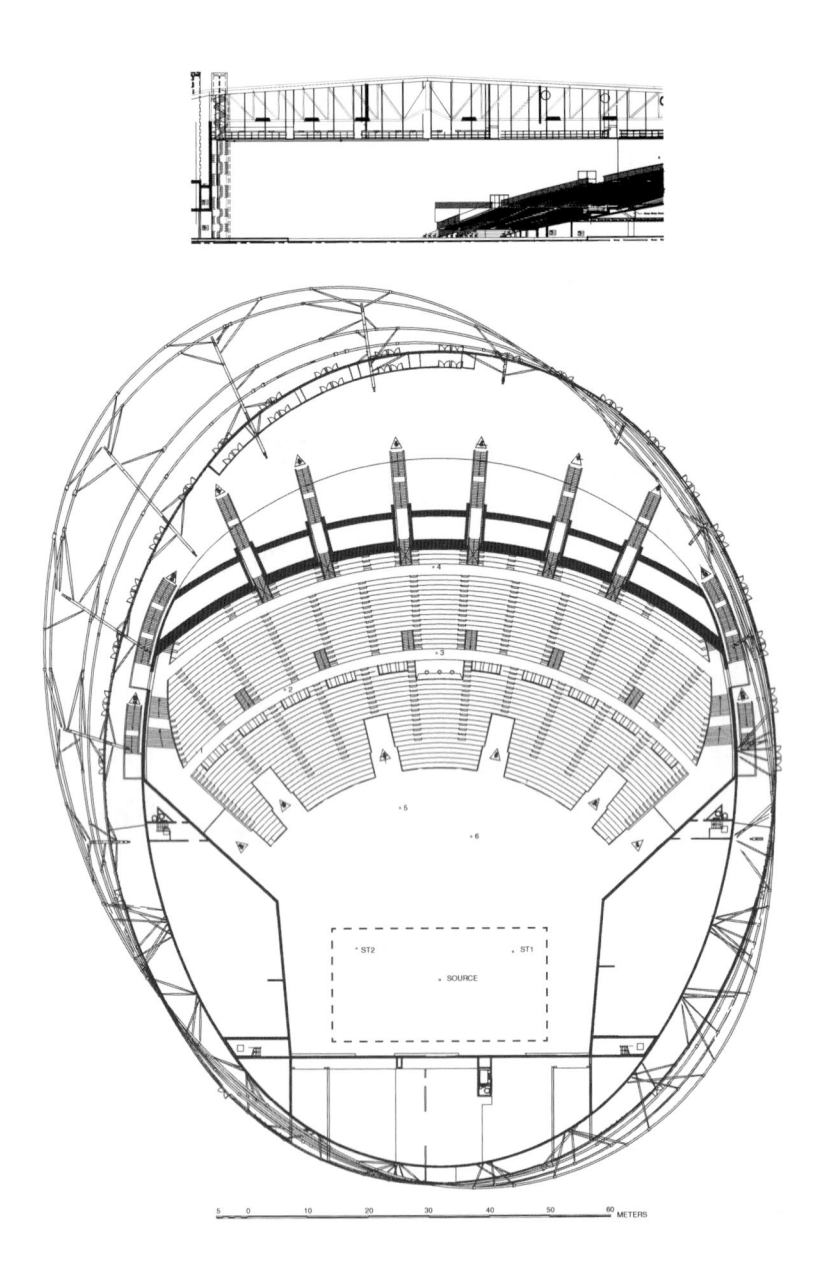

7 Gallery of Halls that Present Pop and Rock Music Concerts

Geometrical data	
Volume	Approximately 85,000 m^3
$L \times W \times H$	$60 \times 60 \times 24$ m

Acoustical data	
Audience area	
$T_{30, 125-2k}$	2.07
EDT_{125-2k}	1.92
$C_{80, 125-2k}$	0.53
$BR_{63 \text{ versus } 0.5-1k}$	1.61
$BR_{125 \text{ versus } 0.5-1k}$	1.28

Materials Used

Audience Area

Floor: Concrete.
Ceiling: Thin trapezoid metal plate.
Walls: 5-cm thick mineral wool direct on concrete; rear wall is diffusive plates.

State of Hall When Measured

Empty as in photo; no additional chairs on floor.

Strasbourg Zenith.

Strasbourg Zenith: "Le grill".

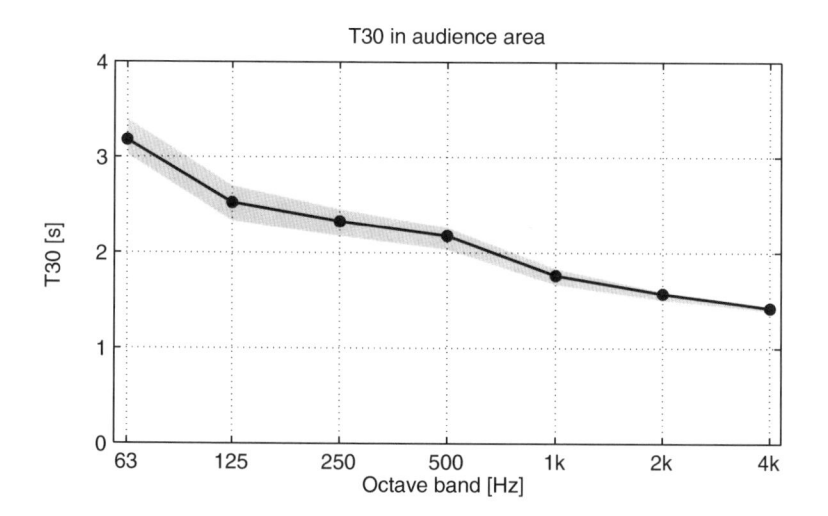

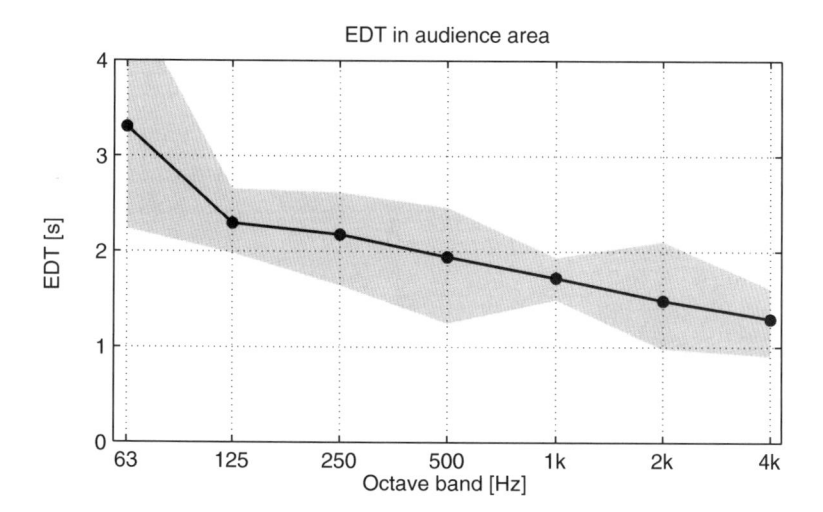

State of Hall When Measured

T30 in audience area

EDT in audience area

Appendix A
Measurements of the 55 Venues Presented in the Gallery in Chapter 7

The aim of this appendix is to document the acoustic behavior of the 55 venues around Europe that were investigated in 2010 and presented in Chap. 7, in terms of what acoustic conditions are achievable and significant differences among the venues. There are no set subjective responses to these halls therefore statistically proven recommendations cannot be made. The venues varied from small clubs to large arenas with volumes up to 600,000 m^3. For pop and rock venues with a volume above 7,000 m^3 there exists no scientific investigation of suitable reverberation times or other parameters. Of the 55 surveyed halls, 31 are larger than 7,000 m^3. Figure 5.7 in this book with recommended reverberation time as a function of bigger volumes originates indeed from the collection of achievable reverberation times in this survey. The results were presented in the scientific paper, "A Survey of Reverberation Times in 50 European Venues Presenting Pop and Rock Concerts," *Forum Acusticum*, 2011, Aalborg, Denmark (Adelman-Larsen and Dammerud 2011). Acoustical and geometrical data of the 55 venues are listed in Appendix A, B and C.

Acoustic conditions using the venues' sound system (those that had one) have not been investigated, in as much as the aim has been to isolate the acoustic responses of the rooms which, as seen, in many cases alone lead to challenging conditions for musicians and sound engineers. A set of objective measures is used to describe the acoustic qualities of the venues. The selection of measures is based on considerations of which characteristics appear relevant. Room acoustic impulse responses were measured using Dirac software installed on a laptop employing a linear sinus sweep of either 5- or 21-s length and a GRAS omnidirectional microphone. The small venues were measured with a Norsonic dodecahedron and a subwoofer, and the larger halls were measured with a d&b PA system consisting of two mid–high frequency speakers (50° vertical and 80° horizontal coverage angle) and two subs, all connected to the same dedicated d&b amp.

Impulse responses of between 5 and 10 relevant microphone positions were recorded in all venues using one constant loudspeaker position approximately at the center of the stage area. In the architectural drawings for each venue presented

Table A.1 Acoustical and geometrical properties of 55 European venues used for pop and rock concerts

Venue	V	L	B	H	r_{max}	H/r_{max}	G_{dmin}	$T_{30,125-2k}$	$EDT_{,125-2k}$	$EDT/T_{30,125-2k}$	$BR_{,125-2k}$	$G_{I,125-2k}$	$G_{dmin}-G_{I,125-2k}$	$Dc_{,125-2k}$	$Dc/r_{max,125-2k}$
1	8500	40.3	18.5	11.7	25	0.47	−8	1.41	1.41	1	1.23	1.8	−9.8	4.4	18
2	8500	32	24	12	21	0.57	−6.4	1.39	1.36	0.98	1.78	1.3	−7.7	4.6	22
3	15000	52	30	9.7	40	0.24	−12	1.47	1.35	0.92	0.89	−0.2	−11.8	5.8	14
4	2800	24	19	6.3	19	0.33	−5.6	0.97	0.97	0.99	0.95	2.7	−8.3	3.1	16
5	1000	15	16	4	17	0.24	−4.6	0.75	0.64	0.85	0.92	3.8	−8.4	2.1	12
6	2800	30	19	4.5	20	0.23	−6	0.89	0.81	0.9	0.99	1.6	−7.6	3.2	16
7	7000	33	19	10.2	22	0.46	−6.8	0.5	0.56	1.11	0.95	−11.3	4.4	6.7	31
8	630	21	11	2.7	21	0.13	−6.4	1.09	0.96	0.88	1.13	10.3	−16.7	1.4	7
9	2800	27	13.5	7.5	17	0.44	−4.6	0.94	0.86	0.92	1.21	2.2	−6.8	3.1	18
10	10000	50	45	13	40	0.33	−12	1.27	1.18	0.93	1.11	0	−12.1	5.1	13
11	1500	25	12	4.8	19	0.25	−5.6	0.72	0.67	0.93	1.8	0.8	−6.4	2.7	14
12	9000	40	18.9	13.4	27	0.5	−8.6	1.09	0.93	0.85	1.27	−1.1	−7.5	5.2	19
13	2400	22.9	15.9	7.3	13	0.56	−2.3	0.73	0.75	1.02	1.35	−0.5	−1.8	3.3	25
14	6000	37	23	7	27	0.26	−8.6	0.97	0.91	0.95	1.34	−0.8	−7.8	4.5	17
15	110000	105	62	29	80	0.36	−18.1	4.15	4.28	1.03	0.91	−1.1	−16.9	9.3	12
16	150000	105	70	26	80	0.33	−18.1	2.46	2.56	1.04	1.05	−5.9	−12.1	14.1	18
17	600000	100	100	85	75	1.13	−17.5	3.71	3.22	0.87	1.67	−9.4	−8.1	23.4	31
18	4200	31.5	24.5	5.4	21	0.26	−6.4	1.48	1.52	1.03	0.75	5.3	−11.7	3.1	15
19	1200	21.9	17.8	3.1	17	0.18	−4.6	0.99	1.04	1.05	0.6	6	−10.6	2	12
20	120000	115	100	25	80	0.31	−18.1	2.48	2.06	0.83	1.1	−4.9	−13.1	12.6	16
21	50000	60	43	20	40	0.5	−12	1.17	1.11	0.95	0.95	−7.7	−4.3	11.8	29
22	20000	58	36	14.4	35	0.41	−10.9	2.35	2.12	0.9	1.03	2.5	−13.4	5.3	15
23	230000	115	80	30	90	0.33	−19.1	2.81	2.92	1.04	0.99	−6.8	−12.3	16.3	18
24	8000	43	23	8.7	31	0.28	−9.8	1.17	1.15	0.98	0.96	0.3	−10.1	4.7	15
25	5000	35	14	6	28	0.21	−8.9	1.15	1.13	0.98	1.33	2	−10.9	3.8	13
26	2600	32	15	5.6	23	0.24	−7.2	0.81	0.78	0.96	0.78	0.6	−7.9	3.2	14
27	250000	115	81	32	85	0.38	−18.6	2.47	2.22	0.9	1.34	−8.2	−10.4	18.3	21

(continued)

Appendix A: Measurements of the 55 Venues Presented in the Gallery in Chapter 7   455

Table A.1 (Continued)

Venue	V	L	B	H	r_{max}	H/r_{max}	G_{dmin}	$T_{30,125-2k}$	EDT_{125-2k}	$EDT/T_{30,125-2k}$	$BR_{,125-2k}$	$G_{l,125-2k}$	$G_{dmin}-G_{l,125-2k}$	$Dc_{,125-2k}$	$Dc/r_{max,125-2k}$
28	500000	150	110	33	100	0.33	−20	2.66	2.49	0.93	1.69	−10.8	−9.2	25.1	25
29	280000	100	95	35	70	0.5	−16.9	2.44	2.03	0.83	1.19	−8.7	−8.2	19.4	28
30	400000	125	115	43	85	0.51	−18.6	2.17	1.83	0.85	1.38	−11.2	−7.3	24.6	29
31	7500	32	22	11	24	0.46	−7.6	1.07	0.88	0.82	1.01	−0.4	−7.2	4.8	20
32	13000	50	20	14.3	40	0.36	−12	1.23	1.1	0.9	1.47	−1.6	−10.4	5.9	15
33	400000	110	105	42	88	0.48	−18.9	4.98	4.4	0.88	1.03	−5.7	−13.2	16.3	19
34	48000	93	32	16	70	0.23	−16.9	1.81	1.67	0.92	1.07	−3.4	−13.5	9.3	13
35	6000	21.5	19	13	15	0.87	−3.5	1.74	1.66	0.96	0.74	5.2	−8.7	3.4	22
36	50000	NaN	NaN	NaN	NaN	NaN	NaN	2.46	2.33	0.95	1.15	−1.2	NaN	8.1	NaN
37	8000	33	26	8.5	25	0.34	−8	1.51	1.43	0.94	1.04	2.8	−10.7	4.2	17
38	2600	31	16	5.3	23	0.23	−7.2	1.68	1.66	0.99	1.01	8.6	−15.9	2.2	10
39	5300	27.6	19.6	9.8	20	0.49	−6	0.95	0.85	0.9	1.36	−0.7	−5.3	4.3	22
40	58000	65	42	21	46	0.46	−13.3	1.47	1.3	0.88	1.73	−6.4	−6.8	11.5	25
41	5000	29	24	8.5	20	0.42	−6	1.08	1.01	0.94	1.47	1	−7	3.9	20
42	1500	21	16	4.3	15	0.29	−3.5	0.96	0.79	0.82	1.21	5.2	−8.8	2.3	15
43	2000	18	16	7.2	13	0.55	−2.3	1.48	1.45	0.98	0.96	8.6	−10.9	2.1	16
44	270000	NaN	NaN	NaN	NaN	NaN	NaN	2.43	2.43	1	1.11	−8.6	NaN	19	NaN
45	150000	73	110	25	60	0.42	−15.6	1.61	1.66	1.03	1.23	−8.8	−6.7	16.3	27
46	700	19	10	3	14	0.21	−2.9	0.94	0.76	0.81	0.94	8.3	−11.2	1.6	11
47	150000	NaN	NaN	20.5	NaN	NaN	NaN	3.53	3.36	0.95	1.11	−3.5	NaN	11.8	NaN
48	150000	NaN	NaN	NaN	NaN	NaN	NaN	2.56	2.41	0.94	1.64	−5.9	NaN	14	NaN
49	5000	29	24	8.5	20	0.42	−6	1.39	1.46	1.05	1.04	4	−10	3.4	17
50	2700	26.7	11.9	9	19	0.47	−5.6	1.13	1.17	1.04	1.23	4.4	−10	2.8	15
51	1000	18	12	5	12	0.42	−1.6	0.9	0.72	0.79	1.54	5.9	−7.5	1.9	16
52	90000	70	70	19	50	0.38	−14	1.8	1.68	0.94	1.07	−6.2	−7.8	12.8	26
53	130000	100	70	24	70	0.34	−16.9	2.07	1.92	0.93	1.28	−6.8	−10.1	14.4	21
54	5800	34	17	10.1	23	0.44	−7.2	1.2	0.97	0.81	0.98	1.9	−9.1	4	17
55	3000							0.64							

in the gallery, ST1 and ST2 are the microphone positions on stage, S denotes the Source position while the measurement point denoted "SE" stands for sound engineer position. B denotes "Balcony" and UB "Under Balcony". All venues were measured without audience but some had temporary chairs set up on the floor area (affecting, for example, measured T at higher frequencies). In some of the large arenas there was no stage set up on the date of the measurement. None of the room responses covered in this work was measured on stage. Although not in all cases fully in compliance with the ISO standard ISO 3382–1:2009, the results are indeed trustworthy. All measured responses were checked during the measuring procedures. In the diagrams of T30, EDT and D50 presented in the gallery for each venue the grey areas denote the minimum and maximum values.

Objective Measures

The priority for this study was to investigate objective measures that assess the acoustic conditions imposed by the room itself, not including a sound system. Such measures can evidently be a background for the consideration of the improvement of the venue's acoustics on the basis of the already found conclusions in Chap. 5. Venues with a high reverberation time at mid–hi frequencies can result in acceptable conditions if a suitable PA system design and precise installation are applied. The following geometric measures were obtained: volume and length, width and height of the venues, denoted as V, L, W, and H. In addition, the maximum distance from the position of the main sound system to a listener within the audience was measured, denoted here as r_{max}. The ratio H/r_{max} was also calculated, because low venues with listeners close to (respectively, far from) the sound system often result in poor conditions.

The following acoustic measures were obtained (125–2,000 Hz): reverberation time T, early decay time EDT, bass ratio BR ($T_{30,125Hz}/T_{30,500-1,000Hz}$), EDT/T_{30}, D_c/r_{max}, G_{late} (G_1), and $G_{d,min} - G_{late}$. D_c is the critical distance and was estimated using Eq. (A.1). G_{late} is the late part of the acoustic measure G (strength) and represents the level of reverberant sound arriving after 80 ms relative to the direct sound. G_{late} was estimated using Eq. (A.2) based on Barron's revised theory (Barron and Lee 1988), and a source–receiver distance r of 15 m. As seen below, G_{late} is affected by T and V. This measure has been found relevant for classical concert hall stages (Dammerud 2009). $G_{d,min}$ is the free-field direct sound level from a point source at the distance r_{max}.

$$D_c = 0.057\sqrt{V/T}, \tag{A.1}$$

$$G_1 = 10 \cdot \log_{10}\left(\frac{31200 \cdot T}{V} \cdot e^{-0.04r/T} \cdot e^{-1.11/T}\right), \tag{A.2}$$

D_c/r_{max} and $G_{d,min}-G_{late}$ are proposed for studying to what degree the reverberant sound may take dominance over the direct sound or on the contrary be inaudible. These measures ignore early reflections.

The smaller halls in this survey are all dedicated to pop and rock, whereas the larger venues are multifunctional and present famous musical acts as well as musicals, sports games, exhibitions, and so on. All halls were chosen from the same criterion: a large number of pop and rock concerts were being held there during the fall of 2010. Of the 55 venues 24 (44 %) are regarded as small with a hall volume within 7,000 m³, and 31 (56 %) are regarded as large with the largest volume being 600,000 m³.

Results

Figure A.1 shows the results for T_{30} versus hall volume for the small and large venues, respectively. In Fig. A.1a the results for three Danish venues also investigated for a specific purpose in Adelman-Larsen and Dammerud (2011) are indicated with shaded diamonds. The dashed lines in Fig. A.1 indicate combinations of T_{30} and V values that result in theoretical values of G_{late} being equal to −3, 4, and 10 dB. Of the small venues, 54 % are within the range of estimated G_{late} of the three Danish venues. For the large venues, estimated values of G_{late} are below −3 dB for 55 % of the venues.

Figure A.2 shows the results for the calculated bass ratio *BR*. The results for the small and large venues are given as circles and squares, respectively. The range of *BR* for the three Danish venues is given as dashed lines in Figs. A.3, A.4, A.5, A.6. From Fig. A.2 we see that a significant portion (47 %) of the venues has higher values of *BR* compared to the three Danish venues. There is no dominance of either small or large venues for high values of *BR*. Four venues show low values of *BR*. These are all small venues.

Figure A.3 shows the results for EDT/T_{30}. The resulting values are generally below 1 but there is no clear difference in the results for the small and large venues. The results for the three Danish venues are among the lowest values.

Figures A.4 and A.5 show the results for D_c/r_{max} and $G_{d,min}-G_{late}$, respectively. The results are similar for these two measures, but for D_c/r_{max} fewer venues have values below and more venues with values above the range for the three Danish venues. Low values of these two measures will indicate dominance of late reverberant sound within a larger portion of the audience area. Again there is no clear difference between small and large venues, except the small venues show the highest values for $G_{d,min}-G_{late}$.

Figure A.6 shows the results for H/r_{max}. Results within the range of the earlier-mentioned three Danish venues are shown in 27 (49 %) venues. The lowest values are dominated by small venues.

Regarding correlations between the objective measures, measured T_{30} is highly correlated with the geometrical measures V, W, L, and H ($r = 0.75$–0.77) and

Appendix A: Measurements of the 55 Venues Presented in the Gallery in Chapter 7

Fig. A.1 Measured reverberation time (T_{30}) as a function of volume (V) of **a** small and **b** large venues

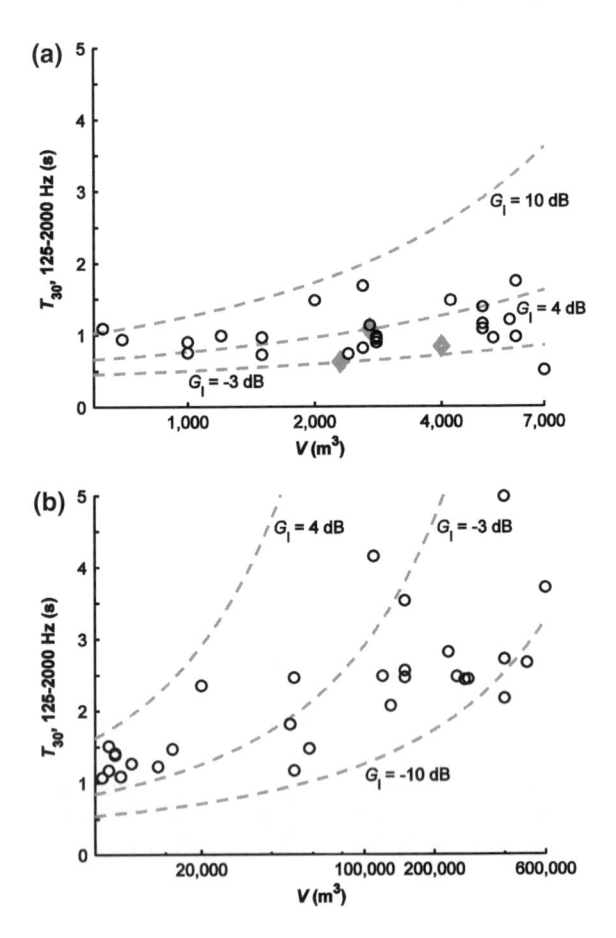

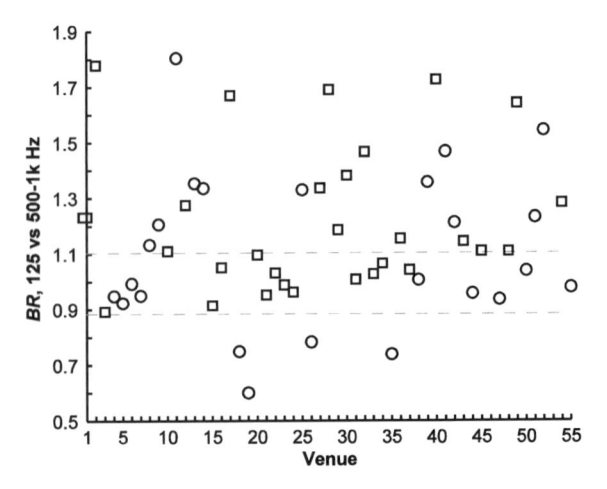

Fig. A.2 Calculated *BR* for the 55 venues

Appendix A: Measurements of the 55 Venues Presented in the Gallery in Chapter 7   459

Fig. A.3 Calculated EDT/T_{30} for the 55 venues

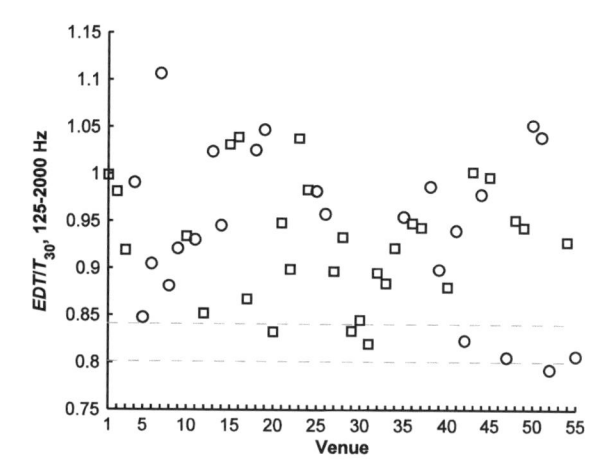

Fig. A.4 Estimated D_c/r_{max} for the 55 venues

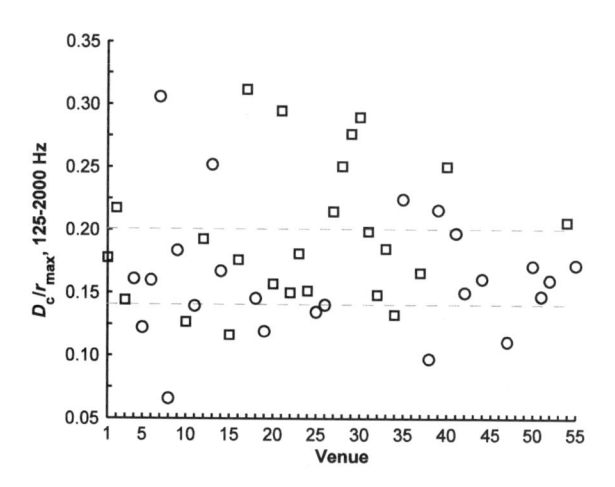

Fig. A.5 Estimated $G_{d,min} - G_{late}$ for the 55 venues

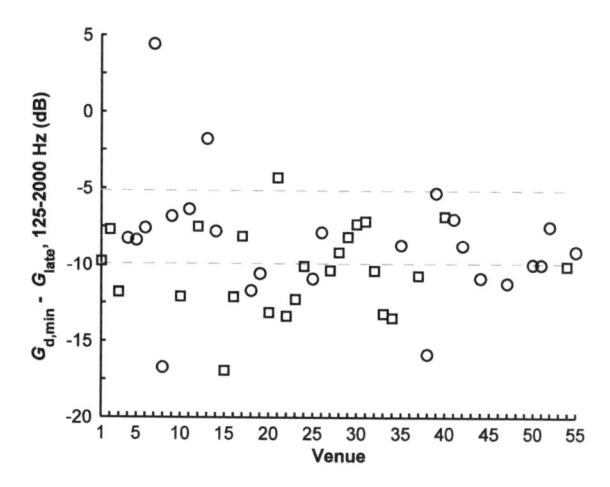

Fig. A.6 Calculated H/r_{max} for the 55 venues

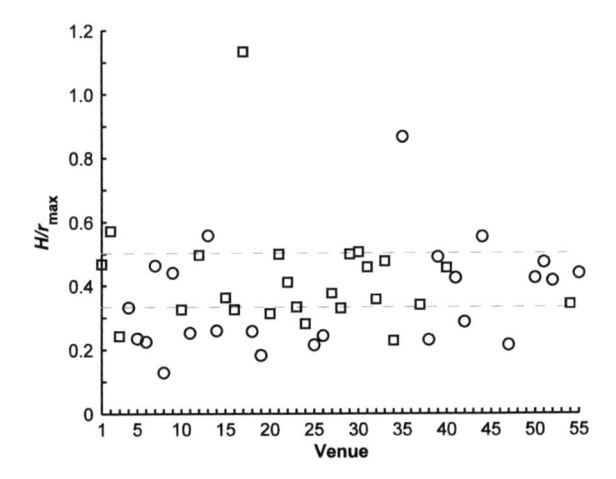

EDT ($r = 0.99$). $G_{d,min}$–G_{late} is not highly correlated with any of the other objective measures apart from T_{30} and EDT ($r = -0.47$ to -0.49). G_l and $G_{d,min}$–G_{late} are not highly correlated ($r = -0.35$), whereas D_c/r_{max} and $G_{d,min}$–G_{late} are moderately correlated ($r = 0.68$). H/r_{max} is not highly correlated with any of the acoustic measures apart from H ($r = 0.60$). All correlations were significant at the 1 % level.

Discussion and Conclusions

For all the objective measures there are significant variations for the venues studied. The variations within the three Danish venues are clearly smaller compared to the 55 venues. The low correlations between the acoustic measures suggest that none of them are clearly redundant, although $G_{d,min}$–G_{late} and D_c/r_{max} show similar but not highly correlated results.

A significant portion of the large venues shows low values of G_{late}, but this may need to be seen in relation to direct sound levels assessed by D_c/r_{max} and $G_{d,min}$–G_{late}. Even a low level of G_{late} is likely to result in unsuitable conditions if the direct sound levels are low (for the large venues).

One 50,000 m³ hall with a T_{30} of only 1.3 s was reported acoustically too dead by sound engineers, a lack of the feeling of being enveloped in sound. All surfaces except the floor are absorptive. In some very large arenas of 2–500,000 m³ impressively low T_{30} around 2 s have been reached though higher at 125 Hz.

Apparently the reverberation time can be too short for both small and large venues, but due to high values of r_{max} very low values of T_{30} may be preferable for the volumes above some 100,000 m³. Even more precise speaker coverage is then necessary because there will be little reflected sound in nondirect sound areas. For volumes below approximately 100,000 m³ and audience capacity below approximately 10,000 listeners, it appears appropriate with a somewhat airy and lively sound. In large arenas a colossal amount of chairs are permanently installed. If these seats are not upholstered the presence of an audience has an enormous impact on the reverberation time. Therefore, recommendations of acoustical measures for that type of venue could be with an occupied audience area.

The results for *BR* suggest that several venues have high values of T_{30} at 125 Hz that will represent a problem. A significant amount of the venues may also suffer from a too-low ceiling compared to the maximum distance to a listener. Regarding EDT/T_{30} a majority of the venues have a value below one. This can indicate that early reflections are not directed towards the listeners. This may be beneficial to avoid strong early reflections from the sound system. As mentioned earlier the reputation of how a venue sounds is not only dependent on beneficial acoustic conditions but also a sound system that suits the geometry and acoustics of the venue. Examples of critical aspects here will be point source versus line array, direct sound interference, directivity, and delay zones. Some European arenas taking part in this survey consistently and independently reported how certain acts were poor soundwise, and other excellent partly because they set up their own PA systems. Other acts were reported to refuse to hook up to preinstalled delay speakers.

Acousticians will never be in a position to ensure perfect PA coverage at every concert inasmuch as mentioned above, that is not even in the hands of the arenas. Acceptable range of T_{30} should in any case be used by the acoustician to achieve what acoustical environment is aimed for primarily by the hall owner who is in charge of the musical program of the venue, or to support the architects in their design ideas for the venue.

References

N.W. Adelman-Larsen and J.J. Dammerud: A survey of reverberation times in 50 European venues presenting pop and rock concerts, *Forum Acusticum*, Aalborg, Denmark (2011).

M. Barron and L.-J. Lee: Energy relations in concert auditoriums, *I. J. Acoust. Soc. Am.* **84** (1988) 618–628.

J.J. Dammerud: "Stage Acoustics for Symphony Orchestras in Concert Halls." PhD thesis, University of Bath, England (2009).

Appendix B

Table B.1 Measured objective acoustical parameters as a function of octave band

Hall	Name	f [HZ]	63	125	250	500	1,000	2,000	4,000
1	AB	T_{30} [s]	2.33	1.66	1.43	1.36	1.33	1.28	1.32
		EDT [s]	1.5	1.43	1.47	1.4	1.4	1.35	1.33
		C_{80} dB	0.87	0.72	1.11	1.49	1.98	2.14	2.12
2	l'Aeronef	T_{30} [s]	2.96	2.07	1.46	1.21	1.12	1.08	0.97
		EDT [s]	2.85	1.71	1.56	1.28	1.12	1.14	1.01
		C_{80} dB	−1.69	1.24	0.07	3.21	3.34	3.15	4.12
3	Alcatraz	T_{30} [s]	1.48	1.38	1.36	1.51	1.57	1.52	1.27
		EDT [s]	1.25	0.95	1.44	1.34	1.53	1.5	1.28
		C_{80} dB	1.18	3.76	1.09	−0.05	0.67	1.04	1.06
4	Apolo	T_{30} [s]	0.9	0.93	0.94	0.9	1.05	1.06	0.96
		EDT [s]	1.24	0.87	0.91	0.91	1.06	1.08	0.92
		C_{80} dB	2.43	5.34	3.87	3.76	2.73	2.37	3.44
5	Apolo la [2]	T_{30} [s]	0.83	0.71	0.71	0.73	0.8	0.82	0.74
		EDT [s]	0.51	0.6	0.7	0.64	0.63	0.62	0.57
		C_{80} dB	5.71	7.04	6.17	6.68	7.36	7.26	8.08
6	Astra	T_{30} [s]	0.87	0.84	0.98	0.88	0.81	0.94	0.93
		EDT [s]	0.56	0.76	0.84	0.82	0.77	0.84	0.82
		C_{80} dB	5.4	6.14	5.78	5.57	6.31	5.08	5.6
7	Bikini	T_{30} [s]	0.37	0.48	0.53	0.52	0.49	0.5	0.47
		EDT [s]	0.6	0.76	0.49	0.48	0.54	0.52	0.5
		C_{80} dB	4.03	8.1	11.34	13.31	12.05	12.23	12.73
8	The Cavern Club	T_{30} [s]	1.17	1.26	1.19	1.17	1.05	0.78	0.66
		EDT [s]	1.21	1.58	1	0.84	0.74	0.65	0.57
		C_{80} dB	0.02	0.23	3.03	4.79	5.43	6.56	7.89
9	Le Chabada	T_{30} [s]	1.33	1.05	0.97	0.81	0.92	0.95	0.8
		EDT [s]	1.27	1.1	0.97	0.76	0.71	0.78	0.57
		C_{80} dB	−1.09	−0.18	2.35	4.7	5.25	5.32	7.53
10	Cirkus	T_{30} [s]	1.63	1.4	1.33	1.3	1.22	1.09	0.91
		EDT [s]	1.07	1.09	1.23	1.27	1.25	1.07	0.93
		C_{80} dB	4.4	3.01	3.56	3.06	3.1	3.89	5.17

(continued)

Table B.1 (continued)

Hall	Name	f [HZ]	63	125	250	500	1,000	2,000	4,000
11	Le Confort Moderne	T_{30} [s]	1.2	1.09	0.75	0.61	0.59	0.58	0.57
		EDT [s]	1.35	0.99	0.64	0.65	0.53	0.55	0.47
		C_{80} dB	−0.09	2.88	6.09	7.36	8.71	8.12	10.09
12	La Coopérative de Mai	T_{30} [s]	1.82	1.29	1.03	0.95	1.07	1.11	0.94
		EDT [s]	1.64	1.06	1	0.82	0.9	0.87	0.74
		C_{80} dB	0.07	1.71	4.02	5.86	5.47	5.08	6.58
13	Debaser Medis	T_{30} [s]	0.95	0.9	0.76	0.66	0.67	0.69	0.65
		EDT [s]	0.95	0.86	0.76	0.7	0.71	0.73	0.63
		C_{80} dB	1.52	2.26	4.78	5	6.41	5.33	6.68
14	Elysée Montmartre	T_{30} [s]	1	1.19	1.01	0.89	0.89	0.86	0.76
		EDT [s]	1.04	0.84	0.93	0.97	0.94	0.89	0.79
		C_{80} dB	2.36	2.74	3.23	4.39	4.34	5.4	6.25
15	Festhalle	T_{30} [s]	3.22	3.96	4.75	4.46	4.21	3.36	2.37
		EDT [s]	2.68	4.11	4.96	4.67	4.18	3.47	2.37
		C_{80} dB	−3.27	−4.84	−5.99	−3.01	−4.04	−2.4	1.77
16	Mediolanum Forum	T_{30} [s]	3.19	2.67	2.35	2.54	2.54	2.23	1.63
		EDT [s]	3.87	2.55	2.91	2.68	2.46	2.21	1.5
		C_{80} dB	−6.95	−1.7	−5.14	−10.21	−12.72	−15.12	−14.18
17	Globe Arenas	T_{30} [s]	5.97	5.28	4.09	3.27	3.06	2.84	2.27
		EDT [s]	4.71	3.43	2.86	3.53	3.39	2.87	2.16
		C_{80} dB	−4.21	−4.13	0.56	2.06	0.29	2.57	5.81
18	Grosse Freiheit	T_{30} [s]	1.4	1.22	1.28	1.54	1.72	1.64	1.38
		EDT [s]	1.37	1.33	1.27	1.5	1.81	1.67	1.38
		C_{80} dB	1.19	−1.64	−1.91	−1.21	−2.32	−1.67	−1.02
19	Kaiser Keller	T_{30} [s]	0.71	0.66	0.81	0.97	1.24	1.26	1.08
		EDT [s]	0.63	0.92	0.87	1.12	1.16	1.11	0.96
		C_{80} dB	5.54	8.67	6.56	4.31	2.43	1.94	2.67
20	Hallen Stadion	T_{30} [s]	3.66	2.75	2.49	2.52	2.5	2.14	1.45
		EDT [s]	2.8	2.16	2.07	2.06	2.09	1.94	1.55
		C_{80} dB	−3.71	−4.62	1.22	0.24	−0.16	0.41	3.7
21	Heineken Hall	T_{30} [s]	2.24	1.13	1.23	1.21	1.16	1.12	0.96
		EDT [s]	1.72	1.17	1.17	1.19	1.09	0.93	0.88
		C_{80} dB	0.45	2.13	0.46	4.12	4	5.02	6.54
22	HMV Hammersmith Apollo	T_{30} [s]	2.34	2.47	2.3	2.41	2.39	2.21	1.8
		EDT [s]	2.68	1.89	2.06	2.28	2.28	2.08	1.77
		C_{80} dB	−1.4	−0.84	−1.41	−1.1	0.01	0.07	1.32
23	Jyske Bank Boxen	T_{30} [s]	2.76	2.77	2.87	2.75	2.86	2.82	2.18
		EDT [s]	2.43	2.87	2.75	2.91	3.06	3.01	1.55
		C_{80} dB	−0.19	−0.6	2.03	2.97	0.67	1.68	3.59
24	Live Music Club	T_{30} [s]	0.89	1.14	1.16	1.15	1.22	1.2	1.14
		EDT [s]	0.88	1	1.08	1.13	1.28	1.28	1.16
		C_{80} dB	3.43	2.91	2.41	3.41	2.37	2.42	3.28
25	LKA Langhorn	T_{30} [s]	1.53	1.43	1.03	1.05	1.11	1.14	0.96
		EDT [s]	1.49	1.38	1	1.02	1.08	1.17	0.91
		C_{80} dB	−1.22	−0.28	3.45	4.29	4.33	3.52	4.8
26	Melkweg	T_{30} [s]	0.67	0.69	0.76	0.84	0.91	0.87	0.82
		EDT [s]	0.72	0.72	0.7	0.77	0.87	0.84	0.79
		C_{80} dB	8.16	3.17	6.55	6.06	7.27	7.1	7.38

(continued)

Appendix B   465

Table B.1 (continued)

Hall	Name	f [HZ]	63	125	250	500	1,000	2,000	4,000
27	MEN Arena	T_{30} [s]	3.38	3.09	2.6	2.3	2.32	2.06	1.53
		EDT [s]	2.48	2.37	2.3	2.17	2.24	2.01	1.56
		C_{80} dB	−2.23	−3.1	1.11	0.26	−1.31	1.04	2.62
28	o2 World Hamburg	T_{30} [s]	5.84	3.83	2.71	2.21	2.32	2.25	1.9
		EDT [s]	4.61	2.97	2.6	2.31	2.31	2.24	1.84
		C_{80} dB	−4.11	−6.24	−5.39	−1.31	−0.27	−0.26	0.11
29	o2 World Berlin	T_{30} [s]	3.94	2.74	2.6	2.32	2.31	2.23	1.76
		EDT [s]	2.93	1.99	1.88	2.09	2.24	1.96	1.64
		C_{80} dB	−4.21	−2.35	−1.26	0.21	−0.34	0.48	1.78
30	o2 Arena London	T_{30} [s]	4.14	2.77	2.15	2.01	2	1.92	1.62
		EDT [s]	3.51	1.85	1.61	1.76	2.06	1.89	1.63
		C_{80} dB	−1.6	−0.28	0.55	−3.22	−4.88	−5.63	−4.95
31	O13 Tilburg	T_{30} [s]	1.87	1.09	1	1.06	1.1	1.11	1
		EDT [s]	1.24	1.1	0.8	0.71	0.88	0.9	0.78
		C_{80} dB	0.27	2.5	5.65	6.29	5.3	4.71	5.38
32	Olympia	T_{30} [s]	1.79	1.61	1.37	1.15	1.04	0.97	0.82
		EDT [s]	1.78	1.32	1.38	1.04	0.89	0.87	0.71
		C_{80} dB	−0.82	1.27	2.34	4.59	4.74	5.53	6.59
33	Palau Sant Jordi	T_{30} [s]	3.93	5.3	5.56	5.57	4.75	3.73	2.85
		EDT [s]	3.53	4.03	5.04	4.92	4.47	3.57	2.52
		C_{80} dB	−4.7	−6.41	−3.86	−3.82	−5.09	−3.9	−3.64
34	Sala Barcelona	T_{30} [s]	1.43	1.88	1.99	1.84	1.69	1.64	1.4
		EDT [s]	1.55	2.06	1.49	1.75	1.55	1.5	1.27
		C_{80} dB	3.55	1.16	1.42	1.02	2.05	2.6	3.16
35	Paradiso	T_{30} [s]	1.09	1.41	1.56	1.88	1.96	1.9	1.57
		EDT [s]	1	1.41	1.45	1.71	1.91	1.84	1.54
		C_{80} dB	0.6	−2.29	−1.55	−2.34	−2.87	−2.65	−1.89
36	Porsche Arena	T_{30} [s]	3.9	2.78	2.56	2.44	2.38	2.15	1.66
		EDT [s]	2.71	2.3	2.41	2.55	2.28	2.14	1.59
		C_{80} dB	−2.56	−2.98	−3.08	−3.73	−2.01	−1.09	0.01
37	Razzmatazz	T_{30} [s]	1.26	1.59	1.55	1.56	1.5	1.36	1.11
		EDT [s]	1.38	1.52	1.4	1.41	1.46	1.34	1.12
		C_{80} dB	−0.52	−0.12	−0.33	1.01	0.85	1.3	2.69
38	Razzmatazz 2	T_{30} [s]	1.74	1.7	1.8	1.72	1.66	1.55	1.27
		EDT [s]	1.54	1.91	1.36	1.74	1.7	1.6	1.26
		C_{80} dB	−1.64	0	−1.19	0.06	−0.72	−0.77	1.53
39	Rockefeller	T_{30} [s]	0.99	1.2	1.08	0.95	0.81	0.7	0.55
		EDT [s]	0.77	0.95	1.01	0.92	0.75	0.64	0.51
		C_{80} dB	3.53	2.29	4.06	5.07	6.94	8.53	11.5
40	Rockhal	T_{30} [s]	3.89	2.21	1.36	1.25	1.32	1.23	1.1
		EDT [s]	3.4	1.61	1.12	1.22	1.32	1.22	1.04
		C_{80} dB	−3.96	−0.91	0.87	1.06	1.75	1.98	1.97
41	Rote Fabrik	T_{30} [s]	1.3	1.43	1.23	1.06	0.89	0.79	0.67
		EDT [s]	1.28	1.33	1.18	0.98	0.85	0.73	0.63
		C_{80} dB	−1.03	0.57	1.13	2.6	4.19	6.09	7.03
42	Rote Fabrik 2	T_{30} [s]	1.28	1.09	1.01	0.88	0.92	0.91	0.76
		EDT [s]	1.19	1.02	0.77	0.7	0.7	0.78	0.64
		C_{80} dB	0.31	3.67	5.21	6.96	6.85	5.81	7.51

(continued)

Table B.1 (continued)

Hall	Name	f [HZ]	63	125	250	500	1,000	2,000	4,000
43	Scala	T_{30} [s]	1.46	1.46	1.4	1.51	1.54	1.49	1.37
		EDT [s]	1.64	1.39	1.45	1.44	1.51	1.46	1.25
		C_{80} dB	0.82	0.81	−0.48	1.77	0.91	1.43	2.36
44	Hans Martin Schleyer	T_{30} [s]	2.35	2.64	2.55	2.27	2.49	2.21	1.57
	Halle	EDT [s]	3.08	2.72	2.46	2.13	2.59	2.23	1.66
		C_{80} dB	−1.24	−5.19	1.16	1.75	−1.3	−1.9	−0.57
45	Oslo Spektrum Arena	T_{30} [s]	2.56	1.92	1.69	1.59	1.53	1.33	1.08
		EDT [s]	2.55	1.96	1.61	1.75	1.6	1.37	1.18
		C_{80} dB	−0.15	0.42	0.91	−0.55	1.57	2	2.04
46	Tunnel	T_{30} [s]	0.78	0.89	0.98	0.97	0.93	0.94	0.88
		EDT [s]	0.57	0.71	0.66	0.79	0.83	0.8	0.76
		C_{80} dB	6.18	6	5.03	5.93	5.72	5.79	6.41
47	Forest National	T_{30} [s]	4.75	3.95	3.58	3.68	3.45	2.98	2.46
		EDT [s]	4.57	3.33	2.93	3.81	3.6	3.1	2.63
		C_{80} dB	−2.97	0.57	−1.38	−3.43	−3.58	−3.47	−2.52
48	Wembley Arena	T_{30} [s]	4.68	3.66	2.5	2.08	2.38	2.16	1.97
		EDT [s]	4.51	3.06	2.35	2.21	2.42	2.04	1.87
		C_{80} dB	−4.2	−7.39	−6.78	−1.56	−1.44	−0.2	−0.51
49	Werk Backstage	T_{30} [s]	1.32	1.42	1.28	1.23	1.51	1.5	1.33
		EDT [s]	1.6	1.44	1.26	1.34	1.66	1.59	1.39
		C_{80} dB	−1.65	−0.93	0.85	0.84	−0.52	−1.01	0.03
50	Zeche	T_{30} [s]	1.21	1.32	1.25	1.13	1.01	0.94	0.82
		EDT [s]	1.17	1.51	1.28	1.12	1.03	0.93	0.8
		C_{80} dB	2.42	−1.17	−1.43	0.87	1.57	2.39	3.11
51	Zeche Carl	T_{30} [s]	1.54	1.2	0.99	0.79	0.77	0.76	0.67
		EDT [s]	1.42	0.92	0.71	0.67	0.58	0.7	0.64
		C_{80} dB	3.23	2.27	5.06	6.22	7.71	6.5	6.55
52	Zenith Paris	T_{30} [s]	1.41	1.84	1.96	1.69	1.75	1.75	1.49
		EDT [s]	1.56	1.83	1.74	1.53	1.69	1.63	1.25
		C_{80} dB	1.25	−1.82	−2.68	1.43	0.69	1.42	3.1
53	Zenith Strasbourg	T_{30} [s]	3.18	2.52	2.33	2.17	1.76	1.57	1.42
		EDT [s]	3.31	2.29	2.17	1.94	1.72	1.49	1.3
		C_{80} dB	−2	−3.31	−1.37	2.21	1.51	3.6	5.69
54	Vega	T_{30} [s]	1.6	1.17	1.16	1.15	1.23	1.3	1.02
		EDT [s]	1.52	0.97	0.93	0.98	0.98	0.99	0.82
		C_{80} dB	−0.22	3.01	4.82	5.92	6.27	4.83	5.54
55	Nosturi	T_{30} [s]	1.34	1.02	0.64	0.54	0.54	0.48	0.41
		EDT [s]	NaN	NaN	NaN	NaN	NaN	NaN	NaN
		C_{80} dB	NaN	NaN	NaN	NaN	NaN	NaN	NaN

Appendix C

Table C.1 Averaged acoustical parameters, capacity and volumes of the measured halls

Hall	Name	Capacity persons	Volume m^3	T$_{30}$, 125–2k [s]	EDT, 125–2k [s]	C$_{80}$, 125–2k dB	BR, 63vs0, 5–1k	BR, 125vs0, 5–1k
1	AB	2,000	8,500	1.41	1.41	1.49	1.73	1.23
2	l'Aeronef	2,000	8,500	1.39	1.36	2.20	2.54	1.78
3	Alcatraz	3,000	1,5000	1.47	1.35	1.30	0.96	0.89
4	Apolo	1,200	2,800	0.98	0.97	3.61	0.92	0.95
5	Apolo la [2]	400	1,000	0.75	0.64	6.90	1.09	0.92
6	Astra	1,000	2,800	0.89	0.81	5.78	1.03	0.99
7	Bikini	1,500	7,000	0.5	0.56	11.41	0.72	0.95
8	The Cavern Club	350	630	1.09	0.96	4.01	1.06	1.13
9	Le Chabada	900	2,800	0.94	0.86	3.49	1.54	1.21
10	Cirkus	1,800	10,000	1.27	1.18	3.32	1.29	1.11
11	Le Confort Moderne	700	1,500	0.72	0.67	6.63	1.99	1.8
12	La Coopérative de Mai	1,500	9,000	1.09	0.93	4.43	1.8	1.27
13	Debaser Medis	950	2,400	0.73	0.75	4.76	1.44	1.35
14	Elysée Montmartre	1,200	6,000	0.97	0.91	4.02	1.13	1.34
15	Festhalle	13,500	110,000	4.15	4.28	–4.06	0.74	0.91
16	Mediolanum Forum	11,000	150,000	2.46	2.56	–8.98	1.26	1.05
17	Globe Arenas	16,000	6,00,000	3.71	3.22	0.27	1.89	1.67
18	Grosse Freiheit	1,250	4,200	1.48	1.52	–1.75	0.86	0.75
19	Kaiser Keller	400	1,200	0.99	1.04	4.78	0.64	0.6
20	Hallen Stadion	13,000	1,20,000	2.48	2.07	–0.58	1.46	1.1
21	Heineken Hall	5,500	50,000	1.17	1.11	3.15	1.89	0.95
22	HMV Hammersmith Apollo	5,039	2,0000	2.35	2.12	–0.65	0.98	1.03
23	Jyske Bank Boxen	15,000	230,000	2.81	2.92	1.35	0.98	0.99
24	Live Music Club	1,500	8,000	1.17	1.15	2.70	0.75	0.96
25	LKA Langhorn	1,500	5,000	1.15	1.13	3.06	1.42	1.33
26	Melkweg	1,500	2,600	0.81	0.78	6.03	0.76	0.78

(continued)

Table C.1 (continued)

Hall	Name	Capacity persons	Volume m³	T₃₀, 125–2k [s]	EDT, 125–2k [s]	C₈₀, 125–2k dB	BR, 63vs0, 5–1k	BR, 125vs0, 5–1k
27	MEN Arena	21,000	2,50,000	2.47	2.22	−0.40	1.46	1.34
28	o2 World Hamburg	16,000	5,00,000	2.66	2.49	−2.69	2.58	1.69
29	o2 World Berlin	17,000	2,80,000	2.44	2.03	−0.65	1.7	1.19
30	o2 Arena London	20,000	400.000	2.17	1.83	−2.69	2.06	1.38
31	O13 Tilburg	2,000	7,500	1.07	0.88	4.89	1.73	1.01
32	Olympia	2,200	13,000	1.23	1.1	3.69	1.63	1.47
33	Palau Sant Jordi	17,960	400.000	4.98	4.4	−4.62	0.76	1.03
34	Sala Barcelona	4,600	48,000	1.81	1.67	1.65	0.81	1.07
35	Paradiso	1,500	6,000	1.74	1.66	−2.34	0.57	0.74
36	Porsche Arena	6,000	50,000	2.46	2.33	−2.58	1.62	1.15
37	Razzmatazz	1,500	8,000	1.51	1.43	0.54	0.82	1.04
38	Razzmatazz 2	700	2,600	1.68	1.66	−0.52	1.03	1.01
39	Rockefeller	1,350	5,300	0.95	0.85	5.38	1.12	1.36
40	Rockhal	6,500	58,000	1.47	1.3	0.95	3.03	1.73
41	Rote Fabrik	1,200	5,000	1.08	1.01	2.92	1.34	1.47
42	Rote Fabrik 2	600	1,500	0.96	0.79	5.70	1.42	1.21
43	Scala	650	2,000	1.48	1.45	0.89	0.96	0.96
44	Hans Martin Schleyer Halle	15,500	270.000	2.43	2.43	−1.10	0.99	1.11
45	Oslo Spektrum Arena	9,700	150.000	1.61	1.66	0.87	1.64	1.23
46	Tunnel	400	700	0.94	0.76	5.69	0.82	0.94
47	Forest National	10,000	150.000	3.53	3.36	−2.26	1.33	1.11
48	Wembley Arena	12,000	150.000	2.56	2.41	−3.47	2.1	1.64
49	Werk Backstage	1,500	5,000	1.39	1.46	−0.15	0.96	1.04
50	Zeche	1,100	2,700	1.13	1.17	0.45	1.13	1.23
51	Zeche Carl	600	1,000	0.9	0.72	5.55	1.97	1.54
52	Zenith Paris	6,238	90,000	1.8	1.68	−0.19	0.82	1.07
53	Zenith Strasbourg	12,000	130.000	2.07	1.92	0.53	1.61	1.28
54	Vega	1,500	5,800	1.2	0.97	4.97	1.34	0.98
55	Nosturi	900	3,000	0.64	NaN	NaN	2.48	1.89

Appendix D
Two Sound Engineers' Statements

Ben Surman, FOH Engineer for Jack DeJohnette, John Scofield, and many others, United States.

For me, the acoustics of a room are of primary concern for a concert, secondary only to the quality of the musicians and their instruments. Amplified music is much harder to manage in spaces with longer reverberation times, especially at faster tempos and with denser instrumentation. The loss of definition in the bass frequencies really blurs the groove and feeling of music that depends on amplified bass and drums.

Kjetil Bjøreid Aabø, Sound Engineer, Musician, Norway.

Here are some points on how I experience different situations, and how I relate to them.

- Make sure the stage sounds as good as possible. If the band has difficulties playing, no one will have any fun.
- Get the balance right. Perfect balance is EVERYTHING. There is no use reaching for the mixer to make your kick drum sound nice and juicy if the guitar or vocal is too low in the mix. Not until you have perfect balance, can you start to do anything regarding PA tuning, acoustics, EQ, and so on. Balance comes first.
- Make sure the fans in front have a great experience. If the room has good acoustics and the PA is set up in a way that gives everyone a similar audio experience, that's great. If not, the fans have "dibs" on the best sound.
- No two concerts sound the same, nor should they. I try to take advantage of the originality of each room I visit, and use it to color the sound on that specific day.
- The important thing is that the audience is having a good time, not whether I have the right reverb unit or not.

Small Clubs

- In small clubs I try to take advantage of actually having walls close by. They color the sound so much.
- Here I feel I can "push" the mix harder than in bigger rooms, making the experience physical as well as audible.

- If I have "problem" areas in the venue, for example, a boomy balcony, I try not to focus on it too much, and prioritize the sound for the fans in front of the stage.

Large Venues

- Often boomy with large reverbs.
- I tend to mic instruments closer to get more proximity.
- Separation is often the key; I give each instrument it's "own" reverb frequency.
- I try to keep the low end under control. Not many instruments should be audible in the subs.

In nice-sounding, reverberant large rooms, such as churches, I try to soften the mix to make it blend more with the acoustics of the room. This is, though, a little bit genre dependent. If the tempo gets too fast, you can get quite a blurry sound.

Dry Rooms/Outdoors

- I try to create the illusion of actually having a room.
- I mic instruments farther away than usual. Ambient mics can be useful.
- Side fills for the band.
- I get the band to stand closer together to get more bleeding between the mics, thus creating more ambience.
- Outdoors I walk around in the audience quite a bit. The sound in the mixing tent is quite different from what the audience is experiencing.

From the Musician's Point of View

- For me it's important that the stage feels solid and heavy.
- I like it when the stage feels warm. Acoustics, temperature, lighting… all contribute to that.
- It's crucial that there are no standing bass frequencies in the stage. This is important both for monitoring and the dynamics in the music.
- Often stages are overloaded with absorption of different types, and this may lead to a really "dead" stage. This, in turn, makes you work harder on your instrument than you usually would, ruining your dynamics, and you still don't get the sensation that the sound really gets "out there " properly.
- If I feel as if I am standing in the same room as the audience, and I can hear what goes on outside of the stage, it is a great advantage. I get a much stronger feel for what the audience is experiencing, and I can play off of that.

Printed by Printforce, the Netherlands